Collins

Revision

NEW GCSE SCIENCE

Science and Additional Science

for AQA A Foundation

Authors: Nicky Thomas
Rob Wensley
Gemma Young

Revision guide +
Exam practice workbook

William Collins' dream of knowledge for all began with the publication of his first book in 1819. A self-educated mill worker, he not only enriched millions of lives, but also founded a flourishing publishing house. Today, staying true to this spirit, Collins books are packed with inspiration, innovation and practical expertise. They place you at the centre of a world of possibility and give you exactly what you need to explore it.

Collins. Freedom to teach

Published by Collins
An imprint of HarperCollinsPublishers
77–85 Fulham Palace Road
Hammersmith
London
W6 8JB

10 9 8 7 6 5 4

ISBN 978-0-00-741600-4

British Library Cataloguing in Publication Data
A Catalogue record for this publication is available from the British Library

Project managed by Hart McLeod

Edited, proofread, indexed and designed by HartMcLeod

Printed and bound in China

Photographs
The Authors and Publishers are grateful to the following for permission to reproduce photographs.
p8 ©Hank Morgan/Science Photo Library; p15 ©Roman Kobzarev/istock.com; p35 © Stuart Kelly / Alamy; p53 ©Alan Williams/Alamy; p58 ©Michael Eichelberger, Visuals Unlimited/Science Photo Library; p103 fig 4 ©Shutterstock; p103 fig 5 ©Shutterstock; p104 ©Shutterstock; p117 ©Michael Eichelberger, Visuals Unlimited/Science Photo Library; p157 left © Klemiantsou Kanstantsin/istock.com, right ©Gabor Izso/istock.com; p162 © Karl Dolenc/istock.com; p164 ©Kevmn, wikipedia

Whilst every effort has been made to trace the copyright holders, in cases where this has been unsuccessful, or if any have inadvertently been overlooked, the Publishers would gladly receive any information enabling them to rectify any error or omission at the first opportunity.

About this book

This book covers the content you will need to revise for GCSE Science and Additional Science AQA A Foundation. It is designed to help you get the best grade in your GCSE Science Foundation Exam.

The content exactly matches the topics you will be studying for your examinations. The book is divided into two major parts: **Revision guide** and **Workbook**.

Begin by revising a topic in the Revision guide section, then test yourself by answering the exam-style questions for that topic in the Workbook section.

Workbook answers are provided in a detachable section at the end of the book.

Revision guide

The Revision guide (pages 6–110) summarises the content of the exam specification and acts as a memory jogger. The material is divided into grades. There is a question (**Improve your grade**) on each page that will help you to check your progress. Typical answers to these questions and examiner's comments, are provided at the end of the Revision guide section (pages 111–123) for you to compare with your responses. This will help you to improve your answers in the future.

At the end of each module, you will find a **Summary** page. This highlights some important facts from each module.

Workbook

The Workbook (pages 148–251) allows you to work at your own pace on some typical exam-style questions. You will find that the actual GCSE questions are more likely to test knowledge and understanding across topics. However, the aim of the Revision guide and Workbook is to guide you through each topic so that you can identify your areas of strength and weakness.

The Workbook also contains example questions that require longer answers (**Extended response questions**). You will find one question that is similar to these in each section of your written exam papers. The quality of your written communication will be assessed when you answer these questions in the exam, so practise writing longer answers, using sentences. The **Answers** to all the questions in the Workbook are detachable for flexible practice and can be found on pages 257–278.

At the end of the Workbook there is a series of **Revision checklists** that you can use to tick off the topics when you are confident about them and understand certain key ideas.

Additional features

Throughout the Revision Guide there are **Exam tips** to give additional exam advice, **Remember boxes** pick out key facts and a series of **How Science Works** features, all to aid your revision.

The **Glossary** allows quick reference to the definitions of scientific terms covered in the Revision guide.

Contents

Diet and energy

A balanced diet

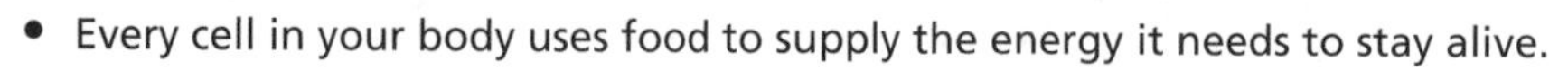

- Every cell in your body uses food to supply the energy it needs to stay alive.
- Everybody needs to eat food containing the right amount of energy to balance their needs.
- If you take in more energy than you use, your body stores it as fat. If you take in less energy than you use, your body raids its fat stores to provide your cells with the extra energy that they need.

Remember!
When writing about balanced diets, you need to check whether you mean nutrient or food. For example, bread is a food. Bread contains the nutrients carbohydrates, proteins and calcium (a mineral ion).

Diet and metabolic rate

- People should eat a balanced diet which contains some of all these different kinds of nutrients:
  - carbohydrates, fats and protein for energy
  - small amounts of vitamins and mineral ions for keeping healthy.
- If your diet is not balanced, you may become malnourished (become too fat or too thin, or suffer from deficiency diseases).
- To lose body mass, people may go on a slimming diet where they eat less. Exercising more also helps. These both lead to more energy being used up than is taken in, and the body is forced to use up some of its stored fat for energy.
- Metabolic rate is the rate at which chemical reactions take place in your cells.
- The greater the proportion of muscle to fat in the body, the higher the metabolic rate is likely to be. It also increases during exercise.
- Metabolic rate can be affected by your genes, which you inherit from your parents.

How Science Works

- Some slimming programmes and products may make claims that are unrealistic. Also it is important to remember that people will put back on the mass they lost if they don't change their eating habits for good.

Remember!
A balanced diet will contain all nutrients in the correct amounts.

Diet, exercise and health

Cholesterol

- Your liver makes cholesterol.
- Cholesterol is needed to make cell membranes, but too much could lead to heart disease.

Diet and cholesterol

- A high level of cholesterol in the blood increases the risk of developing plaques in the walls of the arteries. Sometimes, a clot blocks one of the arteries that take oxygenated blood to the heart muscle. This causes a heart attack – the muscle cannot work, so the heart cannot beat properly.
- Eating saturated fats (those found in animal products) raises blood cholesterol levels. Unsaturated fats, found in plants, seem to lower blood cholesterol level.

A healthy artery has a stretchy wall and a space in the middle for blood to pass through.

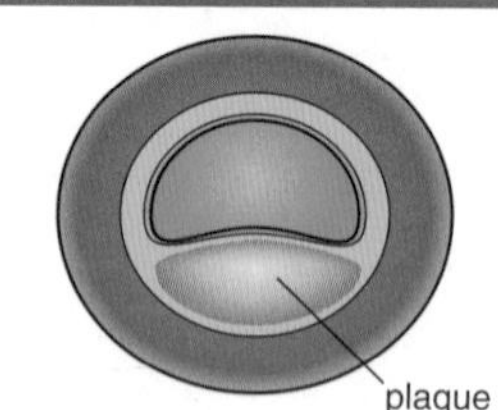

Sometimes, a substance called plaque builds up in the wall. This is more likely to happen if you have a lot of cholesterol in your blood.

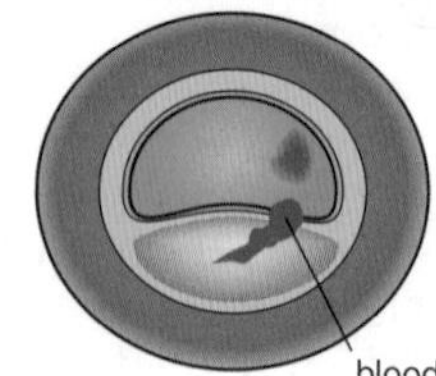

The plaque slows down the blood and a clot may form. A part of the plaque may break away.

Figure 1: How a plaque develops in an artery

- Some people's bodies are better than others at keeping low levels of cholesterol in their blood. They have inherited this from their parents.

Improve your grade

Harry's diet is very high in saturated fats. Suggest two ways that this could affect his health. **AO2 (3 marks)**

Pathogens and infections

Microorganisms

- Microorganisms are living things that are too small for us to see without a microscope.
- They include bacteria and viruses.

Disease

- Microorganisms that cause disease are called pathogens.
- Bacteria can reproduce rapidly inside the body. They may produce toxins (poisons) that make us feel ill.
- Viruses reproduce inside a body cell then destroy it when they burst out. The viruses then invade other cells.
- An epidemic occurs when a wide spread of people have a disease. A pandemic is when the disease affects a whole country or goes worldwide.
- In the 1840s, a doctor called Semmelweis used evidence from the death rates of women to work out that they were dying because doctors were transferring something to them from dead bodies. He made all the doctors wash their hands in chlorine water and, within a very short time, the death rate plummeted. We now know the infection that killed the women was caused by bacteria.

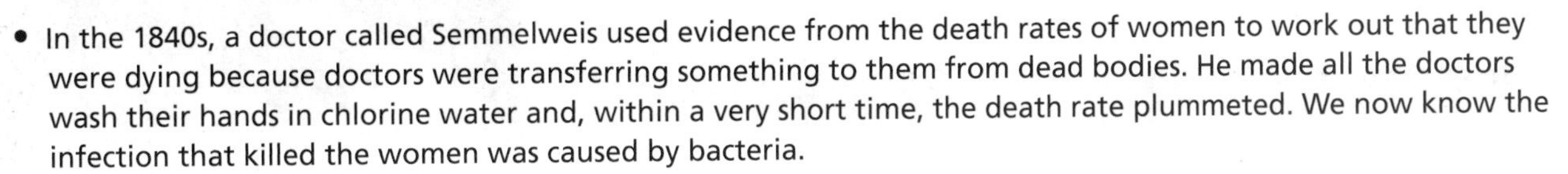

Fighting infection

White blood cells

- White blood cells are part of your immune system. They attack and destroy pathogens that have found their way inside the body.
- Some white blood cells surround bacteria and ingest them. They take the bacteria into their cytoplasm and kill them.
- Other white blood cells make chemicals called antibodies or antitoxins.

Phagocytosis and lymphocytes

- Figure 2 shows how a type of white blood cell, called a phagocyte, can surround and ingest bacteria. This activity is called phagocytosis. Lymphocytes produce chemicals called antibodies.
- The antibodies group round and stick to the pathogen. This may kill it directly, or stick it to other pathogens in clumps so that the phagocytes can destroy them more easily.

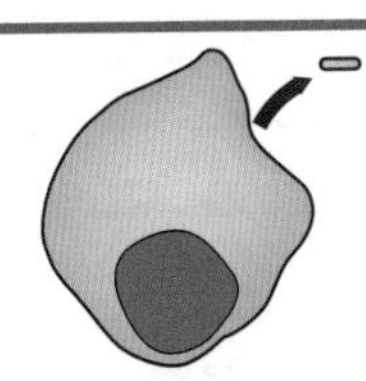

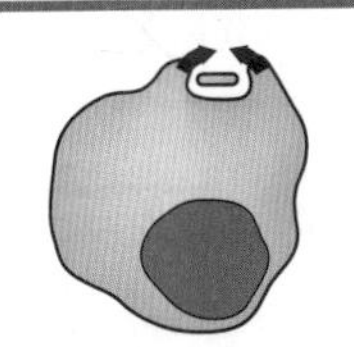

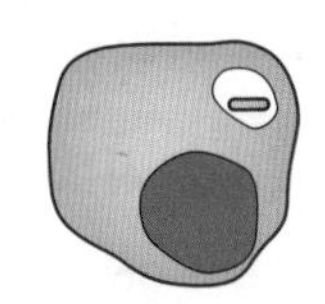

Figure 2: Phagocytosis

- Some lymphocytes make antitoxins that can stick to the toxins given off by bacteria, and destroy them.
- Both antibodies and antitoxins are very specific – each kind only works against a particular pathogen or toxin.

Improve your grade

Describe how phagocytes destroy pathogens. **AO1 (3 marks)**

Drugs against disease

Painkillers

- These are drugs which reduce pain. Examples are aspirin, paracetamol and ibuprofen.
- Painkillers can reduce symptoms but they cannot cure the underlying problem.

Antibiotics

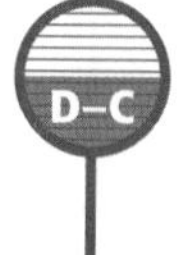

- Antibiotics are drugs that kill bacteria inside your body, without killing your own cells. Examples are penicillin and streptomycin.
- Antibiotics do not all work equally well against all the kinds of bacteria.
- Figure 1 shows how we find the best antibiotic to kill a bacterium. Bacteria are spread onto a jelly. Paper discs soaked in different antibiotics are placed on the jelly and the antibiotics diffuse out. If the antibiotic kills the bacteria, they do not grow around the disc.

Remember!
Antibiotics cannot destroy viruses. Some viral infections can be treated by taking antiviral drugs.

EXAM TIP

You may be shown an image of a test like this and be asked what it shows. The clear jelly shows no bacterial growth: so the bigger this area, the more effective the antibiotic. In this case, antibiotic E is the most effective.

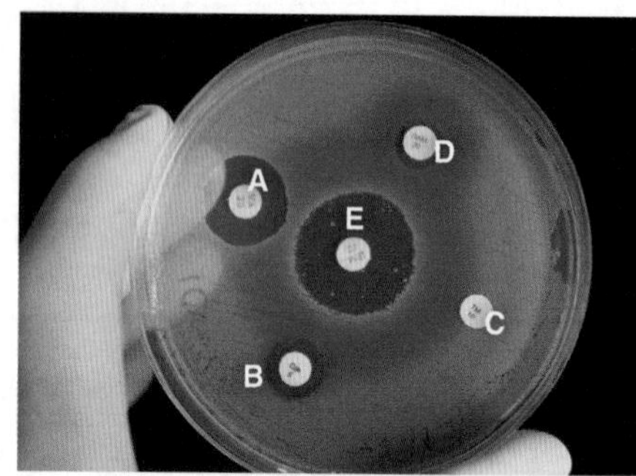

Figure 1: Testing antibiotics

Antibiotic resistance

MRSA

- Many bacteria have become resistant to antibiotics.
- This means that the bacteria are not affected by the antibiotics that used to kill them. One example is MRSA.
- Hospital patients, who are already ill, may not have strong defences against disease. They can pick up MRSA infections easily. These could kill the patient because most antibiotics do not work against it.

Reducing the risk

- Bacteria do not become resistant to antibiotics on purpose. It happens by natural selection. As new antibiotic-resistant strains of bacteria emerge, we have to find new antibiotics to kill them.

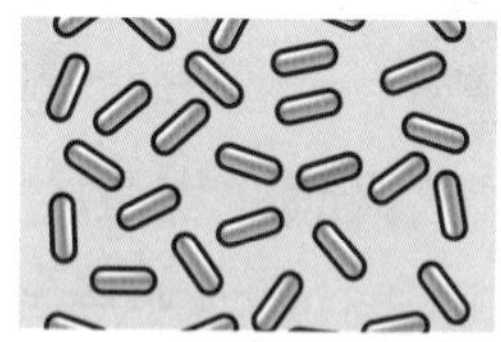

This is a population of bacteria in someone's body. By chance, one of them has mutated and is slightly different from the others.

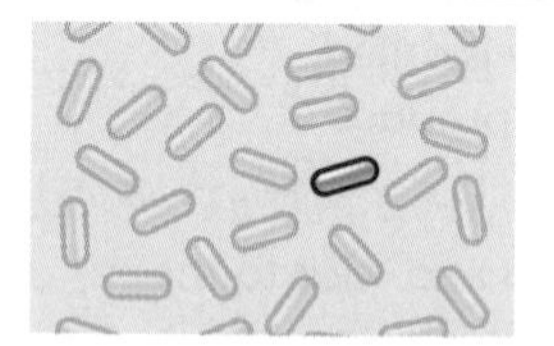

The person takes antibiotics to kill the bacteria. It works – except on the single odd bacterium. This one is resistant to the antibiotic.

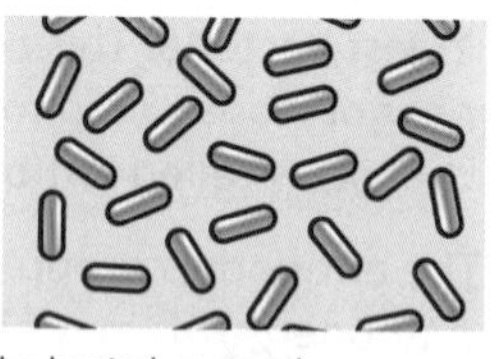

The bacterium now has no competitors and grows rapidly. It divides and makes lots of identical copies of itself. There is now a population of bacteria that the antibiotic cannot kill. This process is an example of **natural selection**.

Figure 2: How antibiotic resistance arises

- To reduce the chance of new strains forming, we need to reduce the use of antibiotics.
- Whenever antibiotics are used, it gives an advantage to any mutant bacterium that happens to be resistant to them.
- If they are not used, then a mutant bacterium does not have any advantage: it is no more likely to reproduce than any other bacterium.

Improve your grade

Suggest two ways that we can stop the spread of MRSA in hospitals. **AO1 (2 marks)**

Vaccination

Immunity

- Certain white blood cells can make antibodies that destroy pathogens. Each pathogen needs a specific antibody to destroy it.
- If you get infected, the correct white blood cells multiply and make the antibody needed to destroy the pathogen. This takes time.
- If you survive, then the next time you are infected, your white blood cells remember how to make the antibody. The antibodies can be made much faster and the pathogen destroyed before it causes harm. You have developed immunity to the disease.

MMR

- In the UK, children have the MMR vaccination which makes them immune to measles, mumps and rubella.
- A small amount of the dead or inactive viruses that cause the diseases are injected into the blood.
- The white blood cells attack them, just as they would attack living pathogens. They remember how to make the antibody, so the child is now immune to the diseases without having to suffer them first.

How Science Works

- In 1998, a group of scientists published an article suggesting that the MMR vaccination might cause autism. Many parents decided not to let their child have the MMR vaccination, even though there was no evidence in the article. Many studies have been carried out since. No one has found any link between the MMR vaccination and autism.

Growing bacteria

Nutrient mediums

- Bacteria and fungi can be grown in a liquid or a jelly containing all the nutrients that they need (a nutrient medium). A jelly called agar is used most often.
- The bacteria (or other microorganisms) that grow on it is called a culture.
- The nutrient medium must be sterile – it must not contain any microorganisms.

Avoiding contamination

- The microorganisms growing on the nutrient medium are called a culture.
- You should use a sterile technique to stop unwanted microorganisms from entering the nutrient medium.
- All equipment, including the medium, should be sterilised before use.
- Metal equipment, such as a wire inoculating loop, can be held in a flame.
- 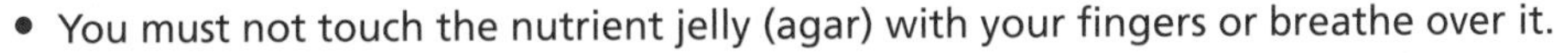You must not touch the nutrient jelly (agar) with your fingers or breathe over it.
- 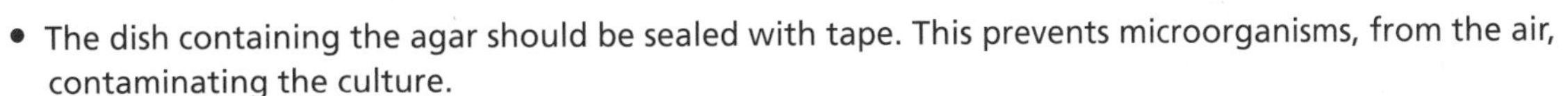The dish containing the agar should be sealed with tape. This prevents microorganisms, from the air, contaminating the culture.
- 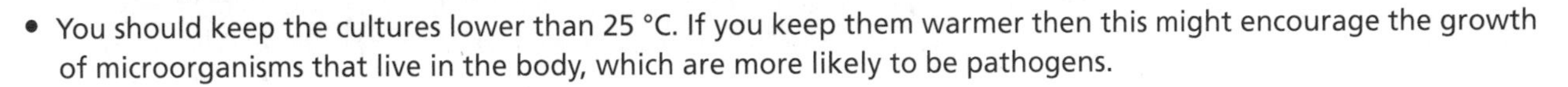You should keep the cultures lower than 25 °C. If you keep them warmer then this might encourage the growth of microorganisms that live in the body, which are more likely to be pathogens.

Improve your grade

You wish to grow some harmless *E. coli* bacteria on some agar jelly. You use an inoculating loop to transfer the bacteria to the jelly. Explain why it is important to hold the inoculating loop in a flame first. **AO1 (2 marks)**

Co-ordination, nerves and hormones

Nerves and nerve cells

- When you go to try to catch a ball, information travels from your eyes along nerves to the brain as fast-moving electrical impulses (signals).
- Impulses then travel from the brain to your arm and hand, telling them how and when to move.

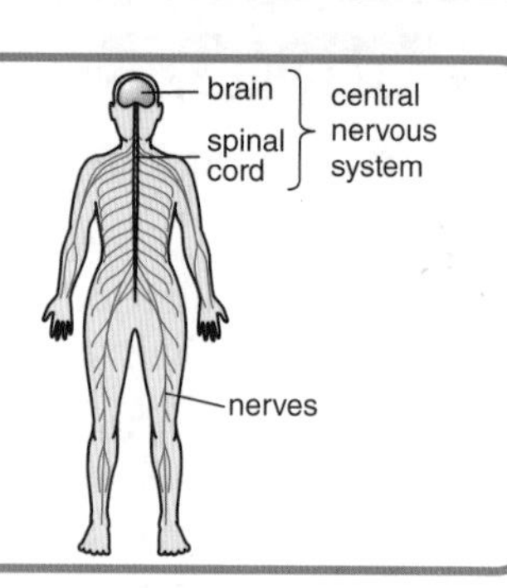

Nerves and hormones

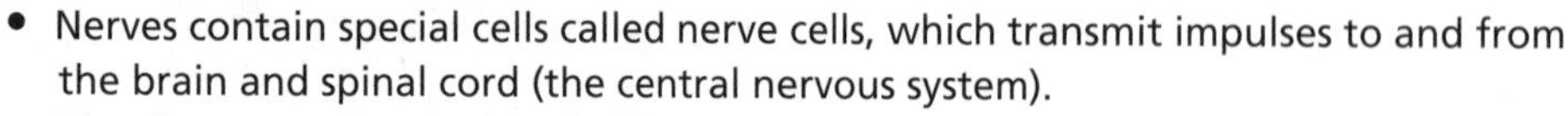

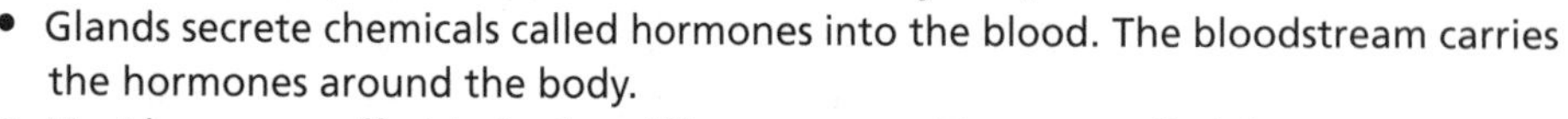

- Nerves contain special cells called nerve cells, which transmit impulses to and from the brain and spinal cord (the central nervous system).
- Glands secrete chemicals called hormones into the blood. The bloodstream carries the hormones around the body.
- Most hormones affect just a few different organs. These are called their target organs.
- An example of a hormone is adrenaline which affects the heart, breathing muscles, eyes and digestive system.

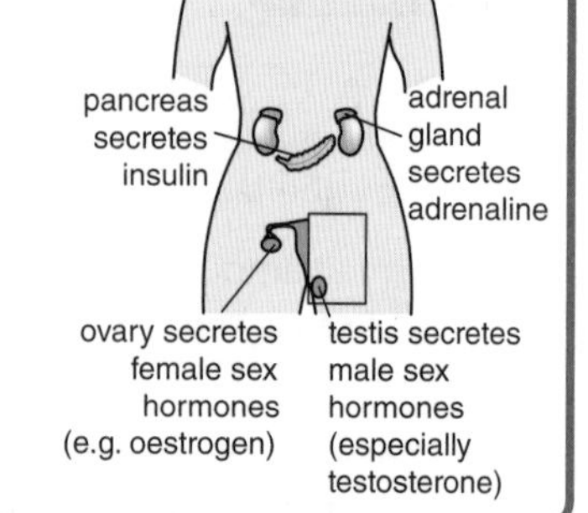

Figure 1: Four glands that secrete hormones

Receptors

Receptors and effectors

- A stimulus is a change in the environment. Examples are light, sound and chemicals.
- Receptors detect stimuli from the environment, and send information to the central nervous system.
- Effectors (muscles and glands) do something in response to the stimulus.

Neurones

- Information is carried in the nervous system as electrical impulses. The cells that transmit these impulses are called nerve cells or neurones.
- The neurones that transmit impulses from receptors to the central nervous system are called sensory neurones.
- The neurones that transmit impulses from the central nervous system to effectors are called motor neurones.

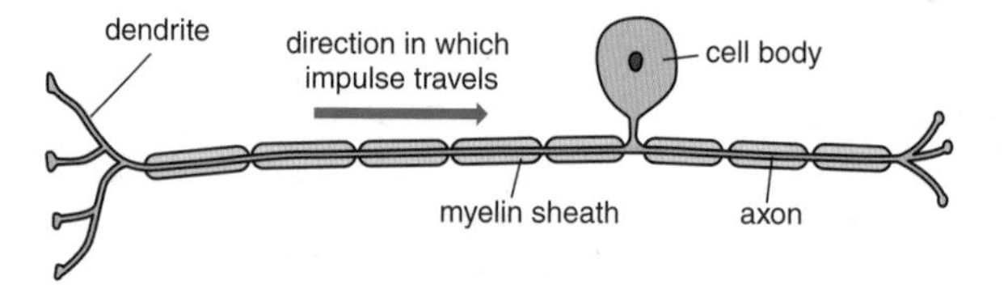

Figure 2: A sensory neurone

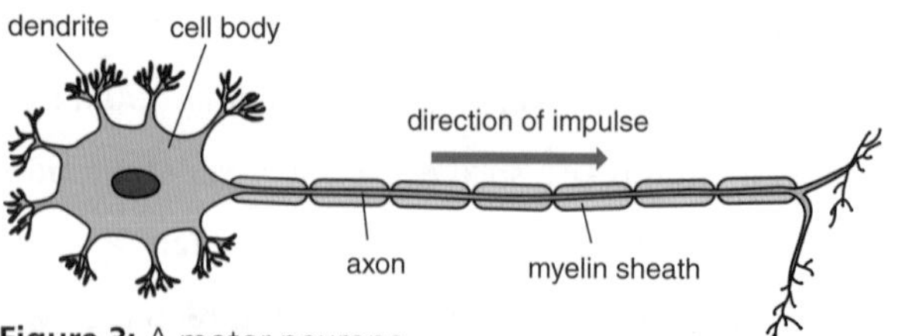

Figure 3: A motor neurone

Remember!

Receptors detect stimuli and send an impulse along a neurone. Effectors (muscles and glands) respond to the stimuli.

EXAM TIP

Make sure you use the correct science words in your answers. Use 'impulses' rather than 'messages'; and 'neurone' if you are talking about a single cell or 'nerve' if you are talking about a bundle of them, as in the spinal cord.

Improve your grade

An injury that results in breaking of the spine may result in the person being paralysed. Explain why. **AO2 (3 marks)**

Reflex actions

Stimulus and response

- A reflex action is a fast, automatic response to a stimulus.
- Most reflexes protect you. For example, if something moves fast towards your face, you blink. This protects your eyes.

Impulse pathways

- A reflex arc is the pathway taken by a nerve impulse as it passes from a receptor, through the central nervous system, and finally to an effector.
- Figure 4 shows the path the impulses take.
- It takes a nerve impulse only a fraction of a second to go along this route. That is why reflex actions are so quick.
- The gaps between neurones are called synapses.

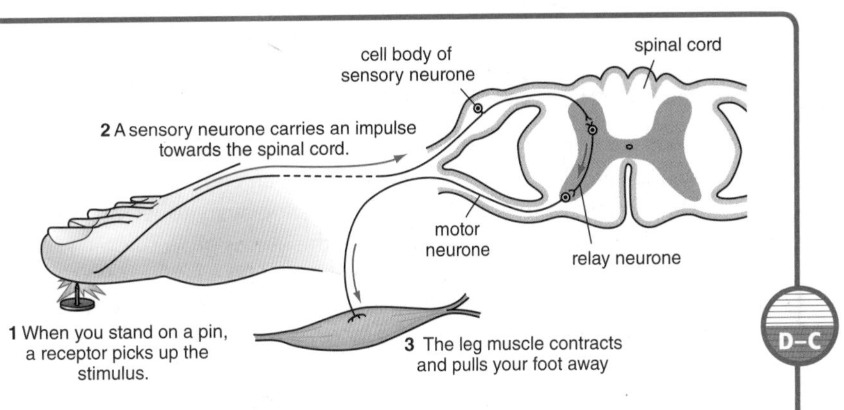

Figure 4: A reflex arc

- Electrical impulses cannot jump across synapses. When an impulse gets to the end of a neurone, it causes a chemical to be secreted. The chemical diffuses across the gap and starts an electrical impulse along the next neurone.

Remember!
Synapses slow down the speed of the impulse.

Controlling the body

Controlling the conditions in cells

- Chemical reactions take place in the cells in your body.
- For them to happen, conditions around each cell must be perfect – and constant (not change). These conditions include: water content, ion (salt) content, temperature, and concentration of sugar in the blood.

Temperature control

- You gain water from food and drink. You lose water in your breath, sweat and urine.
- Your blood has ions dissolved in it, such as those found in salt.
- The kidneys help to keep the balance of water and ions by varying the amount of water and salt excreted from your body in urine.
- Human body temperature needs to be kept at around 37 °C, as this is the temperature at which our enzymes work best.
- The body loses heat by radiation from the skin, and from the evaporation of sweat.
- The body also has mechanisms to keep the concentration of sugar in the blood constant.

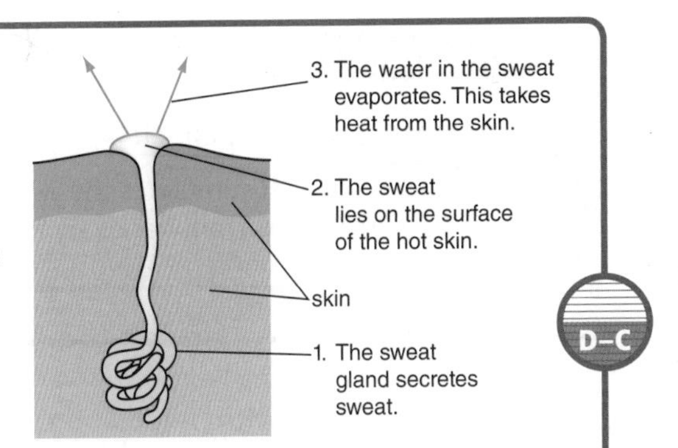

Figure 5: How sweating helps you to lose heat

Improve your grade

James is dancing in a nightclub. He starts to sweat. Explain how sweating helps to cool him down. **AO2 (2 marks)**

Reproductive hormones

The menstrual cycle

- One egg is released from a woman's ovaries about every 28 days.
- Before the egg is released, the lining of the womb (uterus) thickens.
- If the egg is fertilised, the embryo attaches to the lining.
- If it is not fertilised, the lining of the womb breaks down (menstruation).

Hormones and the menstrual cycle

- At the start of the menstrual cycle, the pituitary gland secretes FSH (follicle-stimulating hormone). This causes an egg to mature in one of the woman's ovaries and also stimulates the ovary to secrete oestrogen.
- Oestrogen makes the inner lining of the uterus grow thicker. High levels of oestrogen stop the production of FSH.
- Luteinising hormone (LH) stimulates the release of eggs from the ovary.
- As the level of FSH drops, the ovary stops secreting oestrogen. This cuts off the inhibition of FSH secretion, so the cycle starts all over again.
- The contraceptive pill may contain oestrogen. It stops FSH being produced, so that eggs do not mature.

How Science Works

- When the pill was first used it produced many side effects, for example people put on weight. Nowadays, the pills contain much less oestrogen than they used to (some have no oestrogen at all but contain another hormone called progesterone), so there are fewer side effects.

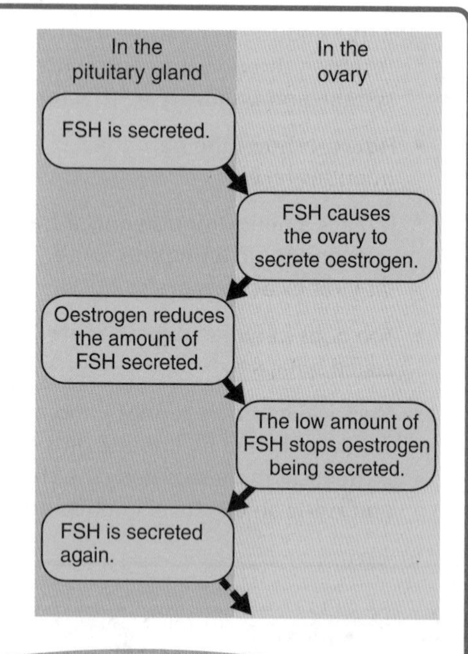

Figure 1: FSH and oestrogen secretion

Remember!

FSH and LH are produced in the pituitary gland which is in the brain. They reach their target organs (the uterus and the ovary) through the bloodstream.

Controlling fertility

Fertility

- Many couples trying for a baby are not successful. They may use fertility drugs that contain hormones.
- The hormone FSH stimulates the woman's eggs to mature in her ovaries. LH stimulates an egg's release. When an egg is released from her ovary into an oviduct, she can conceive in the normal way.

IVF

- IVF stands for 'in vitro fertilisation'.
- The woman is given hormones, such as FSH, to make her ovaries produce several eggs.
- The eggs are removed and are mixed with her partner's sperm for fertilisation to occur.
- One of the embryos is chosen and placed in the woman's uterus. With luck, it will sink into the uterus lining and develop as a fetus.

Improve your grade

FSH is found in fertility drugs. Explain how taking FSH will increase a woman's fertility. **AO2 (2 marks)**

Plant responses and hormones

Responding to light and gravity

- Plants are sensitive to light, moisture and gravity.
- They respond by growing in particular directions. These growth responses are called tropisms.
- Shoots grow towards light and away from gravity. This helps to get plenty of light for photosynthesis.
- Roots grow towards gravity and water – downwards, into the soil. Water is then transported from the roots to the rest of the plant. It also means the root can anchor the plant firmly into the ground.

Phototropism and gravitropism

D–C

- A growth response to light is a tropism called phototropism.
- Auxin is a plant hormone which makes cells in shoots get longer. When light shines onto a shoot, the auxin builds up on the shady side. This makes the cells on that side get longer. So, the shoot bends towards the light.
- A growth response to gravity is called gravitropism or geotropism.
- Auxin tends to accumulate on the lower side of a root. In roots, auxin reduces the rate of growth. So, the lower side of the shoot grows more slowly than the upper surface. This causes the root to bend downwards.
- Gardeners dip the base of a cutting into a powder or gel called rooting hormone, which makes the cutting grow roots.
- Plant hormones are also used as weedkillers. The hormones make the weeds grow very fast and then die. The hormones only affect weeds because they have a different metabolism.

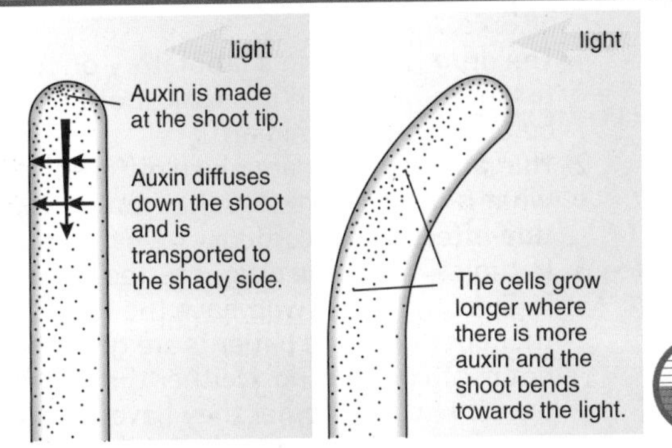

Figure 2: How auxin makes a shoot grow towards the light

Remember!
Growth towards the stimulus is a positive tropism. Growth away is a negative tropism. Shoots display positive phototropism and negative gravitropism.

Drugs

Using drugs

G–E

- A drug is something that changes the chemical processes in the body.
- They have been used for thousands of years. Many come from plants and other natural sources.
- Medical drugs are very helpful. They can help to make people feel better when they are ill, and cure diseases.
- All drugs have side-effects and can be harmful if used wrongly.

Dangers of drugs

- Some people take drugs because they make them feel different. This is called recreational drug use.
- Many recreational drugs are legal but still can be harmful, such as alcohol.

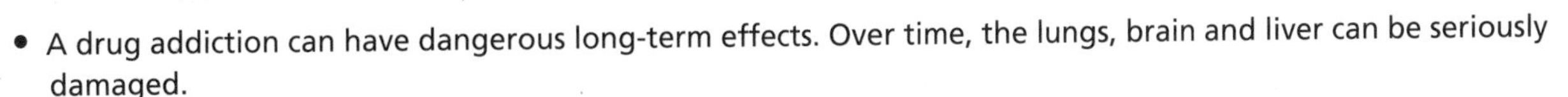

- Some recreational drugs are illegal because of the harm they can cause. Cannabis is an example, it may cause mental illness.

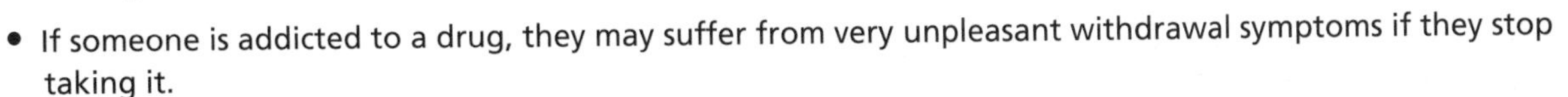

- A drug addiction can have dangerous long-term effects. Over time, the lungs, brain and liver can be seriously damaged.
- If someone is addicted to a drug, they may suffer from very unpleasant withdrawal symptoms if they stop taking it.

Improve your grade

Explain how the hormone auxin brings about a response to light called phototropism. **AO2 (4 marks)**

Developing new drugs

Testing drugs

- Before doctors are allowed to prescribe any new drug, it has to be thoroughly tested.
- Drug trials find out if the drug works and if it is safe.
- Trialling can take years and may not result in a new drug.

Drug trials

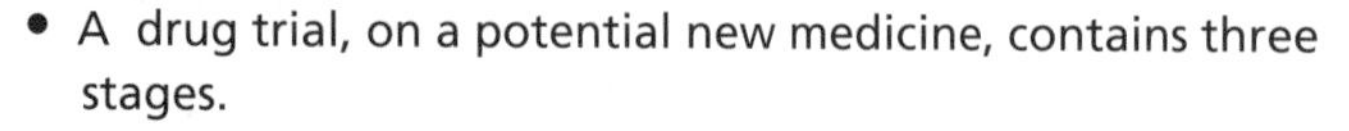

- A drug trial, on a potential new medicine, contains three stages.

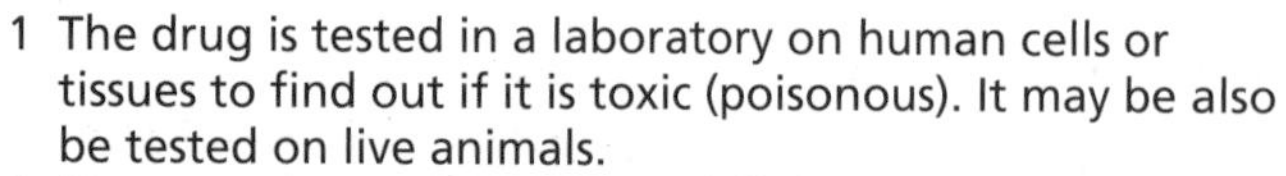

1. The drug is tested in a laboratory on human cells or tissues to find out if it is toxic (poisonous). It may be also be tested on live animals.
2. Human volunteers are given different doses, to find out what is the highest dose that can be taken safely. Any side-effects are recorded.
3. In clinical trials, the drug is tested on its target illness. It is given to people who have the illness, to see if it makes them better. Some patients are given placebos which do not contain the drug. Neither the patient nor their doctor knows whether they have a placebo or the real drug (a double-blind trial). This helps determine whether the drug really works.

How Science Works

- In the 1960s, many pregnant women were prescribed the drug thalidomide to treat 'morning sickness' in pregnancy.
  The drug had been thoroughly trialled as a sleeping pill, but no one had thought to test it on pregnant women. Women who took thalidomide in early pregnancy often gave birth to babies with short arms or no arms.
  Thalidomide was banned worldwide, but it is now being used to treat serious diseases such as leprosy.

Legal and illegal drugs

Cannabis

- People with long-term illnesses, such as multiple sclerosis, claim that cannabis helps them feel better.
- Doctors think that young people who smoke cannabis are more likely to develop a serious mental illness.
- Using cannabis may lead a person to use other, more harmful drugs.

Dangers from recreational drugs

- Alcohol and nicotine cause far more illness and death each year than all the illegal drugs put together.
- People do not worry so much about them because they have been around for so long, and because so many people use them.

Remember!

Recreational drugs, like alcohol and ecstasy, are taken by people because they like the way the drugs change their feelings or behaviour. Other drugs, like aspirin, are medicinal.

EXAM TIP

Make sure you can sort drugs into those that are recreational or medicinal and illegal or legal.

Improve your grade

Why do drugs need to be trialled before they can be prescribed by a doctor? **AO2 (2 marks)**

Competition

Resources

G–E

- All living organisms need resources from their environment in order to stay alive. For example, plants must have:
  - light – to photosynthesise and make food
  - water – to keep cells alive, to transport substances around and for photosynthesis
  - space – for room to put down roots and spread out leaves to capture light
  - nutrients – such as nitrates from the soil.

Competing

- Organisms may have to compete for resources if they are in short supply.
- For example, plants compete for light – the ones that grow tallest win the competition.
- The individuals best at competing are the most likely to survive. Those not good at getting resources are the most likely to die.
- If there are not enough females to go around, then males will compete for a mate.
- Animals may also compete for a territory – a space in which they can find food and a place to breed.

Remember!
Competition does not usually mean that organisms actually fight over resources. They just have to find ways of being better at getting them than others are.

Figure 1: Stags fight over the right to mate with the females in the herd

Adaptations for survival

Resources

G–E

- Plants and animals must have features that allow them to survive in their habitat. These features are adaptations.
- Well-adapted organisms can compete successfully for the things they need.

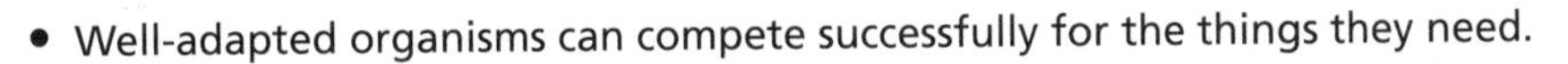

- If they are not adapted for life in a particular habitat, they will either die or move elsewhere.

Living in difficult climates

- Plants that live in dry places usually have:
  - long, wide-spreading roots – the roots grow deep into the soil, to reach water
  - small or no leaves – the smaller the leaf surface area, the less the amount of water evaporating away
  - tissues that can store water.

Remember!
Plants and animals must have features that allow them to survive in their habitat. These features are adaptations.

- Animals that live in dry places must be able to manage without much water. For example, camels' stomachs can hold over 20 litres of water, they can drink very quickly, store water as fat in their humps, and they produce very little urine.
- Desert animals often have large ears. A large surface area helps the animal lose body heat and stay cool.

- Animals that live in very cold places, such as the Arctic, often have thick fur and thick layers of fat. This insulation helps the animal reduce heat loss. They are coloured white, for camouflage against snow.
- Many plants and animals have thorns, poisons and warning colours to deter predators.

Improve your grade

Explain how cacti are adapted to living in the dry desert. **AO2 (3 marks)**

Environmental change

Getting warmer

- The environment is always changing. Global warming means that temperatures around the world are increasing.
- All living organisms must be adapted for survival in a particular environment.
- If that environment changes, they may no longer be able to survive.

Causes of change

- Environmental changes are caused by living and non-living factors. For example:
  - non-living factors include global warming, which has caused rainfall in central Australia to decrease
  - living factors include the introduction of the grey squirrel into Britain, which caused a decrease in the population of the native red squirrel.

Pollution indicators

Pollution

- Producing food and manufacturing goods for an ever-increasing population produces more and more chemical wastes.
- The wastes may be released into the environment. They can be harmful to our health and to wildlife.
- We call harmful chemicals 'pollutants'. Their release causes pollution.

Remember!

Pollutants can cause air pollution (like sulfer dioxide) or water pollution (like sewage).

Measuring changes in the environment

- In the UK, the composition of the air and of the water in rivers and streams; and the air temperature and rainfall, are constantly being measured. This makes sure that any changes can be tracked.
- Oxygen meters measure the concentration of dissolved oxygen in the water. Unpolluted water contains a lot of dissolved oxygen.
- Thermometers measure temperature. Rain gauges measure rainfall.
- Scientists can use the distribution of living organisms to find out about pollution. For example:
  - if there is a lot of sulfur dioxide in the air, many species of lichens will not be able to grow
  - if there is not very much oxygen in a river, there will be no oxygen-loving mayfly larvae in the water, instead there will be just rat-tailed maggots and bloodworms.

Remember!

The higher the number of different lichen species that can grow in an area, the lower the levels of sulfur dioxide in the air.

Improve your grade

Sewage contains a lot of microorganisms. Explain why mayfly larva cannot live in water polluted with sewage.
AO2 (2 marks)

Food chains and energy flow

Energy flow

- You can show how energy passes from one organism to another, in a food chain, like this: grass → antelope → lion
- The energy comes from chemicals in the food that the animals eat.
- The grass in this food chain is a producer. It uses energy from sunlight to produce food. The antelope and the lion are consumers.
- The antelope is a herbivore. Herbivores eat plants. The lion is a carnivore. Carnivores eat other animals.
- The lion is also a predator. Predators kill and eat other animals for food. The animals that they kill are their prey.

Energy wastage

- Green plants capture only a small amount of energy from the light that falls onto them. This is because some light:
  - misses the leaves altogether
  - hits the leaf and reflects back from the leaf surface
  - hits the leaf, but goes all the way through without hitting any chlorophyll
  - hits the chlorophyll, but is not absorbed because it is of the wrong wavelength (colour).

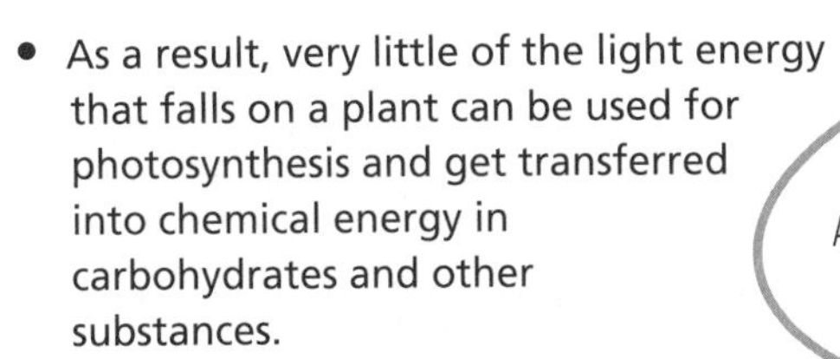

- As a result, very little of the light energy that falls on a plant can be used for photosynthesis and get transferred into chemical energy in carbohydrates and other substances.

Remember!
A food chain shows that energy is passed from producer (plants) to consumers. Consumers gain their energy by eating the organism before it in the food chain.

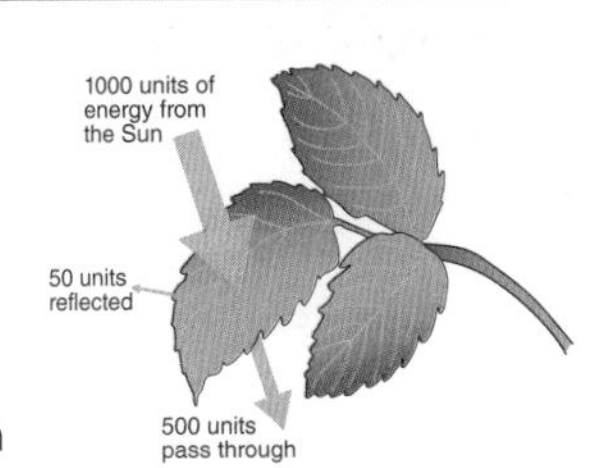

Figure 1: This is what happens to energy from the Sun that falls onto a leaf

Biomass

Pyramids of biomass

- The mass of living material is called biomass. Figure 2 shows a pyramid of biomass drawn to scale.
- The size of each block represents the biomass at each step in a food chain. At each step in a food chain, there is less biomass than in the step before.

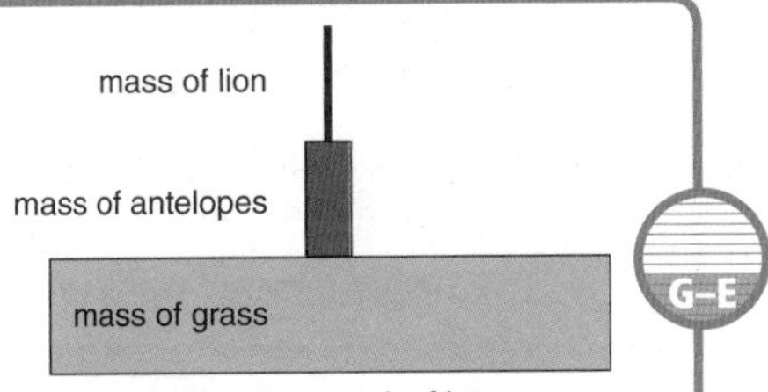

Figure 2: A pyramid of biomass drawn to scale

Energy losses

- The food chain loses energy because:
  - some materials and energy are lost in the waste materials produced by each organism, such as carbon dioxide, urine and faeces
  - respiration in each organism's cells releases energy from nutrients to be used for movement and other purposes, so much of this energy is eventually lost as heat to the surroundings
  - not all of the organism's tissues are eaten, for example the antelope does not eat the roots of the grass as they are under the ground.

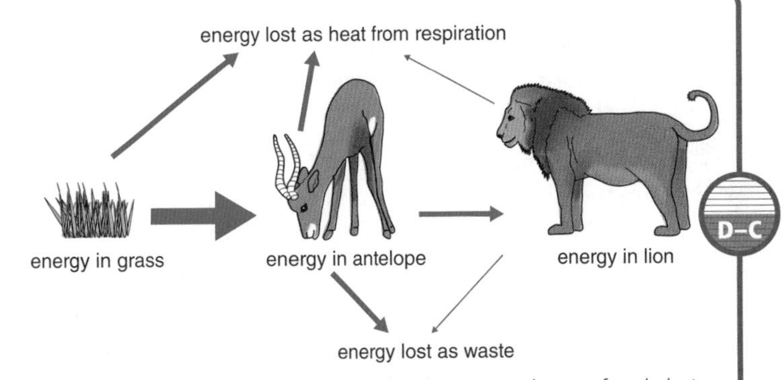

Figure 3: How energy is wasted as it passes along a food chain

Improve your grade

Explain why biomass decreases as you move along a food chain. **AO2 (3 marks)**

Decay

Decay

- Decay happens when microorganisms (bacteria and fungi) feed on a dead body; on waste material from animals and plants; or food.
- The microorganisms produce enzymes that digest the food material. The material gradually breaks down and dissolves.

Speeding or slowing decay

- Most of the bacteria and fungi that carry out decay need:
  - oxygen for aerobic respiration
  - a warm temperature for their enzymes to work at an optimum rate
  - moisture for reproduction.
- Increasing the temperature of microorganisms slows or stops decay.

Remember!

Compost heaps are kept warm, moist and aerated to speed up the decay of plant waste into compost. Compost is high in nutrients and is used to promote plant growth.

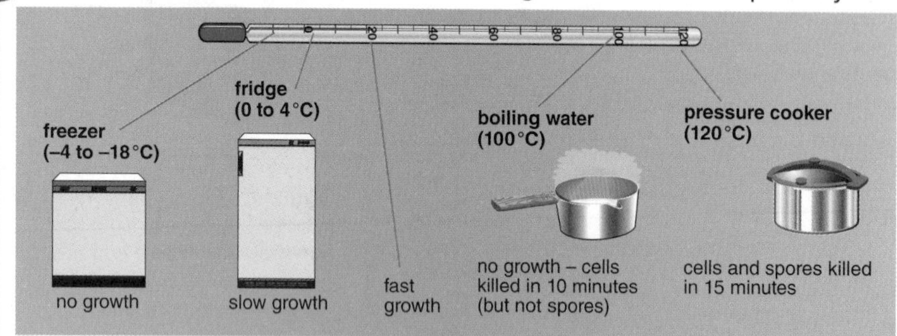

Figure 1: How temperature affects the activity of microorganisms

Recycling

Decay and recycling

- Living organisms take materials from the environment. When they die or produce waste, they are decayed and the material returns to the environment.
- All the organisms that live in one place – a community of organisms – are constantly reusing materials that have been part of other organisms.
- The decay process releases substances into the soil which plants need to grow.

Recycling and food chains

- Figure 2 shows how microorganisms fit into a simple food chain.
- You can see that these decay microorganisms feed on every organism in the chain.
- They will break down most of the waste material that the plants and animals produce, and then their bodies will be broken down by others when they die.

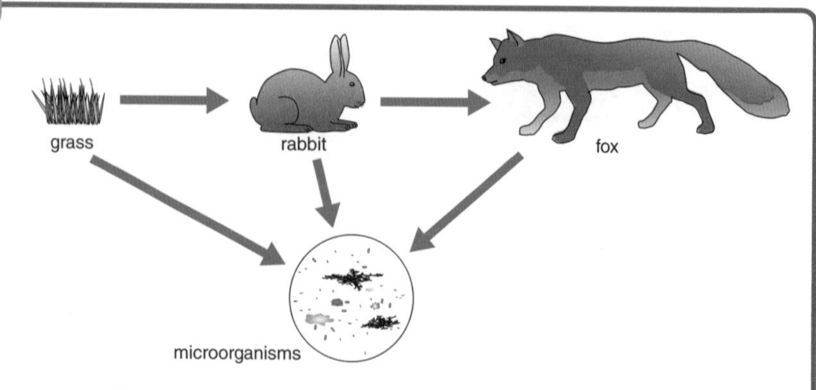

Figure 2: A food chain, including the decay organisms

Improve your grade

Write a simple plan for an investigation to prove that the decay of bread by mould requires moisture.
AO2 (2 marks)

The carbon cycle

Carbon

- Carbon is an element that exists in the air in the gas carbon dioxide (CO_2).
- Processes such as respiration and photosynthesis recycle carbon dioxide between the environment, the dead and the living. This constant cycling of carbon is called the carbon cycle.

Processes in the carbon cycle

- Animals, plants and decomposers all interact with each other in the carbon cycle (Figure 3).
- Photosynthesis converts carbon dioxide into carbohydrates and other food molecules such as proteins: carbon dioxide + water → glucose + oxygen
- When animals eat plants (or another animal) the food goes into their cells and is broken down by respiration: glucose + oxygen → carbon dioxide + water
- Carbon dioxide is returned to the air when the animal breathes out.
- Plants and microorganisms also respire.
- Some dead organisms do not decay. They become buried and compressed, deep underground and change into fossil fuels.
- Carbon dioxide is returned to the air when wood or fossil fuels are burnt (combustion).

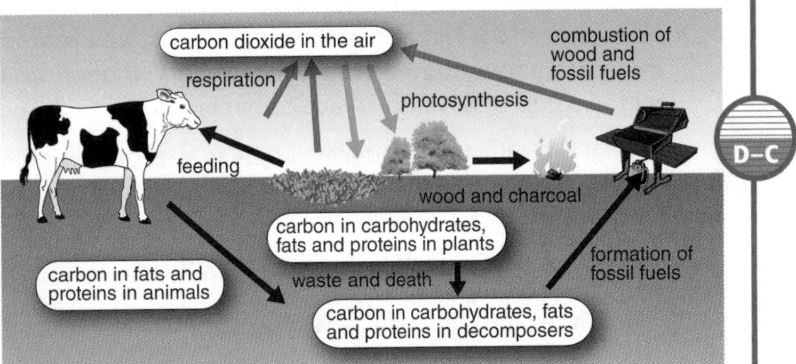

Figure 3: How carbon is recycled in the carbon cycle

Genes and chromosomes

Just like your mum or dad?

- Living organisms look like their parents because they inherit information from them. This information is passed on from parents to offspring as genes.
- Genes are linked together in long chains called chromosomes, which are found in the nucleus of cells.
- One set of chromosomes comes from the mother's egg cell. The second set comes from the father's sperm cell. Eggs and sperms are special cells called gametes.
- An egg and a sperm join together during fertilisation. The new cell that is formed contains genes from the father and the mother.
- We all have different combinations of genes – it makes us all different from one another.

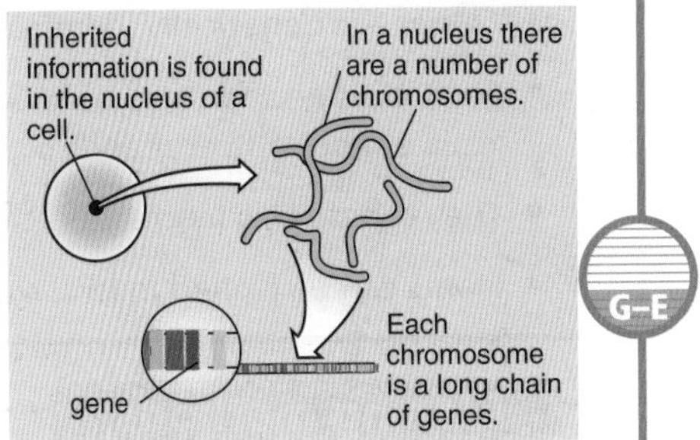

Figure 4: Each chromosome is made up of many different genes

G–E

Chromosome numbers

- Chromosomes are long sections of DNA.
- Most human cells have 46 chromosomes, that is 23 from the gamete of each parent (sperm and egg), which carry about 25 000 genes.
- Each gene contains coded information that controls one characteristic. For example, some genes control hair colour. Other genes control eye colour.
- Most of these genes come in two or more forms. For example, a gene that controls hair colour might have one form that produces brown hair and a different form that produces red hair.

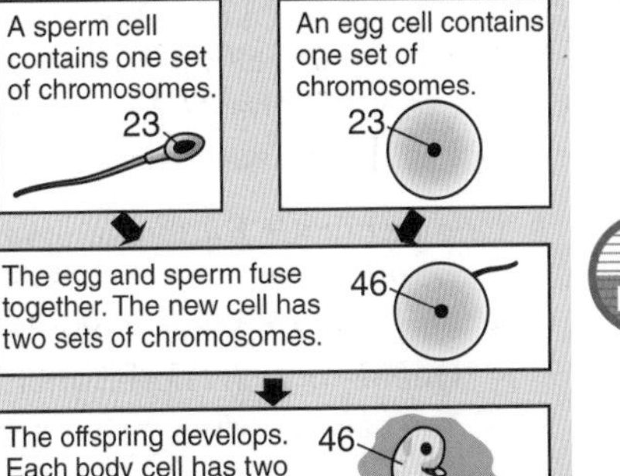

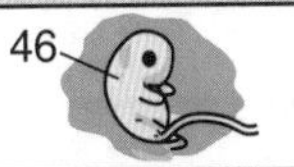

Figure 5: Chromosome numbers in gametes and other cells

D–C

Improve your grade

Matthew and Emma are both very tall. Their one-year-old son, Ryan, is also expected to be tall when he is older. Why is this? **AO2 (2 marks)**

Reproduction

Sexual and asexual

- Most animals and flowering plants reproduce sexually.
- The organisms make sex cells called gametes. In animals, the male gametes are sperm cells and the female gametes are egg cells. In flowering plants, the male and female gametes are made inside flowers.
- Asexual reproduction involves only one parent.

How it works

- In sexual reproduction, gametes and fertilisation are always involved.
- The new cell that is produced by fertilisation is a zygote. It divides repeatedly to produce a little ball of cells. This develops into an embryo and finally into an adult animal.
- Sexual reproduction produces variety in the offspring because each zygote has a different mix of genes from it parents and all its brothers and sisters.

Remember!
Sexual reproduction does not always need two parents. Some plants have flowers that produce both male and female gametes, so they can fertilise themselves.

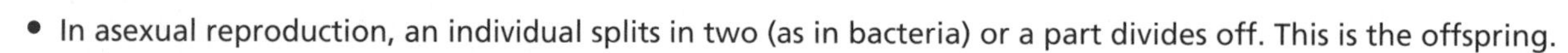

- In asexual reproduction, an individual splits in two (as in bacteria) or a part divides off. This is the offspring.
- There is no variation. The new organisms all have exactly the same genes as their parent, and as each other. They are genetically identical (clones).

Cloning plants and animals

Producing plants

- Clones are genetically identical organisms. Identical twins are an example of natural clones.
- Gardeners and farmers want lots of plants with the preferred characteristics of disease resistance, colour of fruit, shape of flower and so on.
- They can do this by creating clones of plants.
- Plants can be cloned in two ways: by taking cuttings and tissue culture.

Cloning methods

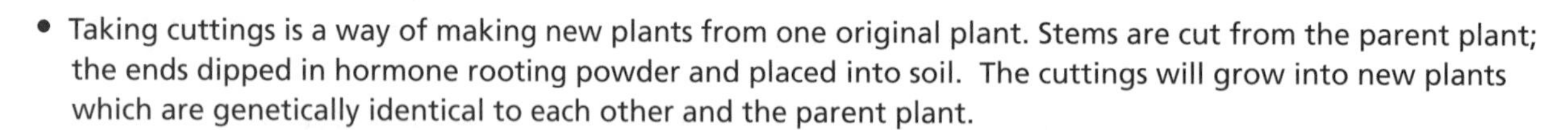

- Taking cuttings is a way of making new plants from one original plant. Stems are cut from the parent plant; the ends dipped in hormone rooting powder and placed into soil. The cuttings will grow into new plants which are genetically identical to each other and the parent plant.

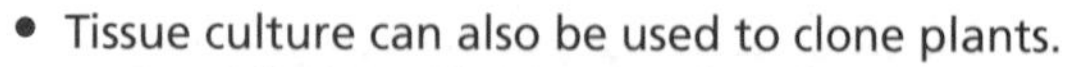

- Tissue culture can also be used to clone plants.
  - A small piece of tissue is taken from a root, stem or leaf of the parent plant. The tissue is then grown on a jelly containing all the nutrients it needs.
  - Everything has to be kept sterile, so this is usually done in a laboratory.
  - Eventually, each tiny group of cells grows into a complete adult plant.

- One technique to clone animals is called embryo transplants. This is sometimes done with farm animals, such as cows.
  - Egg cells are taken from a cow and fertilised with sperm from a bull.
  - One embryo is chosen and split into two (or more) and then each is placed into a host mother.
  - The calves born are clones of each other as they have the same genes.

Improve your grade

Tina has a lavender plant in her garden. She wants to produce more identical lavender plants quickly and cheaply. Outline the method she should use to do this. **AO2 (3 marks)**

Genetic engineering

An example

- Genetic engineering means taking a gene from one organism, and putting it into another.
- Bacteria have been genetically engineered to make human insulin.
- Insulin is a protein which stops blood glucose levels rising too high. If your body does not make enough insulin, you have diabetes. A person with diabetes may need to inject themselves with insulin several times each day.

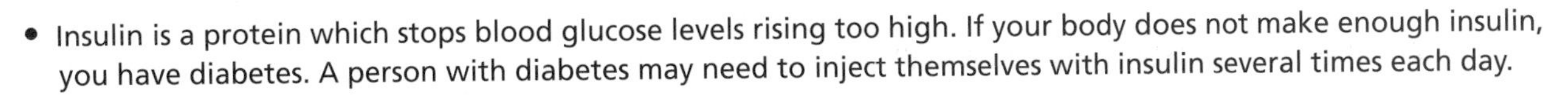

How it is done

- Figure 1 shows how bacteria can be genetically engineered to make human insulin.
- Farmers spray bean fields with herbicides to kill weeds that compete with soya plants. The spray contains a chemical called glyphosate.
- Some soya bean varieties have been genetically engineered to give them a gene that makes them resistant to glyphosate.
- So when a farmer sprays the field with glyphosate, the weeds die but the bean plants do not.

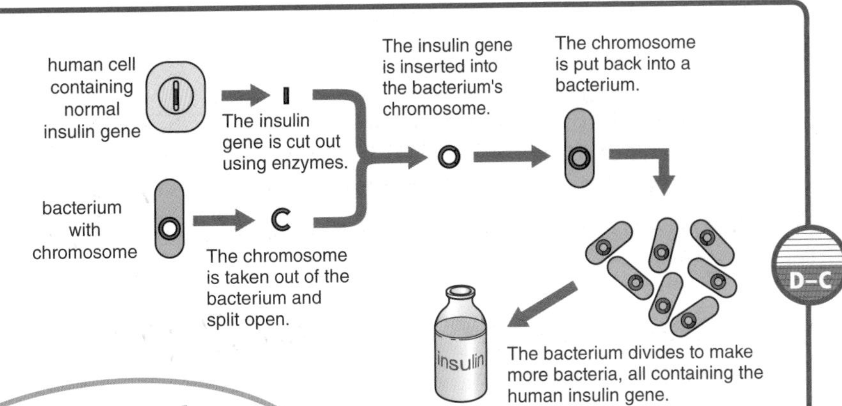

Figure 1: Bacteria have been genetically modified to produce human insulin

Remember!
Crops that have had their genes modified in this way are called genetically modified crops (GM crops).

Evolution

What is evolution?

- Evolution is how living things change over time. It is going on all of the time.
- Life on Earth almost certainly began as very small, single-celled organisms. Since then, over millions and millions of years, new life-forms have gradually evolved from these original ones.
- Throughout history, people have had different ideas about how evolution happens.

Accepting Darwin's ideas

- Jean-Baptiste Lamarck suggested that changes in organisms caused by their environment were passed on to their offspring. We now know that this is not correct.
- Charles Darwin suggested that species gradually changed from one form to another by natural selection. Darwin thought that, in each generation, only the best-adapted individuals survive and reproduce to pass on their characteristics to the next generation.
- Darwin's ideas challenged the established thinking of the day, so his ideas were not accepted at first. They undermined the idea that God made all animals and plants.

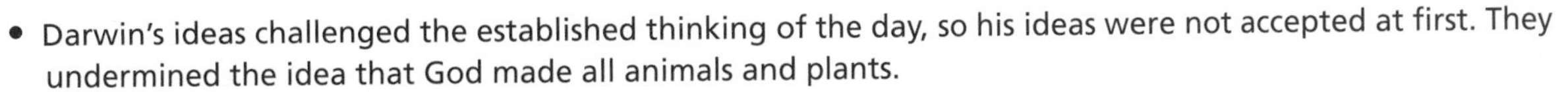

- In the late 19th century, there was not much scientific evidence to support the theories of evolution and natural selection. At that time, no one even knew that genes existed, let alone the way that they are inherited. This was not discovered until 50 years later. However, this is the theory that almost all scientists today believe as there is now a lot of scientific evidence to support it.

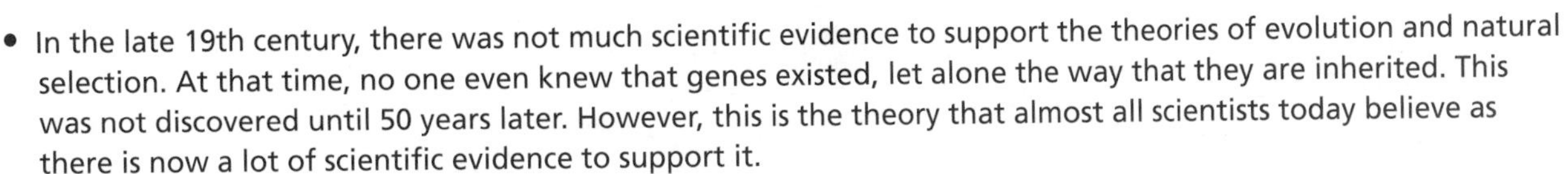

Improve your grade

In 1859, Charles Darwin published a book containing his ideas about natural selection and evolution. Explain why many people at the time did not believe what it said. **AO2 (3 marks)**

Natural selection

Artificial selection

- People breed animals to produce varieties that they want.
- For example, a farmer may choose only to breed from cows that produce a lot of milk. This is called 'artificial selection'.
- Choosing the feature that gets passed along to offspring also occurs naturally. This is called 'natural selection'.

How natural selection works

- This is how natural selection happens.
  - Living organisms produce many offspring.
  - The offspring vary from one another, because they have differences in their genes.
  - Some of them have genes that give them a better chance of survival. They are most likely to reproduce.
  - Their genes will be passed on to their offspring.
- Occasionally, unpredictable changes to chromosomes and genes, called mutations, happen.
- Occasionally, the new form of the gene increases an organism's chances of surviving and reproducing. It is therefore very likely to be passed on to the next generation. Over time, the new feature, produced by this gene, becomes more common in the species.
- The change in colour of the peppered moth from pale to dark is an example of evolution occurring because of a mutation.

EXAM TIP

You may be asked to explain how certain species evolved. Do this by applying the stages of natural selection: variation, competition, survival and reproduction.

Evidence for evolution

How did life start?

- Scientists still do not know how, when or where life on Earth began.
- Evidence, from fossils, shows that there was life on Earth at least 3.5 billion years ago.
- At that time, conditions on Earth were very different from today.

Comparing living organisms

- You can get clues about evolution by looking carefully at organisms that are alive today.
- For example, your arm, a bat's wing and a bird's wing all have the same bones in the same places. Similarities like this suggest that humans, bats and birds are quite closely related and that, long ago, an animal lived from which humans, bats and birds have all evolved (a common ancestor).
- Evolutionary trees like this one (Figure 1) show the pathway along which different kinds of organisms may have evolved.
- Organisms that lived longest ago are at the bottom of the tree.
- Models like this help to show how different groups of organisms might be related, which helps scientists to classify them.

Remember!

There are five main classification groups: bacteria, protoctists, fungi, plants and animals.

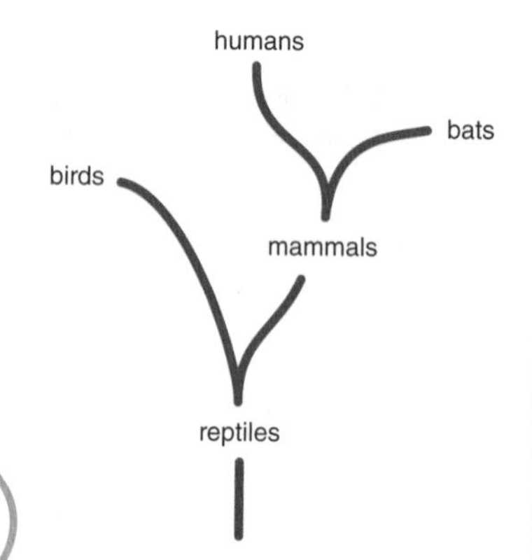

Figure 1: An evolutionary tree showing how humans, bats and birds are related

Improve your grade

The cheetah is the fastest land animal on Earth. It uses its speed to catch its prey. Use natural selection to explain how cheetahs have got faster over time. **AO2 (5 marks)**

Keeping healthy

To lose mass you need to use more energy than you take in. You can do this by eating less and exercising more. **Metabolic rate** is the speed at which energy is used by the body to carry out chemical reactions.

White blood cells attack **pathogens** via **phagocytosis**, **antibodies** and **antitoxins**. Vaccination encourages white blood cells to make antibodies against a pathogen. If the person is later infected, the antibodies will be made much more quickly. We say they are immune.

High blood **cholesterol** can lead to heart disease. It encourages the build-up of **plaque** which can lead to the blockage of arteries. Blood cholesterol level can be influenced by diet and inherited factors.

Antibiotics are drugs which kill bacteria. Some bacteria are becoming **resistant** to anti-biotics. An example is MRSA.

You can grow cultures of bacteria and fungi in liquids or jellies called a **nutrient medium**. It is important that everything is **kept sterile**.

Nerves, hormones and drugs

Sensory neurones carry electrical impulses to the **CNS** from **receptors**. **Motor neurones** carry impulses from the CNS to **effectors** (muscles and glands) where they have an effect.

Hormones control the **menstrual cycle**:

- FSH causes an egg to mature and the ovary to **secrete oestrogen**
- oestrogen causes the uterus lining to thicken and stops the production of FSH
- LH causes ovulation.

Conditions surrounding cells must be **kept constant**. These include water, sugar and salt concentration, and temperature. **Sweating** is one mechanism that the body uses to reduce body temperature when it rises above 37 °C.

Plant hormones called **auxins** cause their shoots to grow towards the light (positive **phototropism**) and roots to grow towards gravity (positive **gravitropism**). Artificial hormones can be used as rooting powder and **herbicide**.

Recreational drugs can be legal or illegal. Many drugs are **addictive** and can be dangerous if abused. Medicinal drugs have to be trialled before they can be prescribed by doctors.

Interdependence and adaptation

Organisms have characteristics that enable them to live in their environment. This is called **adaptation**. The better the adaptation, the more likely it is that the organism will win the **competition** for resources and survive to reproduce.

Energy is wasted at each stage in a **food chain**, which results in less biomass at each trophic **level**. This is represented as a **pyramid of biomass**.

Decay by microorganisms is an important way of recycling nutrients. The **carbon cycle** shows how carbon is moved around the planet.

Pollution levels can be monitored by measuring factors such as temperature and pH; or by studying the distribution of living pollution indicators, such as lichens to show air pollution and invertebrates to show water pollution.

Genes and evolution

Organisms vary because of the **genes** they inherit from their parents. **Asexual reproduction** results in clones. The offspring from **sexual reproduction** are different from their parents.

Genetic engineering means taking a gene from one organism, and putting it into another. This is used to make human insulin from bacteria and to create **GM crops**.

Clones can be created artificially by taking cuttings and carrying out **tissue culture**, **embryo transplants** and **adult cell cloning.**

Scientists believe that all life on earth **evolved** from single-celled organisms that lived 3.5 billion years ago.

Darwin's ideas about **evolution** explained how living things evolved due to variation and survival of the best adapted. Many people did not believe his ideas until 50 years after they were published.

Atoms, elements and compounds

Chemical substances, symbols and formulae

- All substances are made from tiny particles called atoms.
- Atoms of one element are different from the atoms of other elements.
- Compounds are substances that have atoms of two or more different elements joined together.
- Each element has a symbol. For example, C stands for carbon and O for oxygen. Some symbols have two letters, for example Co is cobalt, CO is carbon monoxide.
- A formula shows the types and proportions of atoms in a compound.
- Water (H_2O) has two atoms of hydrogen and one of oxygen.

Composition and structure

- Mixtures can be separated into simpler substances with fewer parts.
- A compound can be broken down into simpler compounds or its elements.
- Elements are substances that cannot be broken down by chemical reactions into simpler substances.

Inside the atom

What is inside an atom?

- At the centre of every atom is a nucleus. It contains protons and neutrons.
- The nucleus is surrounded by electrons that move around it.

Sub-atomic particles

- Sub-atomic particles have both a charge and a mass. You can work out the mass of an atom by adding together the numbers of protons and neutrons.
- Since an atom has equal numbers of protons and electrons, the charges cancel out.
- Neutrons have no charge and are neutral.
- The sum of the number of protons and neutrons in an atom is its mass number.

How are the electrons arranged?

- The electronic configuration of an atom gives the number of its electrons and the arrangement of the shells.
- You can show electrons as dots or crosses.
- Each shell can hold a limited number of electrons.
- The first shell (lowest energy level) can hold up to 2 electrons.
- The second shell (next energy level) can hold up to 8 electrons.
- The third shell (third energy level) can hold up to 18 electrons, but fills up with only eight before the fourth shell is started.

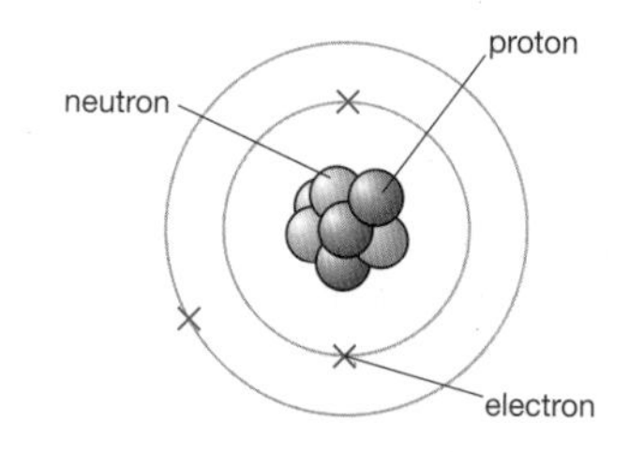

particle	relative mass	charge
proton	1	+
neutron	1	neutral
electron	almost 0	–

Figure 1: The structure of a lithium atom and sub-atomic particle chart

Remember!

In an atom the number of protons and electrons is always identical so the charges cancel out.

Improve your grade

Draw the electronic structures of the following atoms with proton numbers 3, 9, 11, 16, and 20.
AO2 (5 marks)

Element patterns

Patterns in the elements

- Mendeleev arranged the elements using their physical and chemical properties.
- In the periodic table (see page 143), elements with similar properties, such as lithium, sodium and potassium, are in vertical columns – called groups.
- Metals are on the left, non-metals on the right. The red line divides them.
- The block in the middle contains the transition metals.

Reactive and unreactive

- The periodic table (see page 143) lists all the known elements by increasing proton number. It has horizontal rows called Periods, and vertical columns called Groups.
- The Group number tells you how many outer electrons the atom has.
- The number of outer electrons in the atom determines how an element reacts.

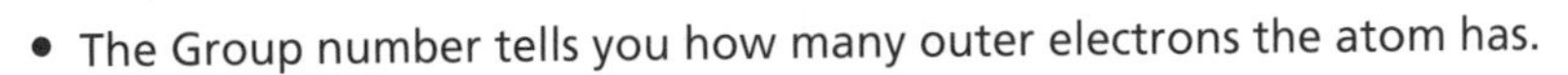

Combining atoms

Two types of bonds

Most elements never exist as single atoms. Instead, two or more atoms join together. A chemical bond holds them together.

- Hydrogen, oxygen and nitrogen molecules, H_2, O_2, N_2, are made from two atoms.
- The atoms are held together by sharing a pair of electrons – one from each atom. This is a covalent bond.
- When non-metals combine with metals, electrons are not shared. Instead some electrons transfer from the metal atom to the non-metal atom. Both atoms are left with an electric charge. Charged atoms are called ions.

G–E

Eight in a shell

- Atoms with fewer than eight outer electrons react in ways that give them a stable group of eight in their outer shell.
- In a water molecule (H_2O), an oxygen atom needs two more electrons to have eight in its outer shell. A hydrogen atom needs one more to become stable (two electrons in the first shell). Oxygen forms two covalent bonds, one with each of two hydrogen atoms. This way all the atoms achieve a noble gas configuration.
- To form an ionic bond in sodium chloride the chlorine atom gains an electron to get an outer shell of eight. It is now a negatively charged chloride ion, Cl^-. The electron comes from the sodium atom, leaving the second shell of the sodium atom as its outer shell. This has eight electrons, so it is stable. The opposite charges attract.

Remember!
Atoms with fewer than eight outer electrons react in ways that give them a stable group of eight in their outer shell. They may share electrons (covalent bonding), or transfer electrons (ionic bonding).

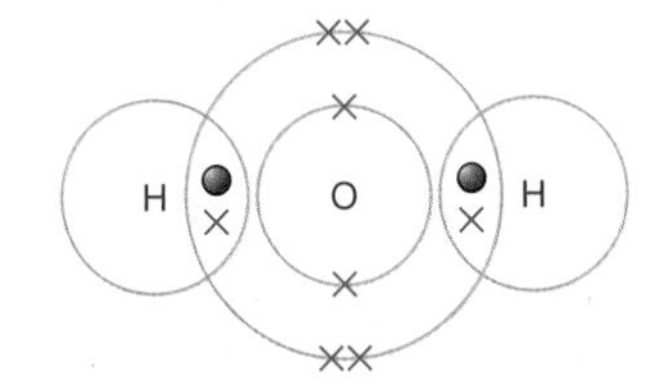

Figure 2: Covalent bonding in water molecules

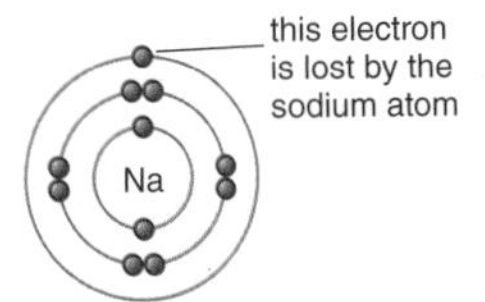

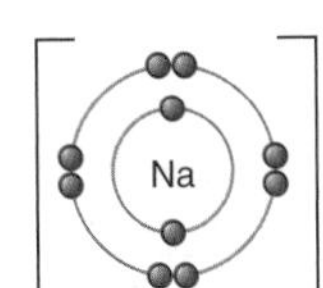

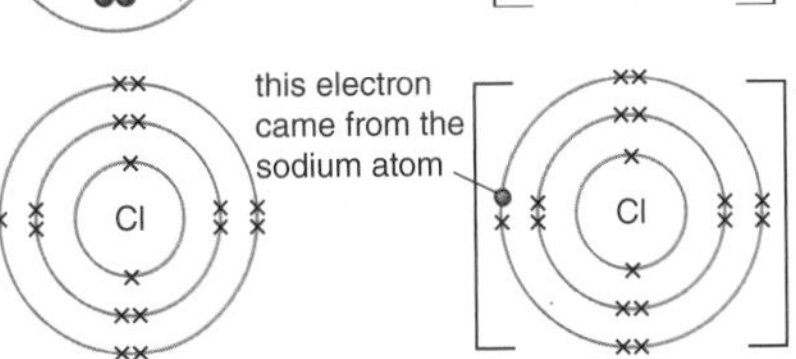

Figure 3: Ionic bonding in sodium chloride

D–C

Improve your grade

Explain why oxygen has a molecule with a double covalent bond and fluorine has only a single covalent bond. You should refer to the number of outer electrons in both atoms in your answer. **AO2 (4 marks)**

Chemical equations

Equations

- All elements have been given symbols to make it easy to understand what happens in reactions.
- A formula shows the elements present in a molecule of a substance.
- A number after a symbol shows how many atoms of that element are in the molecule.
- A word or symbol equation is a quick way to show what has reacted with what (reactants), and what is made (products).

Tracking atoms and molecules

- When methane burns, the atoms in methane (CH_4) and oxygen (O_2) react and turn into carbon dioxide (CO_2) and water (H_2O).
- Figure 1 shows the molecules, but doesn't account for all the reactant atoms as products. Figure 2 shows what happens to all the atoms, it is balanced.
- This is the balanced symbol equation;
  $CH_4 + 2\,O_2 \rightarrow CO_2 + 2\,H_2O$

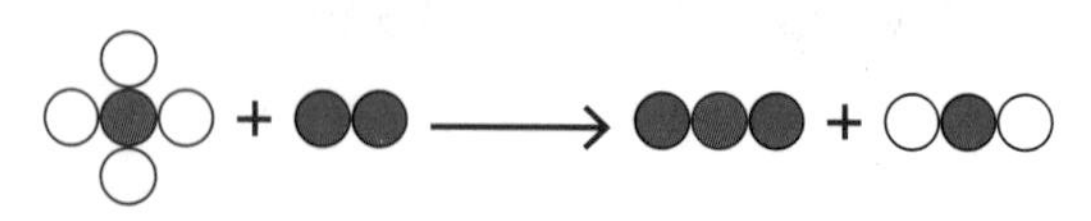

Figure 1: Methane burning with oxygen to form carbon dioxide and water.

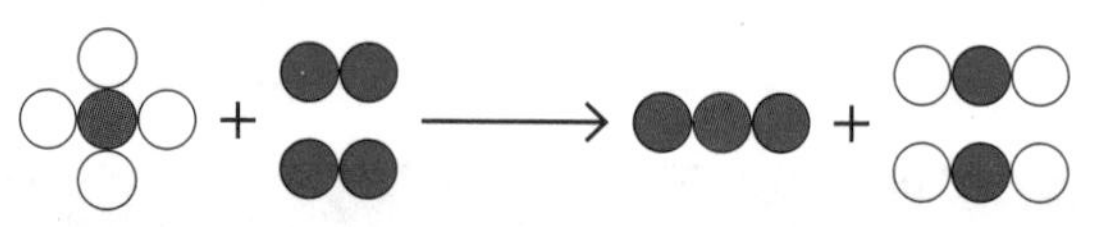

Figure 2: Methane burning with oxygen showing what happens to all the atoms

Building with limestone

Limestone

- Limestone started as the shells and skeletons of sea creatures and coral many millions of years ago, it is mainly calcium carbonate ($CaCO_3$).
- Calcium carbonate reacts with acids.
- Limestone is used for road building; making cement, mortar and concrete; and making iron and glass.
- Limestone buildings can be badly damaged by acid rain.

What are the effects of a limestone quarry?

- Limestone is obtained by quarrying. The table shows some advantages and disadvantages of quarrying limestone.

Advantage	Disadvantage
Jobs for local people in an area with little industry	Damage to the landscape. Loss of wildlife habitats
More, better-paid jobs, so more money to boost local economy	Noise and vibration from blasting, machinery and vehicles
Better healthcare and leisure facilities, as more people move into the area	Dust pollution in the environment
Better transport links, needed for lorries or railways	Traffic congestion and vibration from heavy lorries

Improve your grade

Explain the advantages and disadvantages of quarrying for a metal ore near your local town. Your answer should include two advantages and two disadvantages. **AO2 and 3 (4 marks)**

Heating limestone

Decomposing and slaking?

- Heating limestone strongly decomposes it (breaks it down) to two simpler compounds:

$$\text{calcium carbonate} \rightarrow \text{calcium oxide} + \text{carbon dioxide}$$
$$CaCO_3(s) \rightarrow CaO(s) + CO_2(g)$$

- Other metal carbonates also decompose in this way. It is called thermal decomposition.
- When water is added to calcium oxide it reacts to make calcium hydroxide, slaked lime ($Ca(OH)_2$). Farmers and gardeners spread it on soil that is too acidic.

Cement

- Calcium hydroxide solution (limewater) is used to test for carbon dioxide. If carbon dioxide is bubbled through limewater, it turns cloudy or 'milky' as tiny solid particles of white calcium carbonate form.
- Cement is made by heating limestone and clay. Cement is used widely – on its own, and in mortar and concrete.

- Mortar is made by mixing cement, sand and water. Mortar binds bricks together in brick walls.

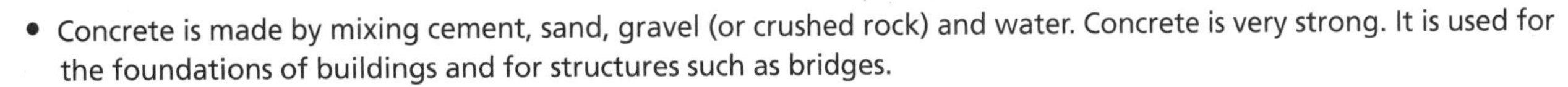

- Concrete is made by mixing cement, sand, gravel (or crushed rock) and water. Concrete is very strong. It is used for the foundations of buildings and for structures such as bridges.

Metals from ores

Metals in the ground

- Some metals occur naturally in the ground, for example gold and silver. They are unreactive.
- Rocks that are mined for metals are called **ores**.
- More reactive metals occur in rocks as compounds.

- Metals low in the reactivity series, such as iron and lead, can be extracted or smelted by heating with carbon.
- Very reactive metals, such as aluminium, sodium and magnesium, need electricity. Electricity is passed through a molten compound of the metal to obtain the metal.

Making use of ores

- Only minerals with enough metal to make it worth extracting are used as ores.
- Some ores are metal oxides. These can be smelted directly.
- At the smelter, the ore is crushed and concentrated, to remove rock with little or no metal.
- Other ores are converted to the metal oxide before or during smelting.
- To convert the metal oxide to the metal, the oxygen must be removed. This is called reduction.
- The metal oxide is reduced by heating it in a furnace with carbon if the metal is below carbon in the reactivity series. Originally, the carbon was charcoal, but now it is coke (a nearly pure form of carbon, from coal).
- Limestone is often added, to remove impurities in the ore forming slag.

Table 2: The reactivity series.

Element	Symbol
potassium	K
sodium	Na
calcium	Ca
magnesium	Mg
aluminium	Al
carbon	C
zinc	Zn
iron	Fe
lead	Pb
hydrogen	H
copper	Cu
silver	Ag
gold	Au

Increasing reactivity (upwards)

Improve your grade

Draw a table to show which of these metals is extracted by carbon and which by electrolysis. **AO2 (2 marks)**
iron, magnesium, aluminium, copper, zinc, lead, potassium, and calcium

Extracting iron

Iron

- Iron is the most widely used metal. We use more of it than all the other metals put together.
- Most iron is made into steel. It is the most important metal, because it is cheap and strong.
- Iron is extracted in a blast furnace from iron ore (Fe_2O_3), coke (nearly pure carbon), limestone, and hot air.

Inside a blast furnace

- **Iron ore** (Fe_2O_3), coke and limestone are fed in at the top of the furnace.
- Molten iron and **slag** collect at the bottom.
- As the hot air is blown through, the coke burns in oxygen to produce carbon monoxide:

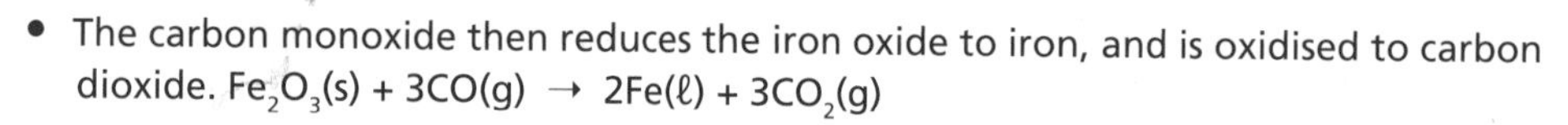

carbon + oxygen → carbon dioxide

$C(s) + O_2(g) \rightarrow CO_2(g)$

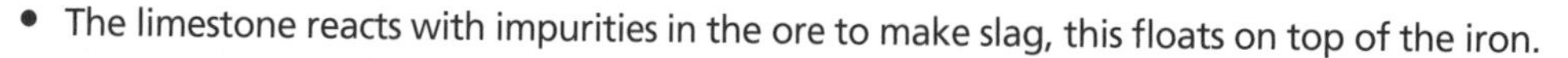

carbon dioxide + carbon → carbon monoxide

$CO_2(g) + C(s) \rightarrow 2CO(g)$

- The carbon monoxide then reduces the iron oxide to iron, and is oxidised to carbon dioxide. $Fe_2O_3(s) + 3CO(g) \rightarrow 2Fe(\ell) + 3CO_2(g)$
- The limestone reacts with impurities in the ore to make slag, this floats on top of the iron.

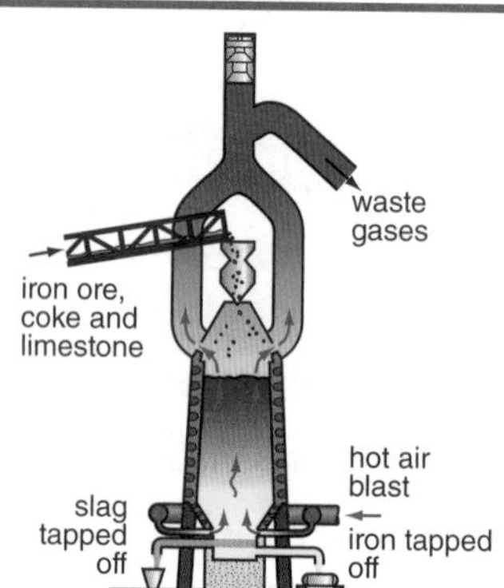

Figure 1: Blast furnace

Metals are useful

Important properties and uses of metals

- Metals are good conductors of heat and electricity; are easily bent into shapes, such as wires, tubes and car body panels; and can be given a shiny surface.
- Many metals we use are transition metals – for example, copper, iron, titanium, and chromium – and are found in the central part of the periodic table.
- Most transition metals are stronger than non-transition metals.
- Titanium is strong and lighter than steel but much more expensive.
- Copper is an excellent conductor of electricity and can be pulled into wires easily, making it ideal for electrical wiring.

Atoms and alloys

- Metal atoms are arranged in giant structures, in regular rows and layers.
- Metals can bend because these layers can slide over each other. The shape can change but the atoms remain bonded together.
- Metals conduct electricity because some of their outer electrons are free to move through the layers.
- To make metals harder they can be mixed to form alloys.
- An alloy is not a compound but is a mixture of elements, usually metals. Steel is an alloy of iron and carbon. Brass is an alloy of copper and zinc.
- The proportions of each element in an alloy can differ. This affects the alloy's properties and uses.
- The more carbon in steel the harder it is. Nine carat gold has more copper in it than 22 carat gold, and is harder wearing.

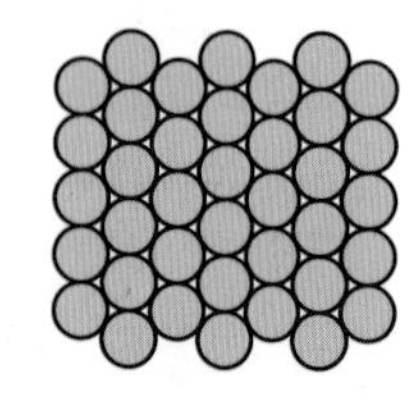

Figure 2: Layer of copper atoms

Improve your grade

Explain the difference between oxidation and reduction. **AO1 (2 marks)**

Iron and steel

Types of iron and steel

- Blast furnace iron is not very pure. It is about 96% iron and 4% carbon.
- Most iron is turned into steel, but not all.
- Some is used as cast iron.
- Steel is an alloy of iron and carbon (up to 2%). The blast furnace iron has most of the carbon removed, and small amounts of other metals are added to make designer steels. Stainless steel is a designer steel.

Remember!
Iron from a blast furnace contains 4% carbon, steel contains only 2% or less carbon. So steel is purer iron than iron!

Designer steels

- Pure iron is too soft for most uses, so steels are used instead.
- Adding other metals to molten steel can give it special properties. The choice of metal depends on how the steel will be used, and therefore the properties required.
- Stainless steel is about 70% iron, 20% chromium and 10% nickel. Stainless steel is very resistant to corrosion and does not rust.

D–C

Table 1 How other metals change steel's properties

Metal added to steel	Improvement to steel properties
Chromium and/or nickel	More corrosion resistance
Manganese	More strength and hardness
Molybdenum and/or tungsten	More strength, hardness and toughness
Vanadium	More strength, less brittle

EXAM TIP
When asked about the use of an alloy always think about the properties the alloy needs to have for the suggested use.

Copper

Why is copper important?

- Copper is a really useful metal, it conducts electricity and heat, and is not very reactive. These properties explain its everyday uses. It is extracted from chalcopyrite ore ($CuFeS_2$).
- Extracting the copper involves several stages.
  - Concentrating: separating copper minerals from unwanted rock
  - Roasting: heating the minerals in air
  - Smelting: heating in a furnace to make impure copper
  - Refining: purifying the impure copper using electrolysis.
  - Both roasting and smelting produce poisonous sulfur dioxide.

G–E

Physical and chemical changes during extraction

- Copper is purified using electrolysis.
- Electricity is passed through a copper sulfate solution. This is called electrolysis.
- Copper atoms in the impure positive electrodes lose electrons. They become copper ions (Cu^{2+}) and dissolve into the solution. The impurities fall to the bottom.
- At the pure copper negative electrodes, copper ions in the solution gain electrons and coat the electrode with copper atoms.

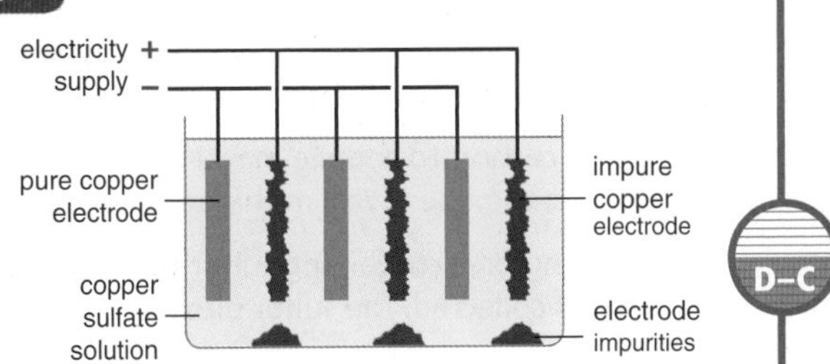

Figure 3: Refining copper

D–C

Improve your grade

Explain how copper is produced at the negative electrode when copper sulfate is electrolysed. **AO2 (2 marks)**

Aluminium and titanium

Properties and uses

Aluminium and titanium have the same properties as other metals, but:

- both are resistant to corrosion (rusting), but titanium has a higher melting point and is stronger than aluminium
- aluminium is used for aircraft bodies and wings, train carriages, window frames, food cans and foil trays, overhead electrical cables, and cooking pans
- titanium is used for military aircraft and missiles, rockets, bicycles, replacement joints such as hips, and laptop cases.

Extraction

- Aluminium is very expensive as it can only be extracted using electricity.
- The ore bauxite (Al_2O_3) is dissolved in cryolite to allow it to melt at 900 °C.
- The molten aluminium oxide then is electrolysed:
  aluminium oxide → aluminium + oxygen
  $2Al_2O_3 \rightarrow 4Al + 3O_2$
- Each aluminium ion needs three electrons to become an atom.
  This is why so much electricity is needed. $Al^{3+} + 3e^- \rightarrow Al$

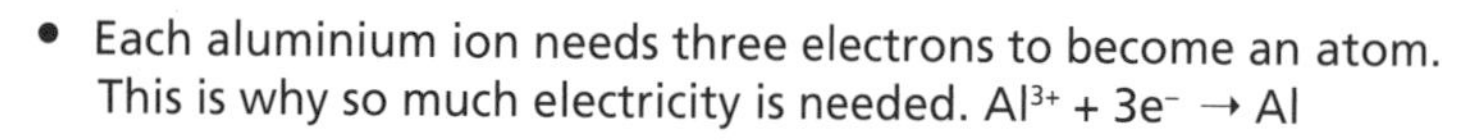

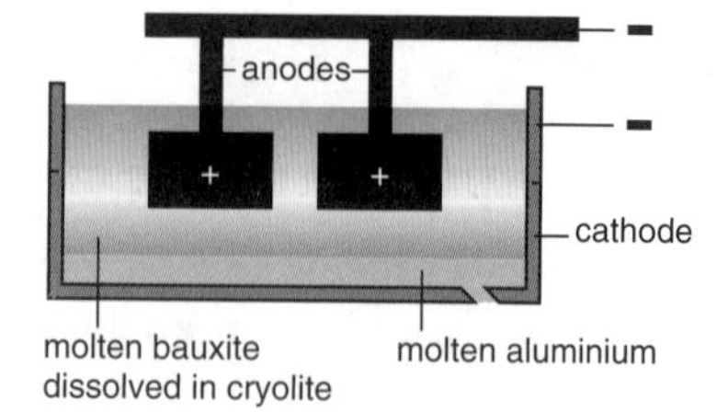

Figure 1: Extracting aluminium by electrolysis

- Titanium cannot be extracted by carbon or electricity. First the ore, rutile (titanium dioxide) is converted to titanium chloride. Then titanium chloride is reacted with magnesium.
- Titanium chloride + magnesium → titanium + magnesium chloride.
  $TiCl_4 + 2Mg \rightarrow Ti + 2MgCl_2$
  This method is very costly which means that titanium is also very expensive.

Metals and the environment

Metal extraction

Obtaining metals causes many environmental problems.

- Extracting metals from ores uses fossil fuels for energy and carbon (coke).
- Ores are often found in environmentally sensitive areas. Bauxite is mined in the Amazon Rainforest in Brazil.
- Opencast mining literally moves mountains. It turns them into holes in the ground, and creates a mountain of waste rock somewhere else.

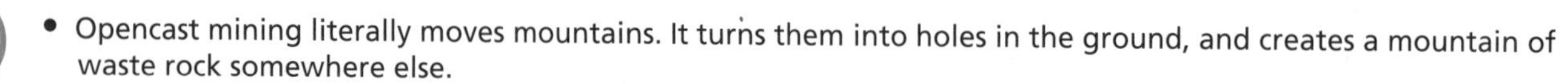

- Ores and fossil fuels are finite resources. There are limited supplies. Once used, they cannot be replaced.
- If metals can be recycled then there will be less need to mine ores.
- A new aluminium can takes twenty times more energy to make than a recycled one.

Effects on the environment

- Mining metal ores destroys the landscape, wildlife habitats, and displaces the local people, changing their way of life.
- Using carbon to reduce metal ores produces carbon dioxide. This adds to carbon dioxide levels in the atmosphere.
- Smelting ores containing sulfur produces sulfur dioxide which causes acid rain. If collected, the sulfur dioxide can be used to make useful sulfuric acid.
- Recycling aluminium and steel saves energy and the environment. Aluminium cans are cleaned and melted before reusing the metal. Iron and steel are added as scrap to steel-making furnaces.

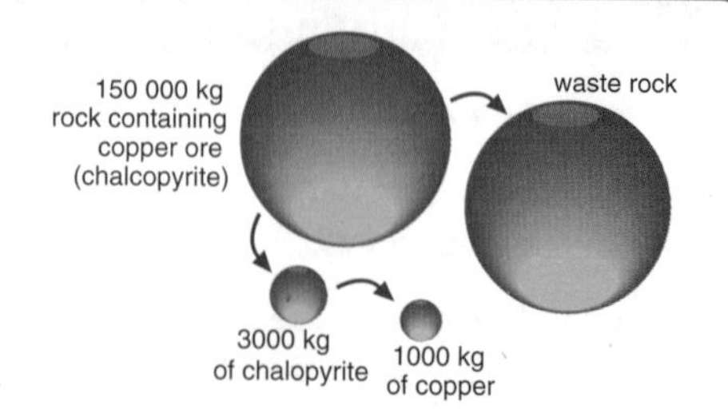

Figure 2: How much waste rock is mined, and later dumped, for each 1000 kg copper produced?

Improve your grade

Explain why titanium and aluminium metals are both reactive and corrosion free. **AO1 (2 marks)**

A burning problem

What happens when a fuel burns?

- Fossil fuels such as coal, oil and natural gas, and biofuels, such as wood and biodiesel, are compounds that contain mainly carbon and hydrogen (hydrocarbons).
- When they burn (combustion), the carbon and hydrogen react with oxygen to make carbon dioxide (CO_2) and water (H_2O) in the form of steam.
  hydrocarbon + oxygen → carbon dioxide + water
- Carbon dioxide contributes to global warming.
- If there is insufficient oxygen, poisonous carbon monoxide (CO) and carbon (as soot) are produced. The soot contributes to global dimming.
- Fossil fuels contain small quantities of sulfur compounds. These burn to form sulfur dioxide (SO_2), a poisonous gas. Sulfur dioxide causes breathing problems and acid rain.

G–E

Global effects

- Global warming. Increasing carbon dioxide in the atmosphere traps more energy from the Sun making the Earth warm up.
- Global dimming. Dust particles in the air prevent sunlight reaching the ground. This reduces the energy available for photosynthesis, and may cool the Earth as well.
- Both may affect the weather patterns on the Earth.
- Acid rain is caused by nitrogen, sulfur and carbon oxides dissolving in rain water. The acid rain formed attacks limestone buildings much more quickly than normal rain. It can also cause serious damage to trees and to aquatic life in affected lakes.

D–C

Reducing air pollution

What are the problems?

- Sulfur compounds can be removed from oil and gas. When burned, these 'cleaner' fuels produce little or no sulfur dioxide.
- Sulfur compounds are not removed from coal – it is too difficult. Instead, the sulfur dioxide produced is removed by limestone.
- Biofuels and hydrogen are sulfur free, so do not produce sulfur dioxide when burnt.

G–E

Alternative solutions

- Vehicles that burn fossil fuels produce poisonous carbon monoxide, and they also produce nitrogen oxides which can cause acid rain.
- To reduce the amount of these compounds released by vehicles all new cars have **catalytic converters** fitted.
- Carbon monoxide is oxidised to carbon dioxide, and nitrogen oxides are reduced to nitrogen.
- Alternative fuels are also being used, such as **biodiesel** made from vegetable oils, which contain almost no sulfur. Ethanol produced by fermentation from sugar, and hydrogen, are sulfur free.

Improve your grade

Suggest why growing plants to use as fuels may cause more problems than it solves. **AO2 (3 marks)**

Crude oil

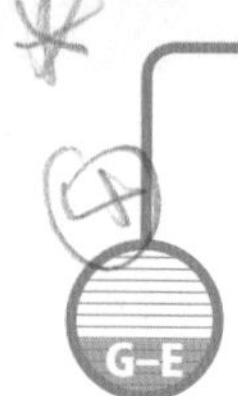

Crude oil fractions

G–E

- Crude oil is a mixture of many different compounds. Most are hydrocarbons – compounds containing only hydrogen and carbon.
- Crude oil is separated by a physical process, distillation.
- Each fraction is a mixture of different-sized hydrocarbon molecules with similar boiling points.
- Fractions with larger molecules boil at higher temperatures.

How does fractional distillation work?

D–C

- When crude oil is heated to about 400 °C, most of it boils and vaporises.
- Vapours consisting of hydrocarbons rise up the column, gradually cooling. When cooled below their boiling point, they condense back to liquid and are run off.
- Hydrocarbons with high boiling points condense first. The lower their boiling point, the higher up the column they rise before condensing.
- Hydrocarbons with different size molecules condense at different levels, separating the crude oil mixture into a series of fractions with similar numbers of carbon atoms and boiling points.

Name of fraction	Carbon atoms per molecule	Uses
petroleum gas	1 to 4	heating, cooking, LPG fuel
petrol	5 to 9	fuel (cars and lorries)
naphtha	6 to 10	to make other chemicals
kerosene	10 to 16	jet fuel, paraffin
diesel	14 to 20	fuel (cars, lorries and trains)
fuel oil	20 to 50	fuel for ships, factories and heating
bitumen	more than 50	tar for road making

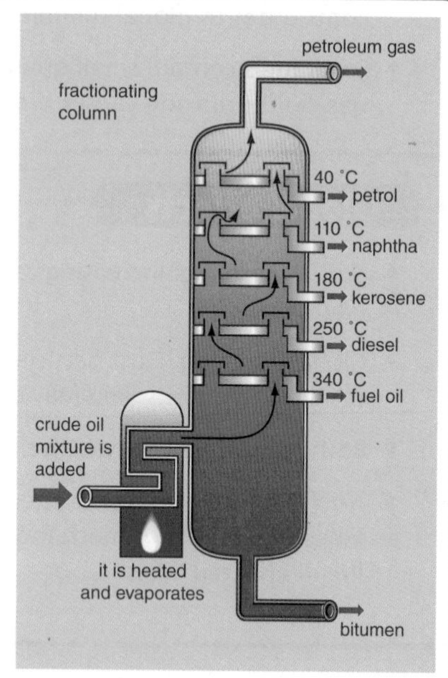

Figure 1: Fractional distillation of oil

Alkanes

What are alkanes?

- Methane belongs to a family of hydrocarbons called alkanes. Their names end in '–ane'
- Methane molecules are made up of one carbon atom joined to four hydrogen atoms. Methane's formula is CH_4.
- The next members of the family are ethane (C_2H_6), propane (C_3H_8) and butane (C_4H_{10}).

Patterns and properties

D–C

- The general formula for alkanes is C_nH_{2n+2} where n is the number of carbon atoms.
- All alkanes have similar chemical properties.

- As the number of carbon atoms increases;
  - molecules become larger and heavier
  - boiling point increases
  - flammability decreases (catch fire less easily)
  - viscosity increases (liquid becomes thicker).

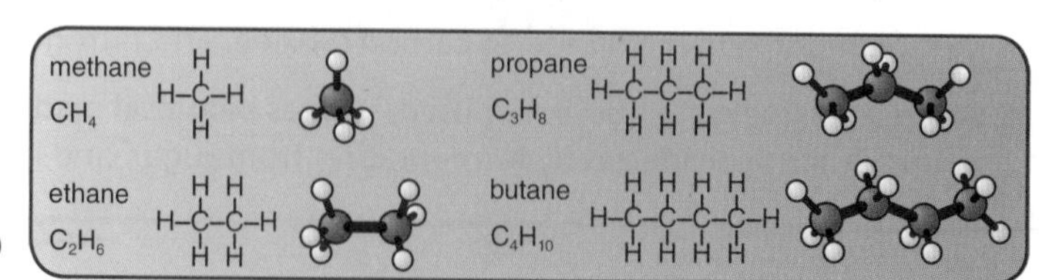

Figure 2: Alkane molecules

Improve your grade

Draw out the structure of the straight chained alkanes with 5 carbon atoms and 7 carbon atoms. **AO2 (2 marks)**

Cracking

What's the problem?

- Crude oils have different amounts of different hydrocarbons, but never enough petrol-sized molecules for our needs. There are too many of the larger hydrocarbons. Fewer people want these fractions, so oil companies break up the larger molecules into smaller ones. This is called cracking.
- Cracking breaks the long alkane chains into shorter lengths, such as petrol.
- A hot catalyst is used, which can be used again and again.
- Cracking also produces another type of hydrocarbon called alkenes.
- These can be made into plastics and other petrochemicals: long-chain alkane → shorter-chain alkanes + alkenes.

How cracking works

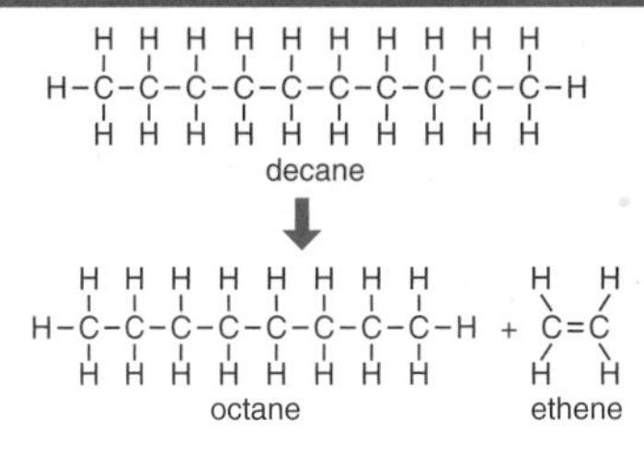

Figure 3: Cracking decane

- Crude oils have different amounts of different hydrocarbons, but never enough petrol sized molecules for our needs.
- Heating the hydrocarbons makes each molecule waggle until a **carbon–carbon** bond breaks. They break in different places, giving a mixture of products. So $C_{10}H_{22}$ could crack to form;
  $C_8H_{18} + C_2H_4$ or $C_7H_{16} + C_3H_6$ or $C_6H_{14} + C_4H_8$ and so on.
- One of the molecules made is an alkane, the other has two bonds (a double bond) shown as C=C between a pair of carbon atoms, and is called an alkene.
- Alkenes have the general formula C_nH_{2n}
- The double bond in alkenes makes them much more reactive than alkanes. This makes alkenes extremely useful.
- Ethene is a particularly important alkene product of cracking. It is the starting point for making polythene and many other plastics.

D–C

Alkenes

The alkene family

- Alkenes are another family of hydrocarbons. Their names start like the alkanes, but end in '-ene' instead of '-ane'.
- All alkenes have a carbon–carbon double bond, shown as C=C.
- In industry, ethene and propene are known as ethylene and propylene.

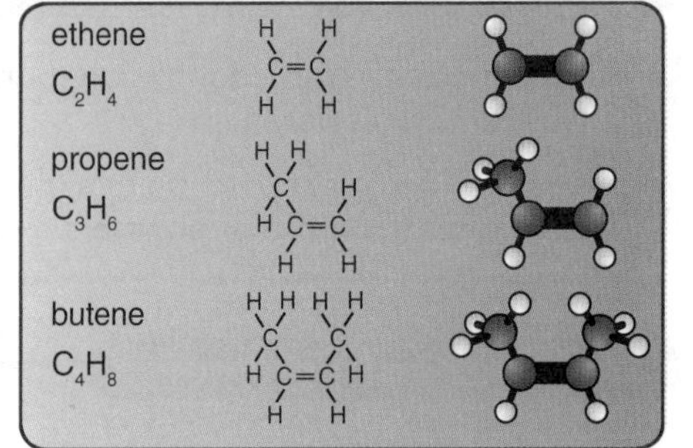

Figure 4: The first three members of the alkene family

Reactive alkenes

- A double bond is just two bonds.
- One of the two bonds can 'open up', allowing each carbon atom to form a bond with another atom. This means that alkenes are reactive compounds.
- Each carbon atom in an alkane already has bonds to four other atoms. So, unlike alkenes, alkanes cannot react by adding extra atoms. Alkanes are saturated – they cannot add any more atoms. Alkenes can, so alkenes are unsaturated.
- Fats are more complex than hydrocarbons, but we also refer to them as being saturated and unsaturated. Polyunsaturated fats have lots of double bonds.

Improve your grade

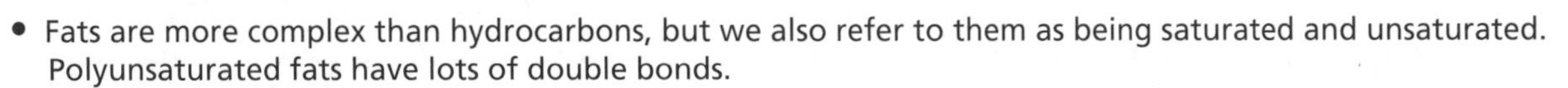
Draw structural diagrams to show how $C_{12}H_{26}$ can be converted to C_3H_6, and another molecule. State which of the two products is unsaturated, and why. **AO1 and 2 (3 marks)**

Making ethanol

Alcohol and ethanol

- Alcohol is the name of another chemical family.
- Alcohols are compounds with a hydroxyl group (–OH).
- Ethanol is a member of the alcohol family, ethanol has two carbon atoms in its molecule. Its formula is C_2H_5OH. It is written like that, rather than C_2H_6O, to show that it has the –OH group.
- Ethanol is made by two methods:
  - from crude oil. Cracking large alkanes produces ethene. Blowing ethene and steam over a hot catalyst makes ethanol.
  - from sugar. Adding yeast to sugar dissolved in water causes fermentation and changes sugar into ethanol.
- Both methods give a solution of ethanol in water. Ethanol is separated by fractional distillation.

Converting into ethanol

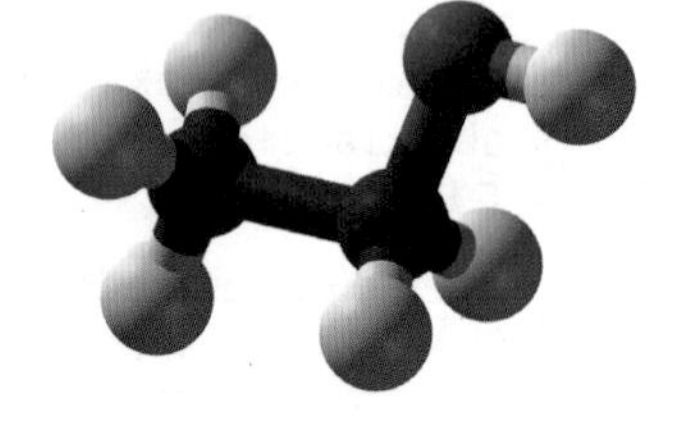

Figure 1: Molecular model of ethanol

- Fermentation happens when micro-organisms, called yeasts, feed on the sugar and convert it into ethanol.
- To manufacture ethanol for use as a fuel, a sugar solution made from sugar cane or maize (corn) is prepared. Yeast is added to cause fermentation: sugar → ethanol + carbon dioxide.

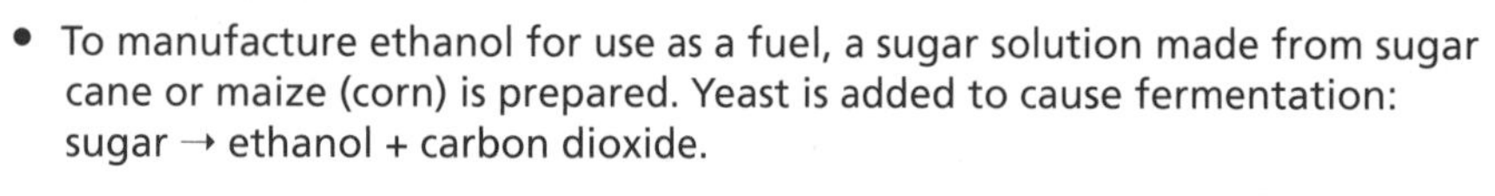

- Fuels obtained from animals and plants are called biofuels, so ethanol made by fermentation is a biofuel.
- As more plants can be grown after harvesting, biofuels are a renewable source of energy. Unfortunately it uses land and crops that could be used to feed people.

Polymers from alkenes

From crude oils to polymers

- Alkene molecules have double C=C bonds. These bonds enable alkene molecules to link together into long chains, known as polymers.
- Polymer molecules contain up to a million atoms joined together like beads in a necklace. Chemists nearly always refer to plastics as polymers.

A variety of polymers

- To make poly(ethene), ethene is heated under pressure. Ethene is known as the monomer. A catalyst sets off a chain reaction. It makes a C=C bond open up and join onto another ethene molecule. That double bond then opens up and joins onto the next, and so on, forming a polymer chain.
- The reaction involves only the C=C double bond. It does not matter what other atoms are attached to the carbons. By changing the other atoms attached to the carbons, it is possible to produce lots of other different polymers with different properties.

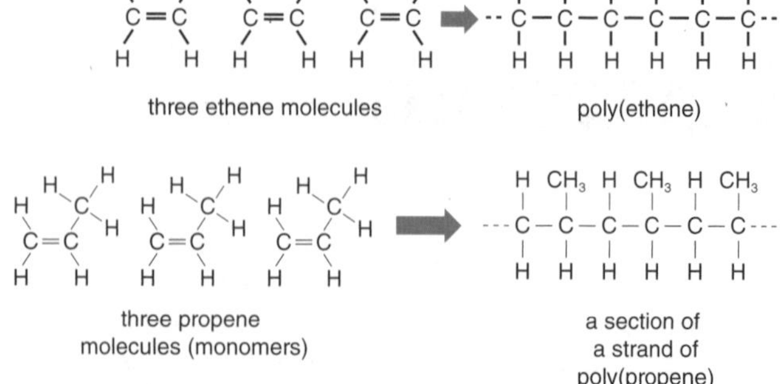

Figure 2: Monomer to polymer

Improve your grade

PVC (polyvinylchloride) is a commonly used polymer. Its monomer has the structure shown here. Show, using three monomer molecules, the structure and bonding of a PVC polymer chain. **AO2 (2 marks)**

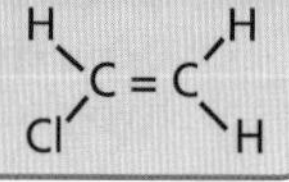

Designer polymers

Special polymers

- Plastic polymers are easily moulded into shape, low density (lightweight), waterproof and resistant to acids and alkalis.
- Properties such as strength, hardness and flexibility vary.
- Polymers can be designed to have the specific properties needed for a particular purpose.

D–C

Polymers for a purpose

- Teflon is a designer polymer. It is slippery and is used to make artificial hips as it allows the joint to move easily. It is also used for non-stick frying pans.
- A surfboard must be tough and light. Many are made of a polymer called polyurethane, produced in the form of rigid foam that has a low density so it floats.

Polymers and waste

A load of rubbish

- Most plastics are not biodegradable. Plastics put in rubbish bins mostly end up in landfill sites.
- Most plastics can be recycled. They can be melted down and used again.
- Recycling plastics means less litter, less buried in landfills and less oil used to make new plastics.

G–E

Figure 3: Recycling information

Sorting out waste

- Most disposable plastic items are labelled with a recycling symbol and code number to identify the polymer.
- Instead of dumping in landfills, household rubbish, including plastics, can be burned in incinerators. The heat produced may be used to generate electricity or heat local buildings. This is wasteful of polymers and can cause air pollution.

Improve your grade

Describe the difference between recycling polymer waste and re-using it by incineration. Give two reasons why recycling is more environmentally friendly than incineration. **AO1 (3 marks)**

Oils from plants

Vegetable oils

- Plants like sunflower, rape, maize, almond and olives are grown for their oils.
- They are crushed to extract the oil, or the oils are dissolved in a solvent.

Oils in food and fuel

- Plant oils store a lot of energy. Oils provide more energy than most other foods.
- Oils from rapeseed, soya beans and other crops are converted into biodiesel fuel.
- Cooking food in oils produces different flavours and textures; the food is cooked faster at a higher temperature than in water. The food absorbs some oil increasing its energy content.
- Oils also contain other nutrients we need.
  - Essential fatty acids (e.g. omega 3 and 6), for the heart, muscles and nervous system to function properly.
  - Vitamins. Seed and nut oils are particularly rich in vitamin E.
  - Minerals. Minerals are compounds of metals and non-metals such as potassium, calcium, iron and phosphorus.
  - Trace elements, but only in tiny amounts. For instance, a single Brazil nut provides the whole recommended daily allowance (RDA) of selenium.

Biofuels

Green fuel

- Fuels produced from animals or plants are called biofuels. These include wood; plant oils; and ethanol made by fermenting sugar.
- Biofuels are greener than fossil fuels as:
  - they are renewable
  - the carbon dioxide produced on burning is the same as the carbon dioxide used in photosynthesis to make them
  - they contain little sulfur so they do not contribute to acid rain.

Remember!

Burning biofuels produces carbon dioxide just like fossil fuels. When burnt the fuel releases the same amount of carbon dioxide as the plants took in when growing, so the level of carbon dioxide in the air isn't increased.

Biofuel issues

Biofuels are made from animal or plant material, for example wood.

- Biofuels contain no sulfur and this advantage means that they do not cause sulfur dioxide pollution.
- Biofuels may be renewable, but they have problems of their own.

  Practical issues:
  - plant oils are very viscous so are hard to use directly in car engines
  - they need more air to burn properly than diesel and petrol
  - few garages sell biofuels as not many people currently use them.
  - Economic and ethical issues;
  - biofuels produce less energy per litre, so a greater volume is needed
  - growing crops needs energy for the machinery, fertilisers and transport, this may be more than is produced by the biofuel crop
  - land used to grow food crops may be used to grow biofuels instead, leading to food shortages and raised food prices.

  Environmental issues:
  - the demand for fuel is so great that huge amounts of land would be needed
  - changing land use can affect the plants and animals that live there, reducing biodiversity.

EXAM TIP

If asked to evaluate the impact of biofuels on the environment or another issue. Give both benefits and problems in your answer to get full marks.

Improve your grade

Explain why biofuels are considered to be carbon neutral. **AO1 (2 marks)**

Oils and fats

A closer look at oils and fats

- Oils and fats are very similar chemicals. Oils are liquid, fats are solid.
- There are two types, saturated and unsaturated. Unsaturated fats and oils have at least one C=C double bond. Saturated fats only have single C–C bonds.
- Animal fats are solid and are saturated fats. Vegetable fats and oils are liquids and are unsaturated.

Unsaturated fats

- Testing an unsaturated fat is similar to testing for an alkene. They both contain double C=C bonds.
- The amount of bromine water decolorised by the fat indicates how many C=C bonds there are in the fat.
- Everyone needs some fats in their diet, for energy and essential nutrients, but too much fat is unhealthy.
- Saturated fats can increase the level of cholesterol in the blood. You should;
  - eat foods rich in polyunsaturates, such as sunflower oil, and monounsaturates, such as olive oil
  - avoid saturated fats, such as lard, limit the amount of fat of all types in your food,
  - eat fish, some fish oils contain omega-3 fatty acids. These have been shown to lower blood pressure.

Emulsions

What is an emulsion?

- Oil and water do not mix together. The oil floats on top.
- An emulsifier helps the oil and water to mix. Food manufacturers use them a lot to make mayonnaise or salad cream, which are emulsions.
- Milk and butter are emulsions. Milk contains fat droplets suspended in water. Butter has water droplets suspended in fat.
- Many cosmetics are also emulsions.

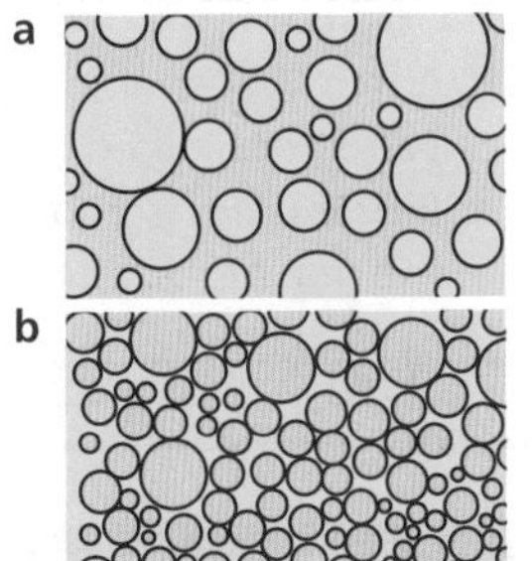

Figure 1: a Milk consists of yellow fat droplets spread out in water.

b Butter has water particles spread through the yellow fat.

Useful emulsions

- Oil and water are immiscible. However they can be made to mix by stirring or shaking, forming an **emulsion**.
- Emulsions are less viscous (less runny) than the oil and more viscous than the aqueous solution.
- Oil-in-water emulsions contain droplets of oil suspended in an aqueous solution.

Improve your grade

Hand creams are oil-and-water mixtures. Explain why adding an emulsifier to the cream helps you spread the cream evenly on your hands. **AO2 (2 marks)**

Earth

Planet Earth

- Earth is made up of layers. Earth's surface, called the crust, is rock, below the crust is the mantle (molten rock), and in the middle is the core (a very hot mixture of iron and nickel).
- Earth has a very thin layer of air round it – the atmosphere, much of the crust is covered by water – the seas.
- People once thought that the centre of Earth is very hot because of heat left over from when Earth first formed, and that the uneven surface was formed by the crust shrinking when Earth cooled down. Neither of these are true.

What is Earth like inside?

- The crust is a thin outer layer of cold, solid rock. Its thickness is between 5 and 30 kilometres.
- The mantle is made of very hot molten rock that flows very slowly by convection currents.
- The core is at the centre; an inner core, in the centre, is solid iron and nickel and an outer core is a molten mixture of iron and nickel.

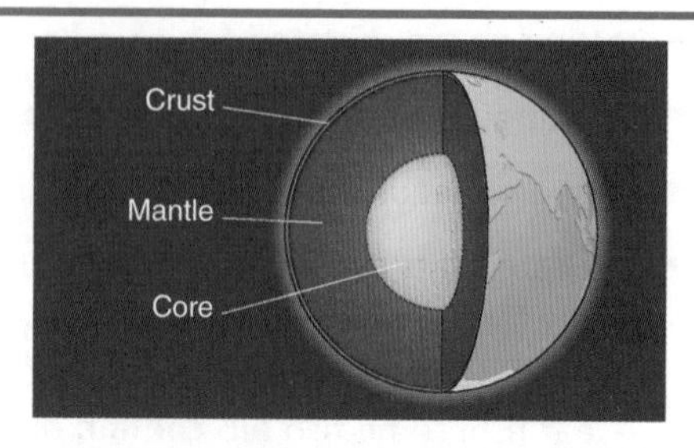

Figure 1: The layers of the Earth

Continents on the move

Pangea, the super continent

- Alfred Wegener in 1915 suggested that all the continents were once joined together in a super continent he called Pangea. He had noticed that South America could fit into the side of Africa.
- Wegener suggested that the continents had drifted apart.

Remember!

Scientists once believed that the mountains and valleys of the Earth had formed because the Earth had been hotter and bigger. As it cooled down they thought it shrivelled up, causing wrinkles which formed the mountains and valleys.

Continental drift

- When Wegener suggested his hypothesis, other scientists rejected it because no one could explain how huge continents could move. Later other scientists found evidence to support Wegener's theory.
- Earth's crust and the semi-solid upper part of the mantle make up the lithosphere.
- The tectonic plates float on the liquid rock of the mantle. Heat from radioactive processes within the Earth drives convection currents in the mantle. These currents carry the floating plates, and the continents very slowly, only a few centimetres a year, in a process called continental drift.

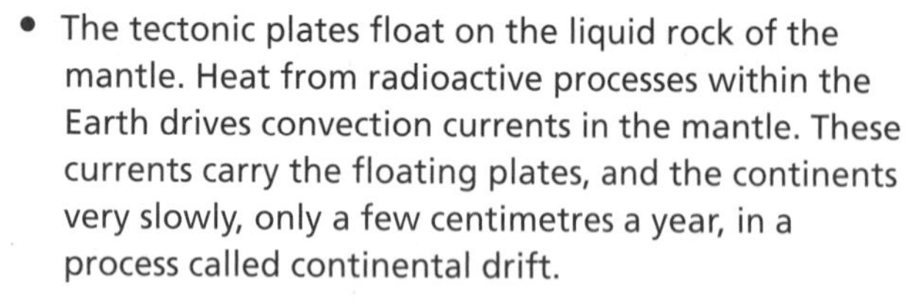

- When two tectonic plates the size of continents collide they push the land upwards, forming mountains. The world's largest mountain ranges are formed where continental plates collide.

200 million years ago 100 million years ago 50 million years ago

Figure 2: Pangea breaking up

How Science Works

- You should be able to explain why Wegener's theory of Continental Drift was not accepted for many years, and give two pieces of evidence in support of it.

Improve your grade

Use the theory of tectonic plates to explain how mountains are formed. **AO1 (2 marks)**

Earthquakes and volcanoes

Earthquakes

- Most earthquakes occur along plate boundaries. Plates grind past one another or collide. Where two plates meet, pressure gradually increases. Suddenly, one plate slips over the other and all the stored energy is released. This causes an earthquake.
- If a big earthquake happens under the sea, the sudden movement causes a huge wave. The wave spreads out in all directions. This is a tsunami.

Volcanoes

- Volcanoes often occur at plate boundaries.
- Magma stays sealed under the crust, often for hundreds of years. Eventually the pressure builds up enough for magma to burst through a vent – a crack or weak spot in the crust. The blast creates a crater, lava and gas pour out, and the familiar cone shape of volcanoes forms.
- Scientists can monitor the movement of tectonic plates. They can tell when pressure is building and an earthquake is possible. However, they cannot obtain hard data about the forces involved, the friction between plates and structural weaknesses in plates. Eruptions are not the same as earthquakes, but the problems of predicting them are: the lack of sufficient valid data.

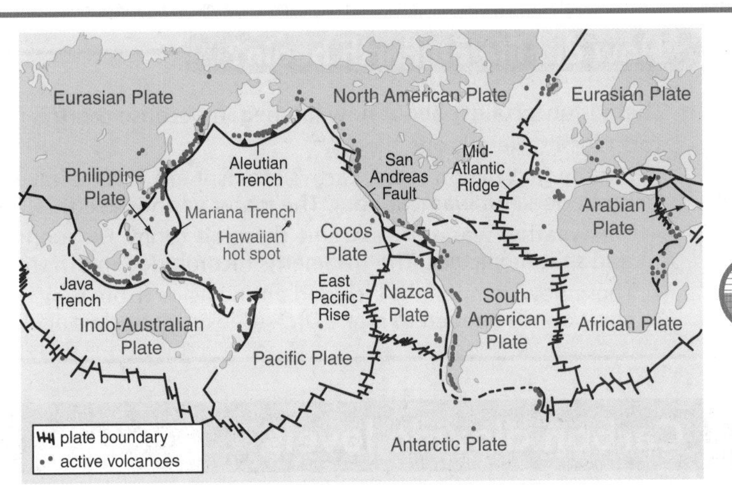

Figure 3: Main tectonic plate boundaries and active volcanoes

D–C

The air we breathe

What's in air?

- Air is mainly oxygen (20%) and nitrogen (80%), with small amounts of other gases. The amounts of the other gases vary according to place, time and day.

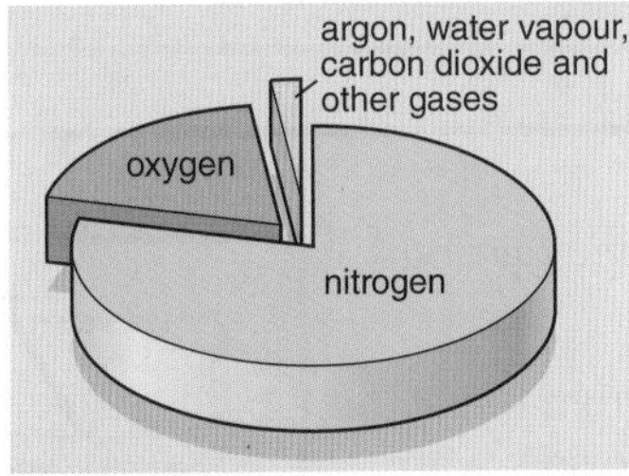

Figure 4: Gases in the air

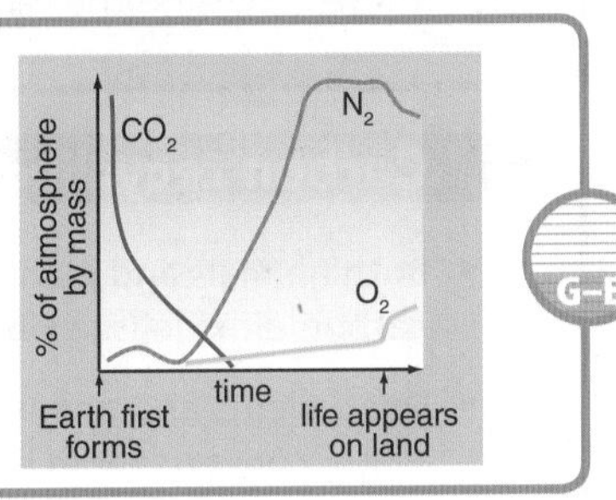

Figure 5: Percentage of nitrogen, oxygen and carbon dioxide in the air since the Earth formed

G–E

How did the atmosphere evolve?

- The Earth was formed about 4 600 million years ago. For the first 1 000 million years there was intense volcanic activity, releasing huge amounts of steam, carbon dioxide, and some ammonia (NH_3) and methane (CH_4).
- The steam condensed and eventually formed the seas and oceans.
- About 3 400 million years ago simple life that could photosynthesise had developed in the oceans and seas. These algae used the carbon dioxide and water and released oxygen. The oxygen reacted with the ammonia to make nitrogen gas.
- About 400 million years ago the atmosphere had enough oxygen to allow land plants and then animals to evolve.
- The atmosphere has stayed the same for about the last 200 million years with 78% nitrogen, 21% oxygen, and small amounts of other gases such as argon and carbon dioxide.

Improve your grade

Sketch a timeline showing how the proportions of water vapour, carbon dioxide, oxygen, and nitrogen have changed since the Earth was formed. **AO2 (4 marks)**

The atmosphere and life

Right for life

- There are many theories about how life developed on Earth, but nobody really knows. We do know that all living things need the right conditions: oxygen, liquid water, and temperatures between 0 and 50 °C. Only Earth has all these conditions.
- Plants developed first, carrying out photosynthesis to make oxygen. They also remove carbon dioxide from the air. This is how the oxygen entered the atmosphere millions of years ago. It is how the concentration of oxygen in the air nowadays stays constant.

How did life on Earth begin?

- There is uncertainty about how life began because there is no evidence. The first primitive life-forms did not form fossils. Here is one theory.
  - For the first billion years, Earth's atmosphere was mainly carbon dioxide, with some methane, ammonia, hydrogen and water vapour. The water vapour eventually condensed to form oceans.
  - The weather was more extreme than it is today. Frequent lightning provided energy to break chemical bonds and split molecules. The fragments recombined in different ways, forming new compounds.
  - These new compounds included amino acids (from which all proteins are built up), sugars and other carbon compounds needed to make DNA (deoxyribonucleic acid). These compounds are the basis of life.

Carbon dioxide levels

Carbon dioxide in the atmosphere

- Carbon dioxide made up most of Earth's early atmosphere. Now it is only about 0.04%.
- It was removed by photosynthesis; dissolving in the oceans; and by marine organisms to make their shells and skeletons. The shells and skeletons have formed limestone and chalk rocks.
- The carbon gas has also been trapped in fossil fuels. Over the last two hundred years people have burnt these, releasing the carbon dioxide back into the air. It is suggested that this has led to global warming.

Carbon recycling

- Both animals and plants take in oxygen and give out carbon dioxide (see Figure 1).
- Dead plants and animals decay. Oxygen from the air or water converts them back into carbon dioxide, with the help of bacteria, fungi, and other organisms.
- Most of the carbon dioxide in the early atmosphere became locked up as fossil fuels or in rocks such as limestone made from animal shells.
- Carbon dioxide is also absorbed by the oceans, removing it from the atmosphere.

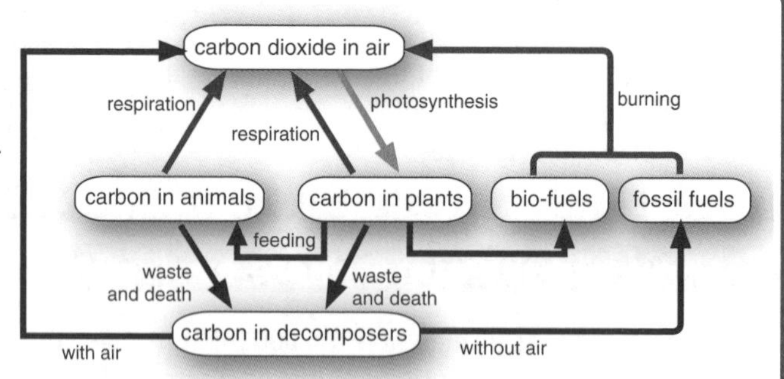

Figure 1: The carbon cycle

How Science Works

- You should be able to explain or evaluate how a particular human activity may affect the atmosphere, for example, increasing the numbers of electric cars will lead to a reduction in carbon dioxide released as less fossil fuel will be burnt.

Improve your grade

Describe how the carbon dioxide in the early atmosphere has been reduced to its current level. **AO1 (2 marks)**

C1 Summary

Fundamental ideas

All substances are made from atoms, which have protons and neutrons in a central nucleus, with electrons in shells around the nucleus.

Electrons occupy specific energy levels or shells but always the lowest available level. Atoms with full outer electron shells (Group 0) are stable.

The atomic number = the number of protons in the nucleus. It is the same as the number of electrons in the atom. The mass number of an element = the number of protons + neutrons.

An element has atoms of only one type. The periodic table is a list of all known elements. It is arranged in horizontal rows, periods; and vertical columns, groups.

Metals with non-metal compounds have ions held together by ionic bonds. Non-metal compounds consist of molecules with covalent bonds.

Limestone and building materials

Limestone is calcium carbonate. When heated, it is thermally decomposed to calcium oxide, an alkali.

Carbonates react with acids to produce carbon dioxide gas, a salt, and water. Limestone is damaged by acid rain.

Limestone is quarried and then used to make cement, mortar, concrete and glass. The quarrying has economic, social and environmental implications.

Metals and their uses

Metals are extracted from rocks known as ores, which contain a large amount of the metal's compounds.

Metals that are more reactive than carbon are extracted by electrolysis. Metals that are less reactive than carbon are extracted by heating them with carbon.

Transition metals have many uses based on their properties. Many pure metals are improved by mixing with others to form alloys.

Crude oil and fuels

Crude oil is made up of hydrocarbon molecules. They can be separated into fractions by fractional distillation using boiling points.

Burning a hydrocarbon fuel produces carbon dioxide and water. It may also produce pollutants.

Most of crude oil's compounds are alkanes (C_nH_{2n+2}). The bonds between the atoms are covalent bonds.

Biofuels, such as biodiesel and ethanol, are produced from plant materials. There are economic, ethical and environmental issues with their production and use.

Other useful substances from crude oil

Hydrocarbons can be broken down (cracked) to produce smaller, more useful molecules.

Alkenes can be used to make polymers such as poly(ethene) and poly(propene).

Alkenes are unsaturated molecules with double bonds represented as C=C.

Ethanol can be produced by reacting ethene with steam or by fermentation.

Many polymers are not broken down by microbes posing waste disposal problems.

Plant oils and their uses

Vegetable oils that are unsaturated contain double carbon–carbon bonds.

Oils do not dissolve in water. They can be used to produce emulsions.

Vegetable oils are important foods and fuels as they provide a lot of energy. They also provide us with nutrients.

Emulsifiers help oil and water molecules to mix.

Changes in the Earth and its atmosphere

The Earth consists of a core, mantle and crust, and is surrounded by the atmosphere.

The Earth's atmosphere has changed over time. It is now 78% nitrogen, 21% oxygen with traces of other gases.

The Earth's crust is cracked into a number of tectonic plates. Convection currents within the mantle cause the plates to move.

Earthquakes and volcanic eruptions occur at the boundaries between tectonic plates.

Energy

What is energy?

- All objects have energy. We can see where energy is stored and what happens when it is transferred (moved).
- For example, a candle stores chemical energy in its wax. When the candle burns, the energy is transferred to the surroundings as light and infrared radiation (heat).

Remember!
Energy doesn't change, but is stored and transferred in different ways.

Storing and transferring energy

D–C

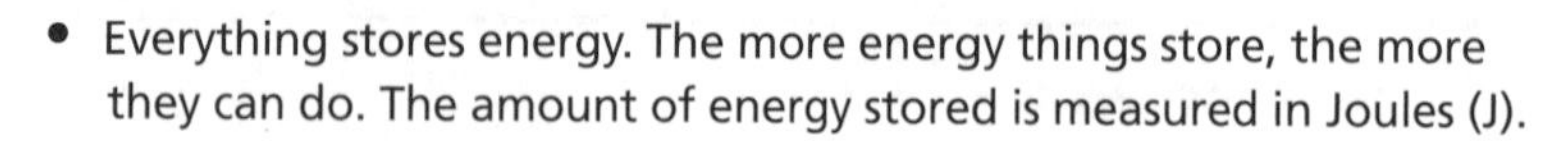

- Everything stores energy. The more energy things store, the more they can do. The amount of energy stored is measured in Joules (J).

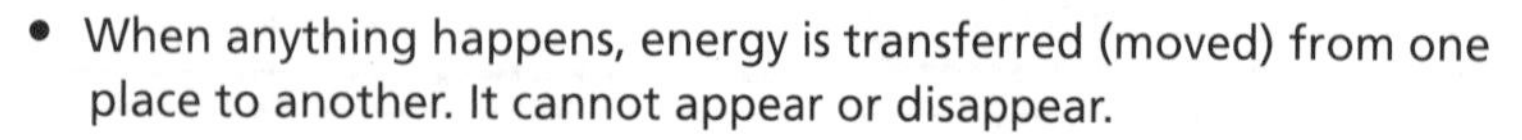

- When anything happens, energy is transferred (moved) from one place to another. It cannot appear or disappear.

Figure 1: Examples of energy transfers

Infrared radiation

Infrared radiation

- All objects emit and absorb infrared radiation. If the objects are the same temperature, black surfaces emit (give out) radiation more quickly than white surfaces. This means black surfaces lose heat and cool down more quickly.

Emission absorption and uses

- All objects emit and absorb infrared radiation.
- Shiny light coloured surfaces absorb infrared radiation slower than black matt surfaces.
- Shiny surfaces reflect more infrared radiation than dark surfaces.
- Hotter surfaces emit radiation faster than cooler surfaces.
- We can design objects to reduce the rate of energy transfer.

Remember!
Objects are absorbing and emitting infrared radiation at the same time. If infrared radiation is emitted faster than it is absorbed, then the object cools down.

How Science Works

- Black surfaces inside an oven emit infrared radiation better than shiny surfaces. This helps heat to transfer to the food cooking inside the oven.

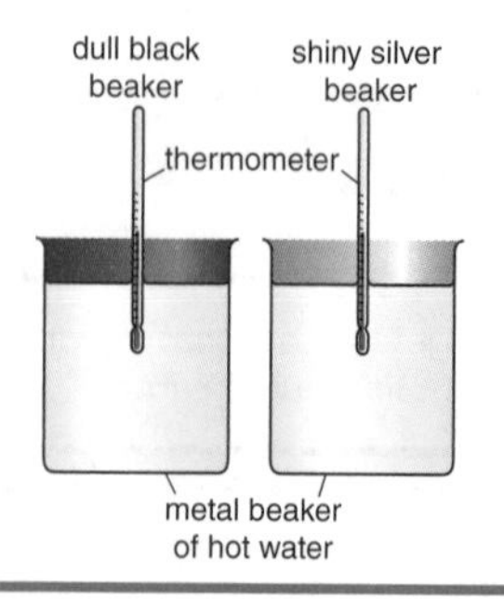

Figure 2: Comparing the emission and absorption of infrared radiation by different surfaces

Improve your grade

Explain whether a kettle of hot water cools down quicker if its outer surface is coloured white or dark green.
AO2 (3 marks)

Kinetic theory

Particle energy

- The three states of matter are solids, liquids and gases. They contain particles arranged differently. Particles gain energy as they change state.

State of matter	Properties	Arrangement of particles
solid	> fixed shape > fixed volume	> vibrating around fixed positions > almost touching
liquid	> takes the shape of container > fixed volume	> moving around > very close to one another
gas	> spreads out > fills the space available	> moving rapidly > moving chaotically

Figure 4: Table of states of matter

increasing energy of particles →

in a solid   in a liquid   in a gas

Figure 3: The energy of particles in each state of matter

Remember!
The particles do not change as they change states, but they behave differently.

Bonds between particles

- Bonds between particles are strongest in solids and weakest in gases.
- Melting is when a solid is heated and changes to a liquid. Particles vibrate more. This breaks bonds between them. The particles can change places.
- Freezing is when a liquid cools and changes to a solid.
- Boiling is when a liquid is heated and changes to a gas. The particles break their bonds. They can move around freely.
- Condensing is when a gas cools and changes to a liquid.

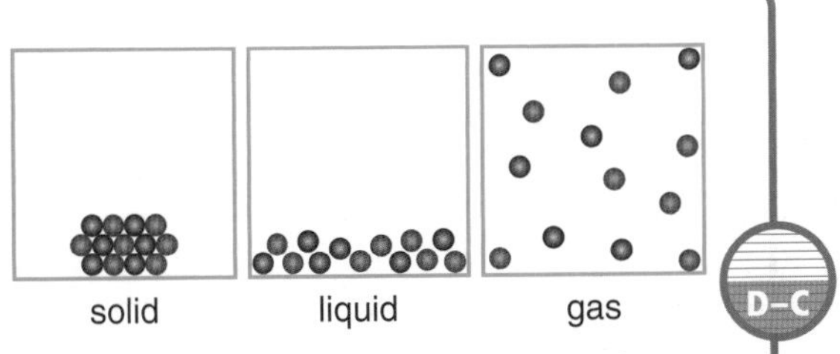

Figure 5: Arrangement of particles in each state of matter

Conduction and convection

Transferring by heating

- Conduction transfers heat energy through solids. Good heat conductors, such as metals, transfer heat well. Insulators, like plastic and cloth, do not transfer heat energy well.
- Convection transfers heat energy in liquids and gases. Warm water or air rises above cooler water or air. It is less dense.

Transferring the energy

- If one end of a solid is heated, particles vibrate more and shake their neighbours. Conduction transfers energy from one particle to another through the solid.
- If one part of a liquid or gas is warmed, the particles vibrate more taking up more space. The warm region of gas or liquid expands, becoming less dense and rising above cooler denser regions. This is convection.
- Convection currents spread heat through liquids or gases that are heated from the base, or cooled from the top.

Figure 6: The dye traces the convection current as the water warms

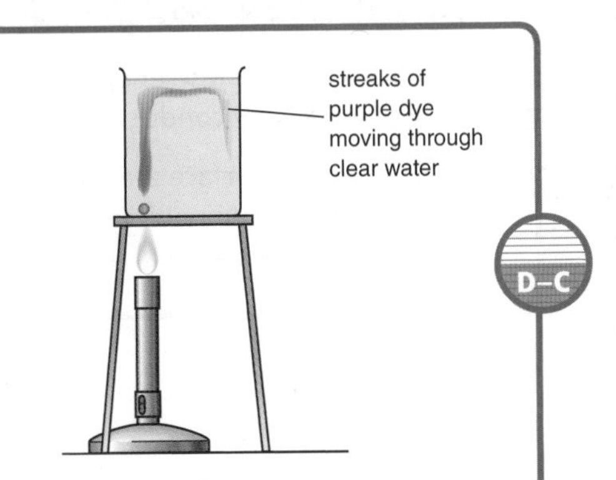

Improve your grade

Explain why convection can take place in a liquid but not in a solid. **AO2 (3 marks)**

Evaporation and condensation

Evaporation and condensation

- Evaporation is when a liquid changes to a gas at temperatures below the boiling point.
- Condensation is when a gas cools and changes back to a liquid.

Evaporation and environment

- Evaporation happens because particles in liquids have different energies. They are held together with bonds.
  - Particles with high energy break the bonds. These particles become particles in a gas.
  - The liquid loses energy and cools.
  - Its surroundings also cool down because the liquid is cooler and absorbs energy from its surroundings.
- Evaporation is quickest if:
  - it is warm (more particles have enough energy to break bonds linking them),
  - the liquid has a large surface area,
  - it is windy (so vapour above the liquid's surface does not become saturated).

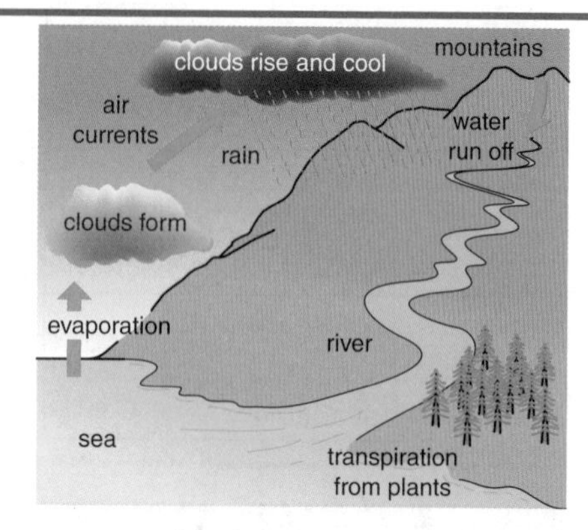

Figure 1: The water cycle

Remember!
Evaporation occurs at the surface at low temperatures, but boiling occurs throughout the liquid at boiling point.

Rate of energy transfer

Energy transfer

Objects transfer energy to and from the surroundings at their surface. Energy is transferred quicker if it:

- is made from a good conductor
- has a large surface area
- has a dull, dark colour
- is much hotter or cooler than its surroundings.

Conduction, infrared radiation and convection

- Hotter objects lose energy more quickly than cooler objects. When hot objects cool, energy is transferred to the surface by conduction and from the surface by convection or radiation.
- The ratio of surface area to volume affects the rate of heat loss. If the surface area is larger, energy is transferred more quickly from the object. Animals in hot countries, such as elephants, have large ears to cool down quicker. Animals in cold areas often grow large. This reduces their surface area compared to their volume.

Improve your grade

Explain why a towel dries quicker on a windy summer day. **AO2 (3 marks)**

Insulating buildings

Home insulation

- Buildings are insulated to stop them losing heat. Less energy is needed so less fossil fuels are burned, which protects the environment.
- Materials have different U-values. A good insulator has a low U-value because it is hard for energy to flow through it.
- Double glazing has a low U-value. The layer of gas between the panes of glass is a good insulator.
- Newer houses have two walls with a small gap (cavity) between them. Cavity walls can be filled with foam or insulating material.

How much can we save?

- Payback time is the time taken to save as much money as it costs to install an energy saving measure.

Method of insulation	Installation cost (£)	Annual saving (£)	Payback time (years)
loft insulation	240	60	4
cavity wall insulation	360	60	6
draught proofing doors and windows	45	15	3
double glazing	2400	30	80

Figure 4: Payback time for different energy saving measures

- Payback time in years = installation cost ÷ annual saving

Specific heat capacity

Absorbing energy

- When energy is absorbed, an object heats up. Some materials have a bigger temperature rise than others. Specific heat capacity is the energy needed to raise the temperature of 1 kg of a material by 1 °C.
- Water has a higher specific heat capacity than oil. You need less energy to heat a kilogram of oil compared to water.

Calculating absorbed energy

- Energy transferred to or from a material (J) = mass (kg) × specific heat capacity (J/Kg °C) × temperature change (°C).
- For example, the specific heat capacity of water is 4200 J/Kg°C. The energy absorbed by 250 g of water heated from 20 °C to 100 °C is 0.25 × 4200 × 80 = 84 000 J.

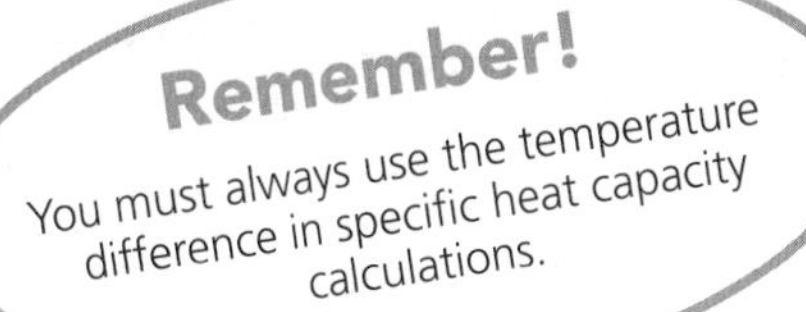

Improve your grade

The specific heat capacity of copper is 390 J/Kg °C. Explain whether copper heats up quicker than the same mass of water when they are put in a hot place. **AO2 (3 marks)**

Energy transfer and waste

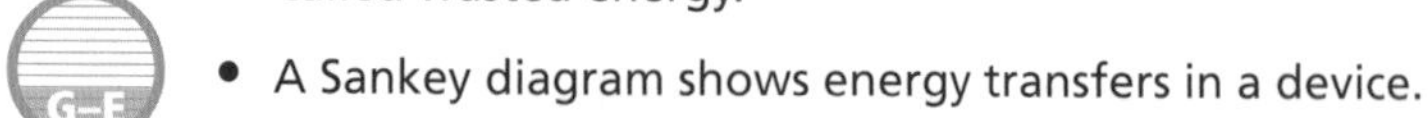
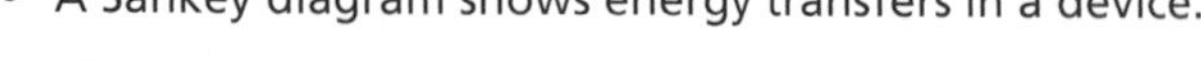

Sankey diagrams

- The Law of Conservation of Energy says energy cannot be created or destroyed when it is transferred. All energy is usefully transferred, dissipated or stored. The energy that spreads to the surroundings is called wasted energy.
- A Sankey diagram shows energy transfers in a device.
- The energy input is shown at the left side of the arrow.
- The arrow splits. Each section shows the output energy form.
- The width of each part of the arrow shows the proportion of energy it represents.

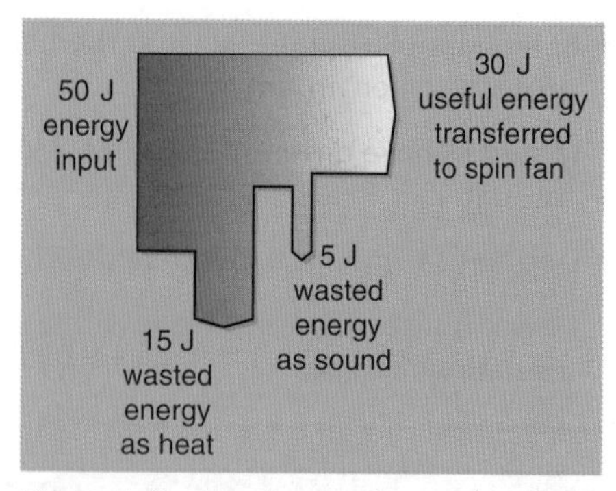

Figure 1: Sankey diagram for an electric fan

Wasted energy

D–C

- Use the Law of Conservation of Energy to work out the missing values on a Sankey diagram. The total energy output (useful energy and wasted energy) must equal the total energy input.

Remember!

All energy must be accounted for during a transfer. It may be stored, usefully transferred or dissipated to the surroundings.

Efficiency

Being efficient

- Efficient devices do not waste much energy.
- $\text{Efficiency} = \dfrac{\text{useful energy out (J)}}{\text{energy input (J)}}$ The answer is always a decimal less than 1.
- Power is the rate of energy transfer. Efficiency can be described as useful power out ÷ total power in.
- Efficient devices cost less to run. They use less energy to do the same work as a less efficient device.

Efficiency and wasted energy

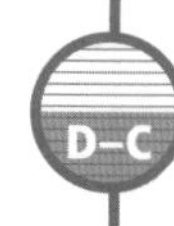

- Convert decimals to percentages by multiplying by 100. Efficiency is always less than 100%.
- You can use a Sankey diagram to calculate efficiency. Efficiency = useful energy ÷ energy input

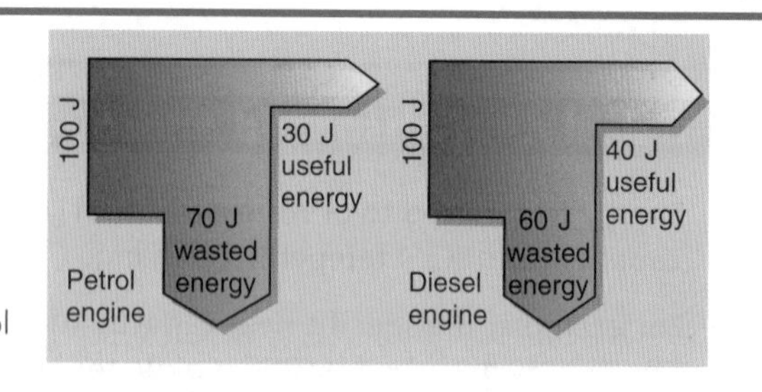

Figure 2: These diagrams show the petrol engine's efficiency is 30% and the diesel engine's efficiency is 40%

Improve your grade

Explain what this Sankey diagram shows in as much detail as possible.
AO2 (3 marks)

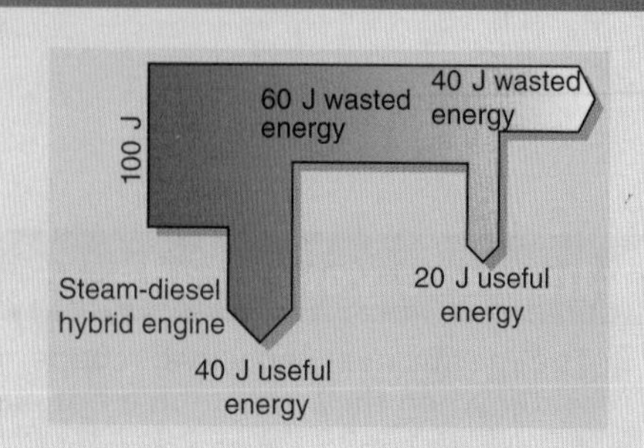

Figure 3

Electrical appliances

Using electricity

- Electrical appliances transfer the energy supplied by electricity into something useful.
- Batteries change chemical energy into electricity when the battery is part of a circuit.
- Batteries do not store as much energy as the mains can supply. Heaters need so much energy they rarely use batteries.

G–E

Alternatives to electricity

- Mains electricity is convenient, safe and pollution free at the point of use. However, generating electricity in power stations causes pollution.
- Alternatives to electricity include gas or biomass for cooking, lighting and transport.
- Some appliances work using solar power or energy stored in springs instead of batteries.

Energy and appliances

Appliance power

- Power is the rate that energy is transferred. It is also the rate that something does work. Power is measured in watts (W) or kilowatts (kW).
  For example, a 60 W light bulb transfers 60 J of energy every second.

How much energy is used?

- Calculate the energy used by an appliance in joules using:
  energy (J) = power (W) × time (s).
- More energy is used if the equipment is more powerful, or is turned on for longer.
- Use seconds in your calculations, for example a 100 W bulb uses 30 000 J in 5 minutes (300 seconds).

EXAM TIP

Check you know the right units for time, power and energy. You will lose marks if you write **j** instead of **J** for example, or if you leave them out.

Improve your grade

Two different bulbs are switched on for 10 minutes. Calculate the energy transferred by each one.

a a 60 W filament bulb

b a 10 W energy efficient bulb **(AO2 4 marks)**

The cost of electricity

Energy from electricity

- Electrical energy is not used up – the energy is transferred to electrical appliances by the current, and transferred in the appliances to useful output. Some appliances transfer energy more quickly than others.
- Transferred electrical energy is measured in kilowatt-hours (kWh), if power is measured in kilowatts and time is measured in hours.
  Energy (kWh) = power (kW) × time (hours)

How much does it cost?

- A kilowatt-hour is calculated using power (in kW) × time (hours).
  – For example, a 0.1 kW light bulb switched on for 15 hours uses 1.5 kWh (0.1 kW × 15 h).
- The cost of using electricity is the number of kilowatt-hours used × cost per kilowatt-hour.
  – For example, if each kilowatt hour costs 12p, the cost of using the light bulb was 18p (1.5 kWh × 12p/kWh).

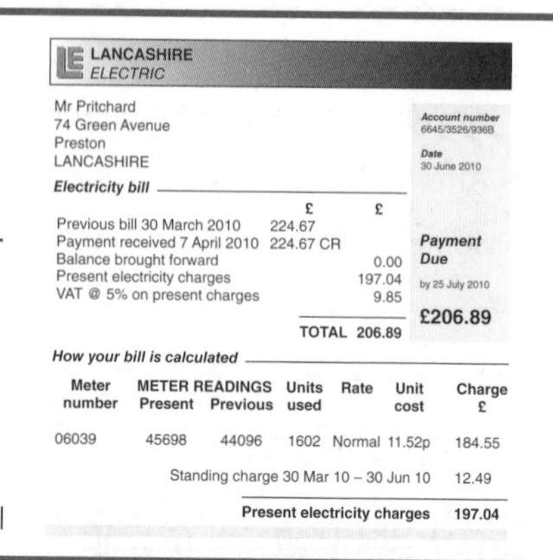

LANCASHIRE ELECTRIC

Mr Pritchard
74 Green Avenue
Preston
LANCASHIRE

Account number 6645/3526/936B
Date 30 June 2010

Electricity bill

	£	£
Previous bill 30 March 2010	224.67	
Payment received 7 April 2010	224.67 CR	
Balance brought forward		0.00
Present electricity charges		197.04
VAT @ 5% on present charges		9.85
	TOTAL	206.89

Payment Due by 25 July 2010 **£206.89**

How your bill is calculated

Meter number	METER READINGS Present	Previous	Units used	Rate	Unit cost	Charge £
06039	45698	44096	1602	Normal	11.52p	184.55
	Standing charge 30 Mar 10 – 30 Jun 10					12.49
	Present electricity charges					**197.04**

Figure 1: A typical electricity bill

Power stations

How does a power station work?

- Fossil fuels (coal, gas, oil) are non-renewable and will run out. They will last longer if we use them more slowly or if we find new supplies.
- In a power station, fossil fuels are burned, or nuclear fuels (uranium, plutonium) undergo fission. Water is heated, changing to steam. Steam drives turbines, which spin generators, generating electricity.

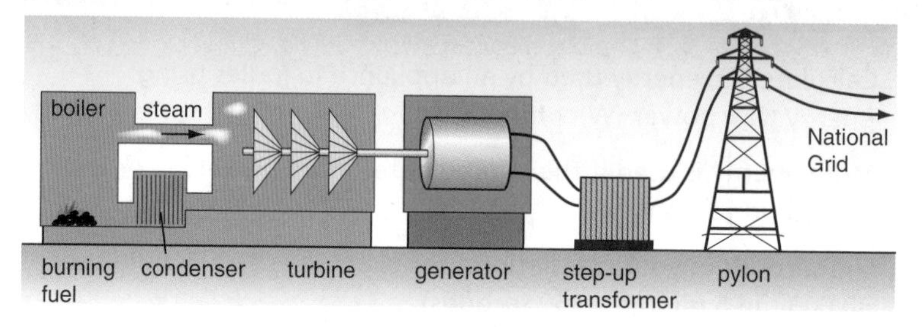

Figure 2: How a power station generates electricity

- A step-up transformer increases the voltage of the electricity. Very high voltages are needed to transmit electricity efficiently through the National Grid.

Energy changes

- Many power stations are about 35% efficient.
- In gas fired power stations, hot gases also drive turbines. Gas power stations are 60% efficient.
- Power stations using waste energy directly for heating are 70–80% efficient.

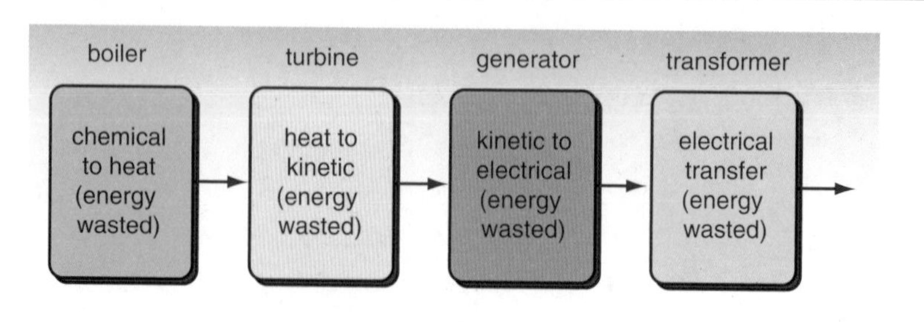

Figure 3: Energy changes taking place in a power station

Improve your grade

Describe the energy changes taking place in these parts of a coal fired power station: the burning fuel; the boiler; the turbine; the generator. **AO1 (4 marks)**

Renewable energy

Three renewable resources

Renewable energy resources will not run out. They include steam, wind and water.

- Steam drives turbines in biomass power stations. Biomass is organic waste that is burned.
- When the wind blows, it spins blades on a wind turbine. This turns a generator.
- Water can be trapped behind hydroelectric dams. When it falls through pipes, it makes a turbine spin.

Four more renewable energy sources

- These renewable resources spin turbines directly:
  - Tidal barrages are walls built across river mouths. When the tide goes in or out, water flows though pipes in the barrage driving turbines.
  - Waves drive turbines in small wave generators.
- Steam drives turbines in geothermal power stations:
  - Heat from rocks deep underground changes water to steam in geothermal power stations.
- Photovoltaic cells change sunlight to electricity directly. This is solar power.

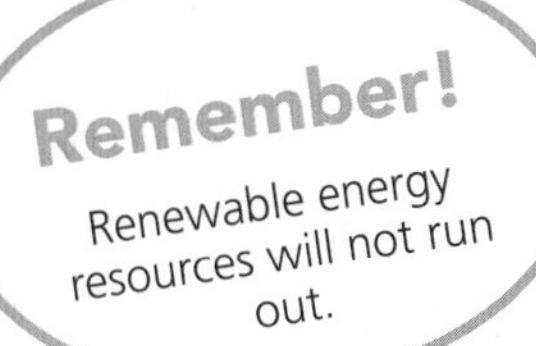

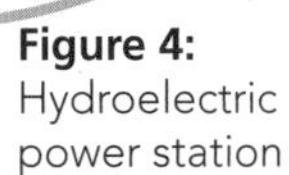

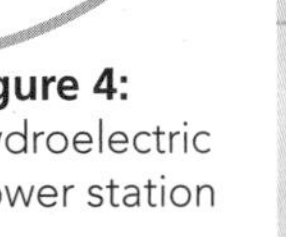

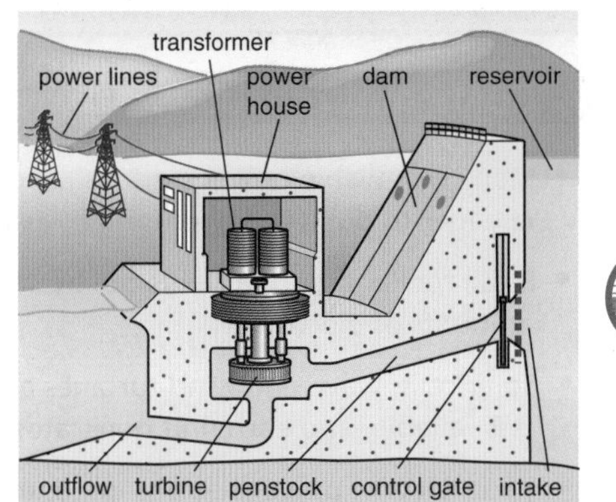

Figure 4: Hydroelectric power station

D–C

Electricity and the environment

Comparing power stations

Type of energy source	Advantages	Disadvantages
Fossil fuels	Produces large quantities of energy reliably	Produces polluting gases and solid waste. Mining damages the environment.
Nuclear fuels	Does not produce greenhouse gases.	Creates radioactive waste that must be safely stored.
Hydroelectricity	Renewable.	Causes large scale flooding. Produces greenhouse gases.
Biomass	Renewable. Reduces need for landfill sites.	Produces greenhouse gases.
Wind power	Renewable. Does not produce polluting gases.	Can affect wildlife. Must be sited in windy areas.
Tidal power	Renewable. Does not produce polluting gases.	Causes large scale flooding and changes water flow.
Geothermal	Renewable.	Can release toxic gases from below the Earth's surface.
Solar	Renewable. Does not produce polluting gases.	Large scale use involves large land areas and can affect wildlife.

G–E

More effects of generating electricity

- Air pollution can be visible (smoke) causing health problems or invisible (for example, greenhouse gases) causing global warming and climate change.
- Power stations should be close to where they are needed. Fossil fuel power stations must be near water, and transport links.

Remember!
Make sure you know at least one advantage and disadvantage for each source of energy.

Improve your grade

Explain whether a hydroelectric power station or a coal fired power station is best for a city located near the coast.
AO3 (5 marks)

Making comparisons

Which power station?

- Renewable power stations can be less efficient than fossil-fuel power stations. Efficiency is less important in renewable power stations because renewable energy will not run out.
- Other important factors are the:
  - building costs and running costs
  - reliability of the energy resource
  - ability to change the amount of energy generated.

How Science Works

- You should be able to compare the benefits and drawbacks of different energy sources in different situations.

Costs and reliability

- Many renewable power stations have high capital costs and low running costs. Fossil fuel power stations are cheaper to build but operating costs will increase as the supplies of fossil fuels fall.
- Fossil fuel and nuclear fuels can be stored, increasing their reliability. A single power station generates large amounts of electricity.

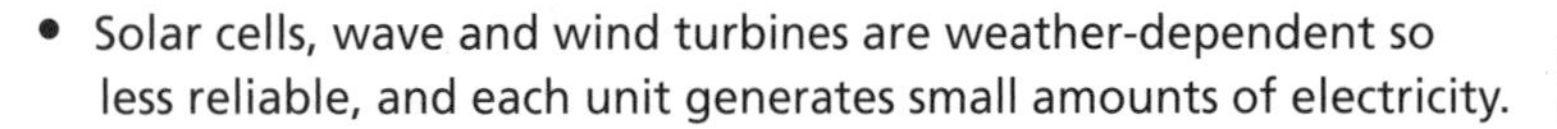

- Solar cells, wave and wind turbines are weather-dependent so less reliable, and each unit generates small amounts of electricity.
- Hydroelectric power stations generate electricity quickly when needed but the reservoir levels must be maintained, sometimes by pumping water back to the reservoir.
- Nuclear power stations have the longest start up and shut down times, followed by coal then gas power stations.

EXAM TIP

If you are asked about the environmental impact of an energy source, your answer must describe environmental damage (for example flooding, disrupting river flow, emission of greenhouse gases) and not a general disadvantage (for example expensive or unreliable).

The National Grid

What is the National Grid?

- Electricity is transmitted from power stations to homes and businesses through the National Grid. This is a network of cables that links power stations with properties.
- Step-up transformers at power stations increase the voltage to 400 000 V.
- Step-down transformers in substations reduce the voltage. Electricity is supplied at 230 V in homes.
- In the home, some plugs include transformers to reduce this voltage even more, for example for mobile phone rechargers and printers.

High voltages

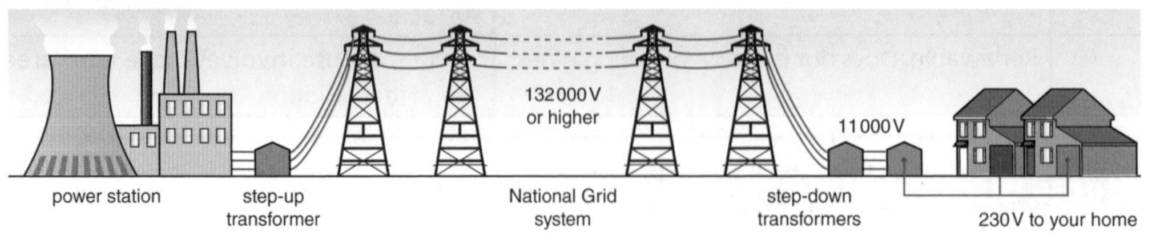

Figure 1: The National Grid

- Power (W) = voltage (V) × current (amp)
- Increasing the voltage in power cables reduces the current. Wires heat up less so thinner cables are needed and less energy is wasted.

Remember!
You should be able to explain what each part of the National Grid does.

Improve your grade

Explain whether a coal-fired power station or a hydroelectric power station is best able to cope with surges in demand during the day. **AO3 (3 marks)**

What are waves?

Types of waves

- Mechanical waves transfer energy through a medium. Examples include sound, and waves in water and along ropes.
- Electromagnetic waves transfer energy through a vacuum and a medium. Examples include radio waves, microwaves, light, and ultraviolet radiation.

Transferring energy

- Waves transfer energy from a source without transferring matter.
- Longitudinal waves oscillate in the same direction that the energy travels, and include sound waves.
- Transverse waves oscillate at right angles to the direction the energy travels, and include electromagnetic waves and water waves.

Remember!
A wave only transfers energy. The substance the wave travels through does not travel with it.

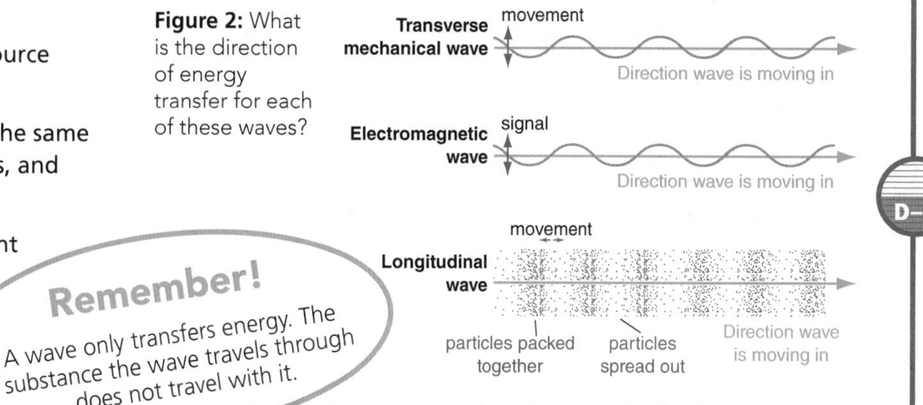

Figure 2: What is the direction of energy transfer for each of these waves?

D–C

Changing direction

Reflection and diffraction

- The normal is an imaginary line drawn at right angles to a surface.
- Waves can be reflected (bounce off a surface). The angle between a reflected ray and the normal is the same as the angle between an incoming ray and the normal.
- Waves can be diffracted (spread through a gap or round an obstacle). Diffraction is greatest when the wavelength is about the same size as the gap or obstacle.

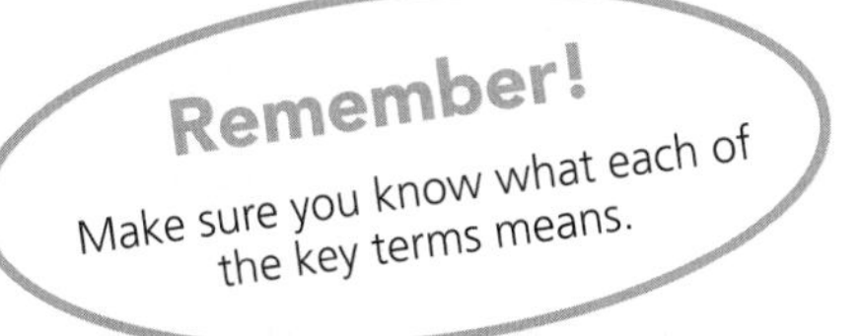

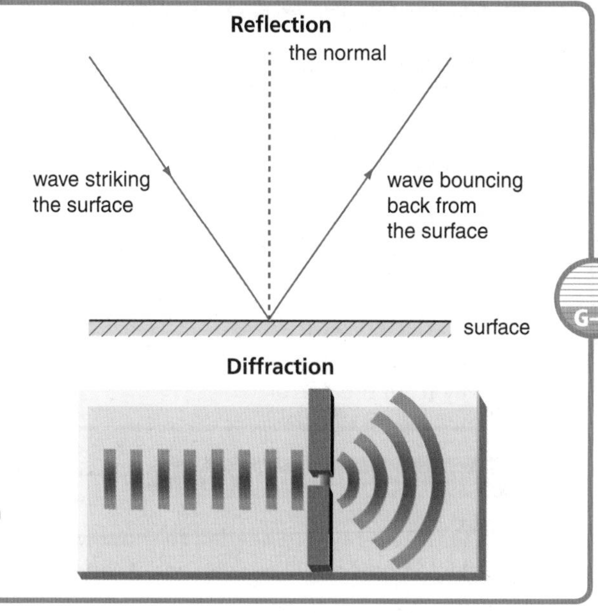

Figure 3: Reflection and diffraction

G–E

Refraction

- Waves can be refracted (change direction at a boundary). Waves refract because they change speed in different materials. Light travels fastest in a vacuum.
- If a wave moves from air into glass
  - it travels slower in the glass
  - the wave changes direction (refracts) towards the normal.

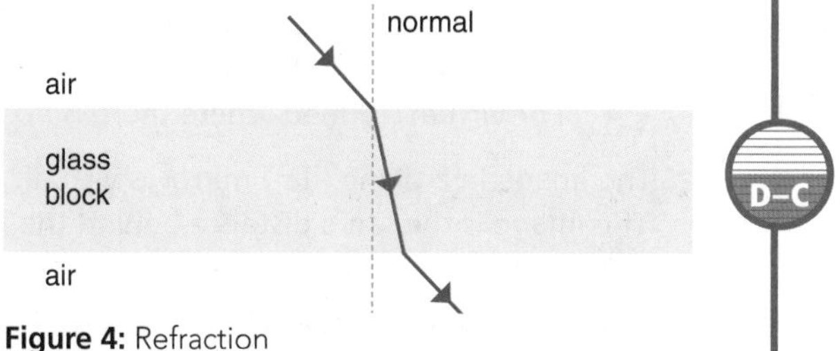

Figure 4: Refraction

D–C

Improve your grade

An echo is a reflected sound wave. Explain why you can hear echoes only in certain places, and why you may hear more than one echo. **AO3 (3 marks)**

Sound

Looking and listening

- Vibrating objects create sound waves. Sound waves are longitudinal mechanical waves and cannot pass through a vacuum.
- All sound waves can be reflected (echoes), refracted or diffracted.
- Loud sound waves have larger amplitude than quiet sound waves.

Sound is a wave

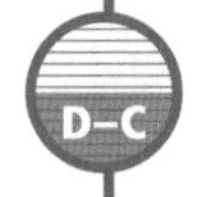

- Frequency measures the number of cycles per second and is measured in Hertz (Hz). Humans hear sounds between 20 Hz and 20 000 Hz.
- High pitched notes have a short wavelength and high frequency; low pitched notes have a long wavelength and low frequency.
- The higher the frequency of a wave, the more energy it can carry.

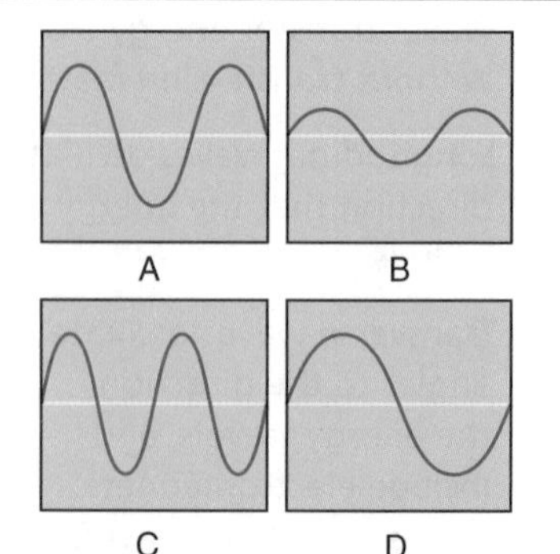

Figure 1: Sound B is quieter than sound A; sound C is higher pitched than sound D

Light and mirrors

Describing light waves

- Ray diagrams show how one or more rays of light travel when they are reflected or refracted.
- The normal is an imaginary line at right angles to the surface of the mirror.
- The incident ray is the ray travelling towards the mirror.
- The reflected ray is the ray travelling away from the mirror.

Remember!
The angle of incidence and the angle of reflection are measured by comparison with the normal.

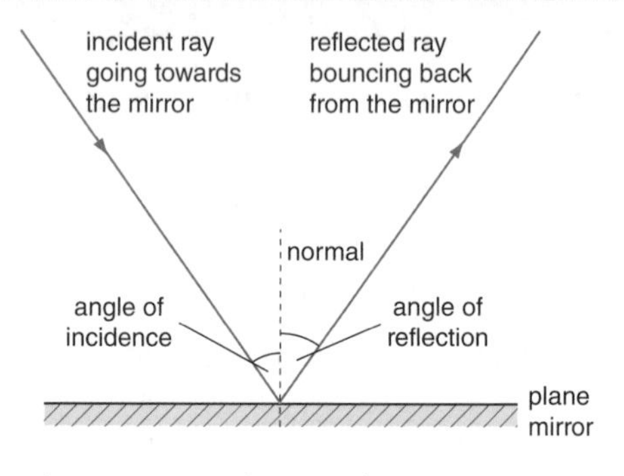

Figure 2: A ray diagram showing a reflected ray

Using curved mirrors

- At least two rays are needed to show where the image of a reflected object will be seen and its appearance.
- Compared to the object, images can be;
  - upright or inverted (upside down)
  - magnified (bigger), diminished (smaller) or the same size
  - laterally inverted (left and right sides reversed)
  - real or virtual (formed where there is no light from the object).
- The image in a plane (flat) mirror is virtual, upright and laterally inverted. The image is the same distance behind the mirror as the object is in front of the mirror.

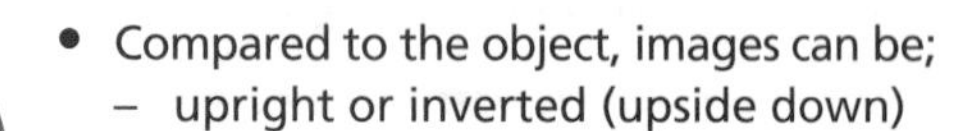

Figure 3:

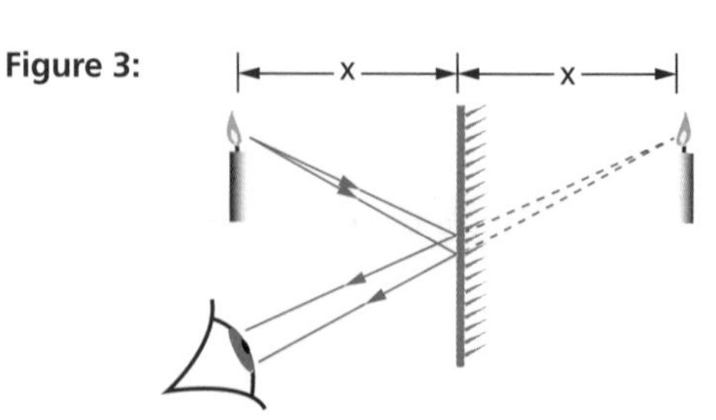

Improve your grade

Draw traces to show two sound waves. One sound is lower-pitched and twice as loud as the other sound. Label the amplitude and wavelength on each trace. **AO1 (4 marks)**

Using waves

Invisible waves

- Electromagnetic radiation is transverse waves that travel at the speed of light. It transfers energy.
- Radio waves, microwaves, infrared, and visible light are electromagnetic waves used in communication systems.
- Radio waves are used for radio and TV broadcasts.
- Microwaves are used in mobile phone networks and satellite communication, as the atmosphere does not absorb microwaves.
- Infrared radiation comes from anything warm. It is used in remote controls.
- Visible light allows us to see and is used in photography.

More about radiation and its uses

- Long wavelength radio waves have the lowest energy. They are used for local and national radio broadcasts.
- Short wavelength radio waves transfer more energy. They are used for TV broadcasts
- Sources of visible light include the Sun, flames and lamps.

Figure 4: These aerials detect short wavelength radio waves

The electromagnetic spectrum

Electromagnetic radiation

- Electromagnetic radiation is a continuous spectrum of transverse waves carrying energy. In a vacuum, all electromagnetic waves travel at the same speed, the speed of light.

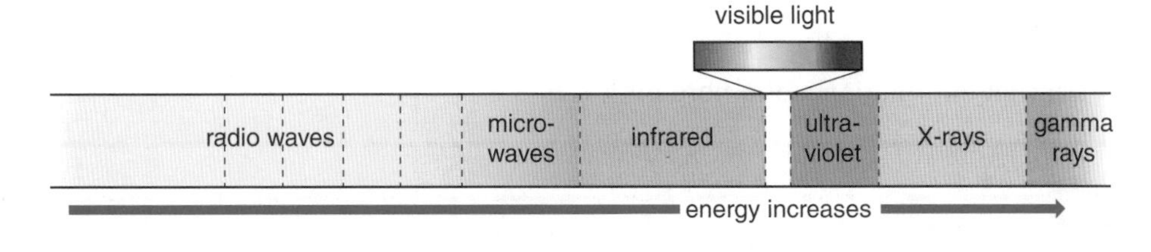

Figure 5: The electromagnetic spectrum

Energy and wavelength

- The speed (or velocity) of the wave is calculated using: speed (m/s) = frequency (Hz) × wavelength (m). The speed of electromagnetic radiation is 300 000 000 m/s.
- The shorter the wavelength of an electromagnetic wave, the higher its frequency. Electromagnetic waves with a short wavelength and high frequency carry most energy.
- Electromagnetic wavelengths range from about 10^{-15} m (gamma rays) to about 10^{4} m (radio waves).

Remember!
Although electromagnetic waves have some properties the same, waves that have a different frequency will carry different amounts of energy.

lower energy
wavelength

higher energy

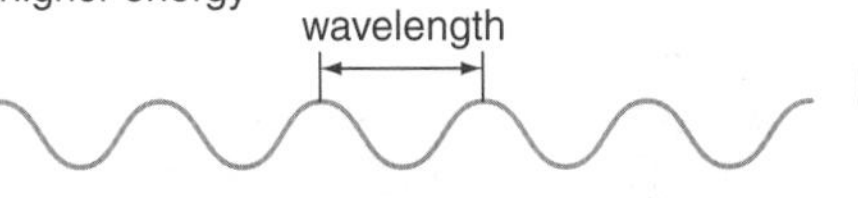

Figure 5: Wavelength frequency and energy are linked

Improve your grade

Calculate the speed of radio waves with a wavelength of 10 000 m and frequency of 30 000 Hz. **AO2 (3 marks)**

Dangers of electromagnetic radiation

How dangerous is electromagnetic radiation?

G–E

- Radio waves are harmless.
- Microwaves are strongly absorbed by water, passing through skin and heating up cells.
- Infrared radiation can cause burns.
- Very intense visible light can damage cells in the eye's retina.
- Ultraviolet radiation can cause sunburn in minutes, and prolonged exposure or repeated sunburn can cause skin cancer.
- X-rays and gamma rays can kill or damage cells, or cause burns.

Remember!
High frequency electromagnetic waves are more harmful than low frequency electromagnetic waves.

Protecting from the dangers of radiation

- People are exposed to very low levels of microwave radiation over long periods of time from mobile phones.
- Microwaves do not cause cancer, but may warm cells when they are absorbed.
- Many studies have looked for evidence that mobile phones cause harm but have not found evidence of any serious risk. The levels of microwaves that a user is exposed to can be reduced by improving the shielding, reducing the intensity of the signal or reducing the time of usage.

Telecommunications

Sending information long distances

G–E

- Electromagnetic waves allow signals to travel long distances very rapidly.
- Radio waves are used for radio and TV communications.
- Microwaves are used for satellite TV broadcasts, sat-nav and mobile phone networks.
- Infrared and visible radiation transmit information along fibre optic cables.

The technological age

D–C

- Microwaves communicate with satellites. Microwaves are not absorbed by the atmosphere. Signals are transmitted to the satellite. The satellite transmits them to another place on Earth.
- Bluetooth and WiFi networks use low energy radio waves to send signals short distances.
- Visible light and infrared signals travel long distances in optic fibres without being absorbed. Optic fibres form part of the telephone network, also used for Internet and email.
- Interactive TV and TV remote controls use infrared radiation.

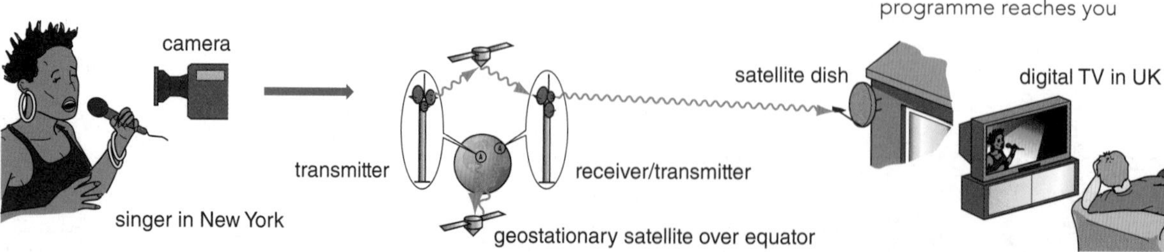

Figure 1: How a satellite TV programme reaches you

Improve your grade

Explain which type of electromagnetic wave is the best choice for satellite communications. **AO2 (3 marks)**

EXAM TIP
You should be able to choose the most suitable type of electromagnetic wave to use in different situations.

Cable and digital

What is digital and cable?

- Analogue signals can have any value, but digital signals are pulses that only have two values: on or off.
- Digital signals are higher quality than analogue signals because:
  - there is less interference between different signals
  - the signal quality is not affected by distance
  - signals can be made stronger without losing information.

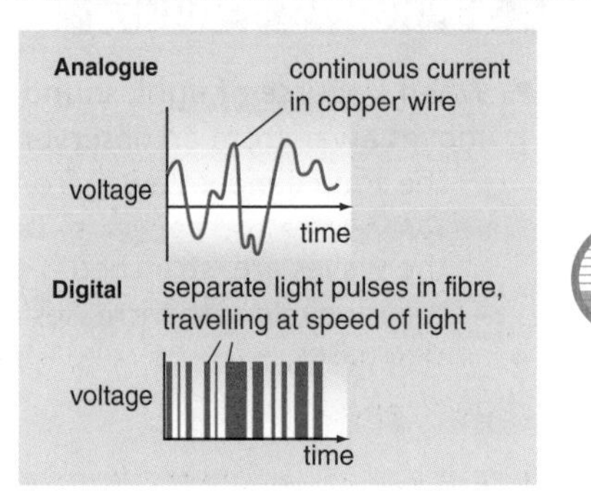

Figure 2: Analogue and digital signals

G–E

Why use digital?

- Digital signals have better quality than analogue signals.
- Digital signals do not interfere with each other
- Digital signal quality is not affected by distance as much as analogue signals
- Digital signals can be strengthened without losing information
- Fibre optic cables are thin flexible strands of very pure glass.
- Optical fibres transmit information fast with good quality and carry many signals.

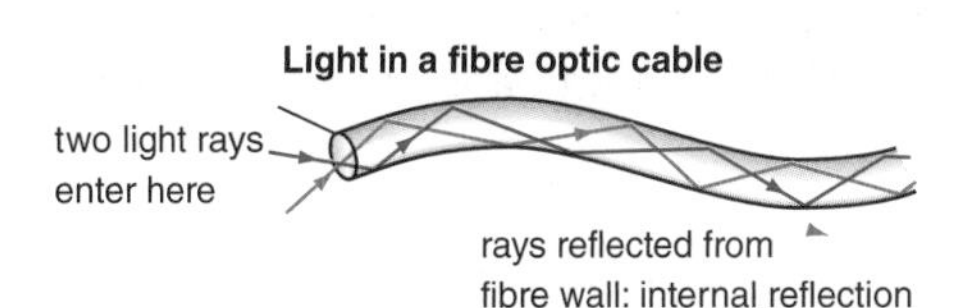

Figure 3: Signals travel through fibre optic cables by total internal reflection

D–C

Searching space

Telescopes

Optical telescopes magnify visible light from distant objects. Radio telescopes detect radio signals from distant objects.

Looking further

- Telescopes can be based on Earth or in space.
- Advantages of Earth-based telescopes compared to space-based telescopes are:
  - reduced costs to manufacture
  - easier to maintain, repair and update.
- Advantages of space-based telescopes compared with Earth-based telescopes are:
  - clearer images as there is no atmospheric interference from moisture, dust, pollution or weather patterns
  - the full spectrum of electromagnetic waves from objects in space can be detected.

Radiation	Objects 'seen' in space
gamma ray	neutron stars
X-ray	neutron stars
ultraviolet	hot stars, quasars
visible	stars
infrared	red giants
far infrared	protostars, planets
radio	pulsars

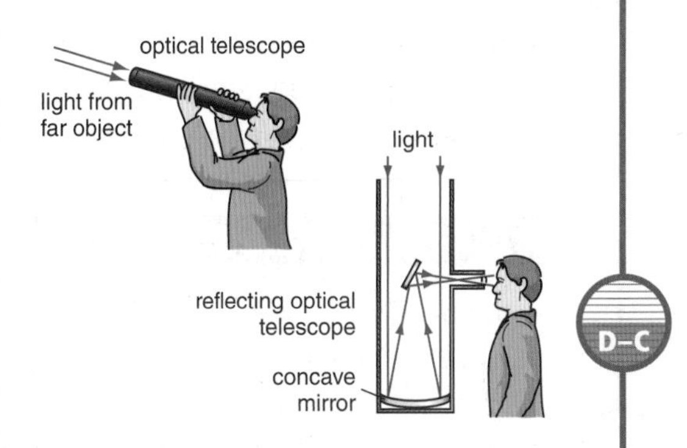

Figure 4: Reflection telescopes use curved mirrors to focus light from stars

D–C

Remember!
All types of electromagnetic waves travel at the same speed throughout space.

Improve your grade

Explain two advantages of using space-based telescopes. **AO2 (4 marks)**

Waves and movement

Changing sounds

G–E

- When a source of light, sound or microwaves moves away from an observer:
  - the source moves away from the waves it produces
  - the waves are stretched
  - their wavelength increases and frequency decreases.

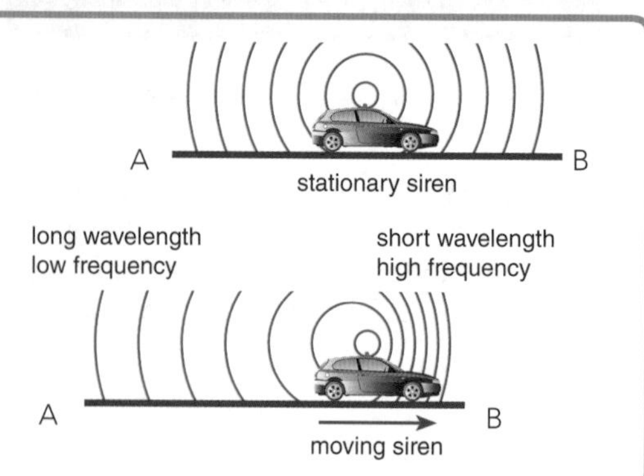

Figure 1: When the siren is stationary both people hear the same sound. When it moves, the person at B hears a higher pitched sound than at A

Doppler and light

D–C

If a source of light or sound moves towards an observer,

- the wavelength appears to decrease
- the frequency appears to increase.

The change in observed wavelength and frequency is called the Doppler effect. It is greater if the source moves faster.

- Red light has a longer wavelength than blue light. Light from galaxies moving away from Earth appears redder (longer wavelength). This is called the red shift.

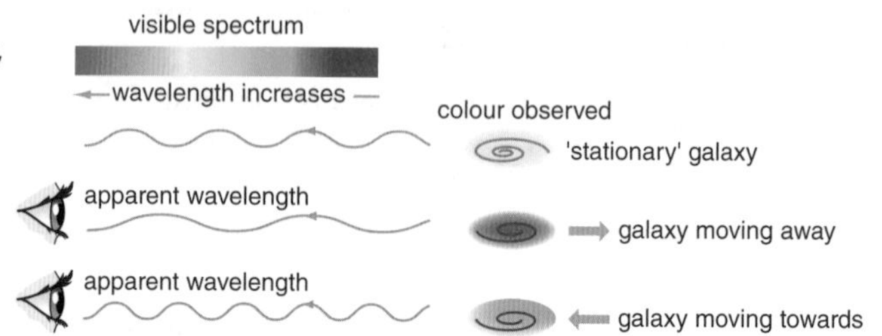

Figure 2: Distant stars and galaxies move fast enough for the colour of their light to change

Origins of the Universe

Red-shift and Big Bang

Light from many galaxies is shifted towards the red end of the spectrum. It appears to have a longer wavelength than light from our galaxy. This is the red-shift.

- Radiation detected from galaxies, in all directions, has a red shift. This is because the galaxies are moving away from us.
- More distant galaxies have a larger red shift. They are moving away faster.
- The Big Bang theory says that all matter and energy in the Universe was initially in one very small point. This point expanded rapidly outwards.

CMBR and the Big Bang theory

D–C

- We believe that the universe began about 14 billion years ago. Matter and space expanded violently and rapidly at that time and is still expanding.
- The red-shift is evidence that the Universe is still expanding, supporting the Big Bang theory.
- Cosmic microwave background radiation (CMBR) is electromagnetic radiation filling the universe. It comes from radiation present shortly after the beginning of the universe.
- The Big Bang theory is currently the only theory that can explain the existence of CMBR.

Remember!

Evidence for the Big Bang includes evidence the Universe is expanding (red shift) and the radiation remaining from the Big Bang (CMBR).

Improve your grade

Explain the evidence we have that supports the Big Bang theory. **AO3 (6 marks)**

P1 Summary

The transfer of energy by heating processes

Factors that affect the rate at which energy is transferred

All objects absorb and emit infrared radiation. The temperature and colour of the surface affect how quickly this happens.

Kinetic theory can be used to explain different states of matter.

U-values measure how effective materials are as insulators.

Solar panels use the Sun's radiation to warm water.

The energy needed to warm 1 kg of a material by 1 degree Celsius is its specific heat capacity.

Conduction, convection, evaporation and condensation transfer energy and involve particles.

Convection transfers heat in liquids and gases; conduction transfers heat most effectively in solids.

The rate of heat transfer depends on surface area, volume, material and type of surface it is in contact with.

Energy transfers, efficiency and electrical energy

Energy can be transferred, stored or dissipated. It cannot be created or destroyed.

The efficiency of a device is the useful energy output ÷ total energy input.

Electrical appliances carry out different energy transfers.

The amount of energy transferred depends on the equipment's power and time it is used for.

Generating and distributing electricity

Electricity is generated in power stations where heat from a fuel or from volcanic rocks changes water to steam. This spins turbines and generators.

Energy from water (hydroelectricity, tides and waves) and wind spins turbines directly. Energy from the sun produces electricity directly.

The use of energy resources has an impact on the environment, including producing greenhouse gas emissions, pollution, flooding and waste products.

The National Grid distributes electricity from power stations to consumers. High voltages in cables reduces energy losses.

Waves, communication and the Universe

Waves transfer energy and can be reflected, refracted and diffracted. Their speed is calculated using wavelength × frequency.

Electromagnetic waves are transverse waves that travel at the speed of light in a vacuum and form a continuous spectrum.

The Big Bang theory is that the Universe began from a very small initial point. The red shift is evidence that the Universe has continued expanding since then.

Radio waves, microwaves, infrared and visible light are used for communication.

Waves are reflected, and this is how an image in a mirror is formed.

Sound waves are longitudinal waves. The loudness and pitch of a note depend on the wave's amplitude and frequency.

Animal and plant cells

Cells

- All living organisms are made of cells.
- Animals, like humans, are made up of animal cells. Plants are made up of plant cells.
- Animal and plant cells have many similarities, but also some differences.
- Every one of the cells in the human body has been produced from the division of one single cell – the zygote.

Cell organelles

- The different parts of a cell are called organelles. Each has a particular function.
- Figures 1 and 2 show the organelles you can see with a powerful light microscope.

EXAM TIP

Make sure you remember which organelles are found in both plant and animal cells and which are found only in plant cells. Figures 1 and 2 show you.

D–C

ribosomes, where protein synthesis takes place

mitochondrion, where energy is released in respiration

cytoplasm, where metabolic reactions controlled by enzymes take place

cell membrane, which controls the passage of substances in and out of the cell

nucleus, containing chromosomes made of genes. Controls the activities of the cell

Figure 1: An animal cell

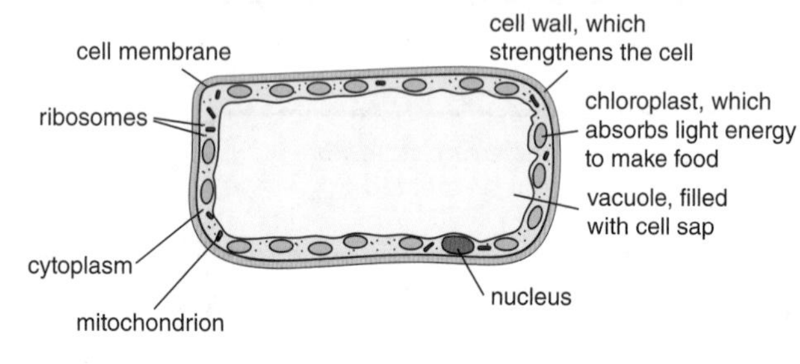

Figure 2: A plant cell

Microbial cells

Microorganisms

- Microorganisms are living things that are too small to see with the naked eye.
- They include yeast, bacteria and some kinds of algae.
- Most microorganisms are made of a single cell.

Yeast, algae and bacteria

- Yeast is a single-celled fungus. Its cells have cell walls but they are not made of cellulose like plant cell walls. Fungi cannot photosynthesise as they have no chloroplasts.
- Algae are simple, plant-like organisms. Their cells are similar to plant cells.
- Bacteria do not have a nucleus. Their genes are in the cytoplasm.

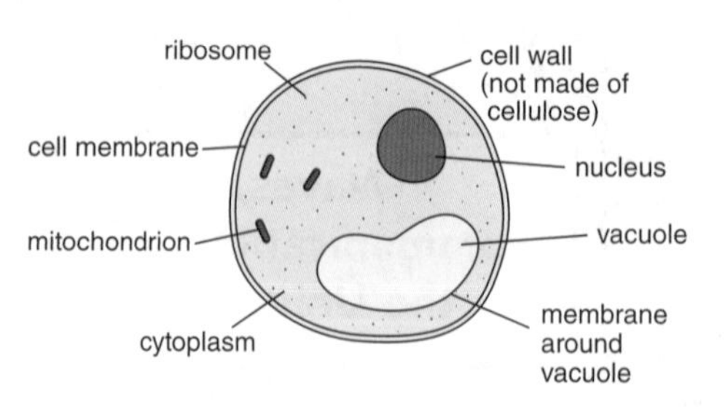

Figure 3: The structure of a yeast cell

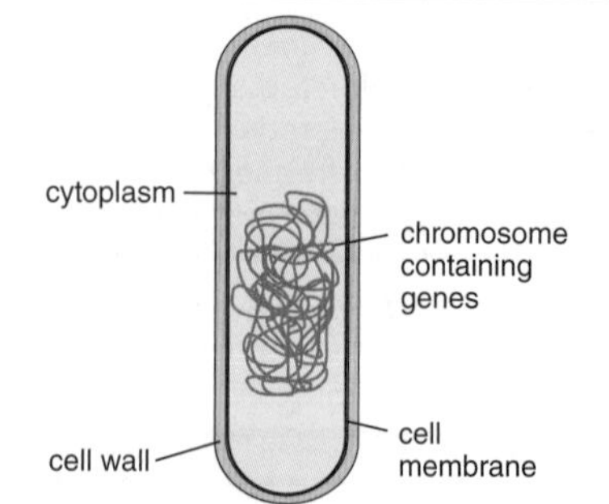

Figure 4: The structure of a bacterial cell

Improve your grade

The image below was taken of some cells using a light microscope. State what type of cells they are and give reasons for your answer. **AO2 (2 marks)**

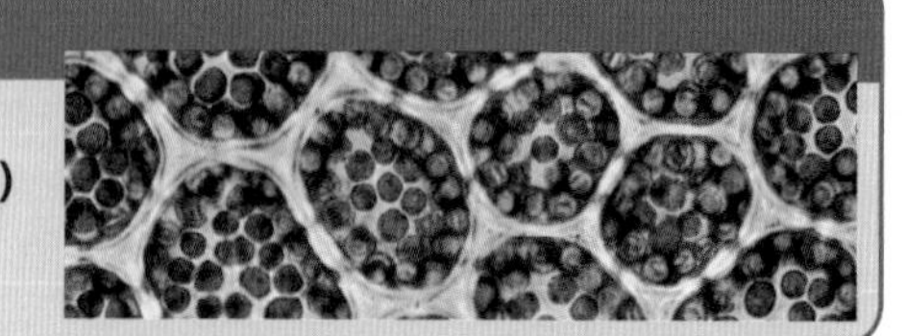

Diffusion

Moving particles

- Scents are substances that you can smell. A scent consists of molecules of gases that spread out in the air.
- Molecules naturally spread out from a place where there are a lot of them into a place where there are fewer. They spread from a high concentration to a lower concentration. This is diffusion.
- Particles are always moving. Any particles that can move around freely can diffuse. This includes particles in gases and liquids.

Remember!

The speed of diffusion can be increased by increasing the difference in concentration (the concentration gradient) and increasing the temperature. (The faster that particles move around, the faster they will diffuse.)

Cells and diffusion

- Diffusion is the spreading of the particles of a gas, or of any substance in solution, resulting in a net movement from a region where they are of a higher concentration, into a region where they are in a lower concentration.
- Most cells need oxygen so that they can respire. Oxygen diffuses into the cells from a higher concentration outside to a lower concentration inside.
- The concentration of oxygen inside a cell is kept low because the cell keeps on using it up.

Specialised cells

Different cells for different functions

- In a multi-cellular organism, cells are specialised. Their structure is adapted to suit their function.
- Life starts as a little ball of unspecialised cells. During development into an embryo, the cells change. This is differentiation.
- Examples of specialised cells are egg and sperm.

Examples of specialised cells

- The human body contains hundreds of different kinds of specialised cells.
- Red blood cells, goblet cells and ciliated cells are just three of them.

EXAM TIP

Cells are measured in a unit called a micrometre, symbol μm. There are 1000 micrometres in one millimetre.

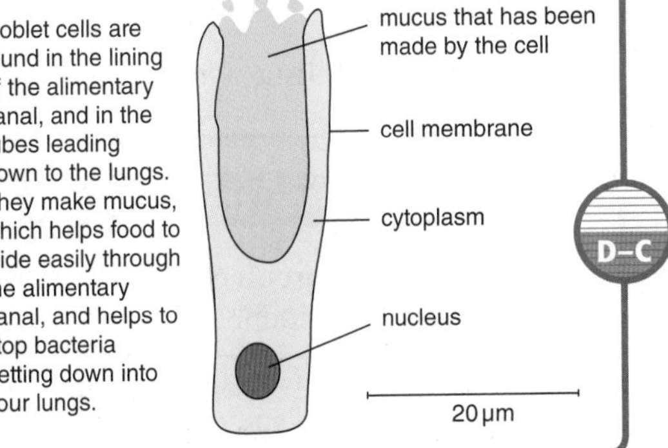

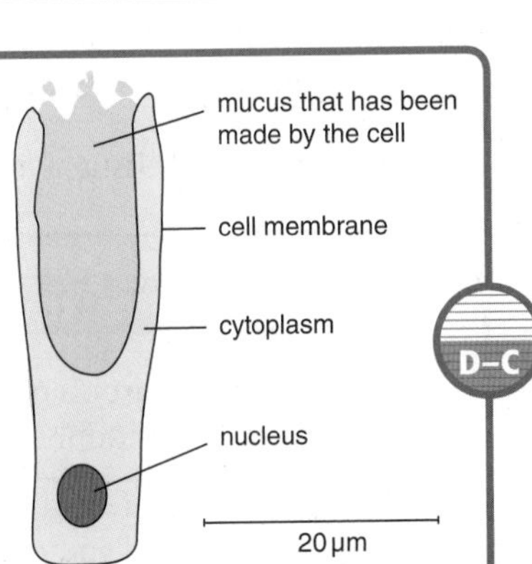

Figure 5: Goblet cells

Improve your grade

Explain how a sperm cell is specialised. **AO2 (4 marks)**

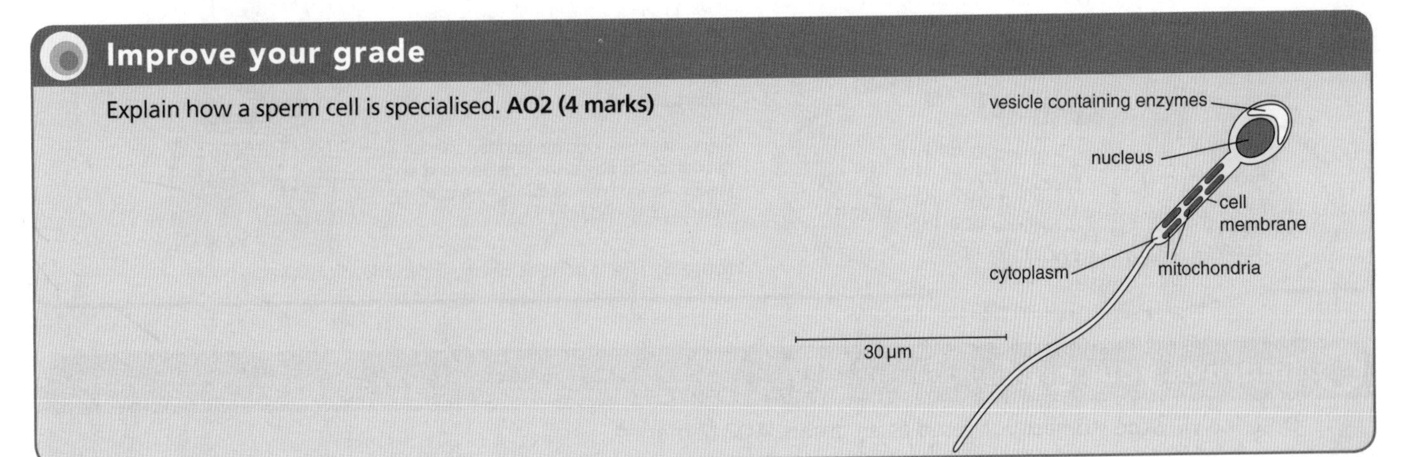

Tissues

What is a tissue?

- In an animal or a plant, a tissue is a group of similar cells that all work together to carry out a particular function.
- Epithelial tissue is made up of epithelial cells. It covers surfaces in the body and protects the cells underneath it. The upper layer of your skin is epithelial tissue.

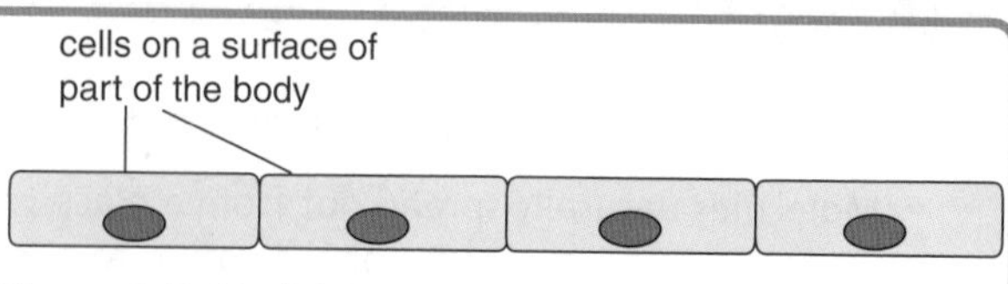

Figure 1: Epithelial tissue covers surfaces in the body. The tissue protects the cells underneath it.

Structure and function of tissues

- The cells that make up tissues are adapted for particular roles.
- Muscular tissue is specialised to produce movement.
- Glandular tissue is made up of cells which secrete (release) useful substances such as enzymes or hormones.

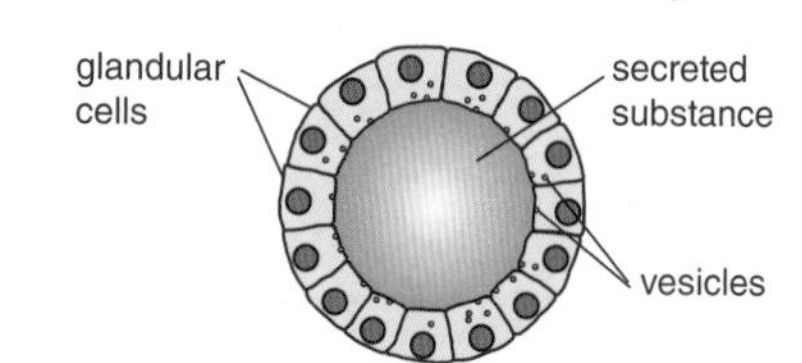

Figure 2: Glandular cells

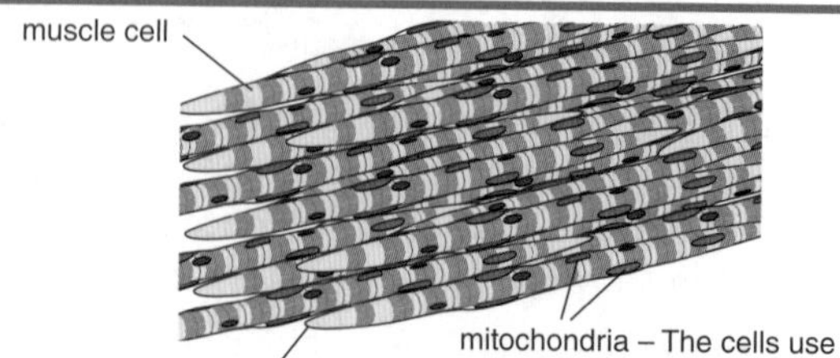

Figure 3: Muscular tissue

Animal tissues and organs

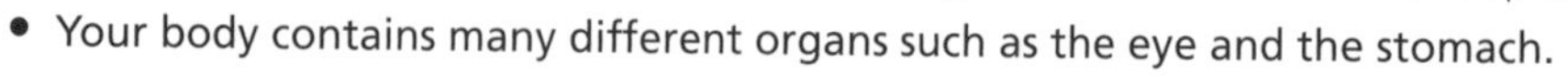

Organs

- Your body contains many different organs such as the eye and the stomach.
- Organs are made of tissues. Each organ contains many different tissues. For example, the wall of the stomach contains:
  - glandular tissue, which secretes digestive juices to break down the food
  - muscular tissue, which contracts and relaxes to mix up the contents of the stomach
  - epithelial tissue, which covers the inside and outside of the stomach wall.

Figure 4: The tissues that make up the wall of the stomach

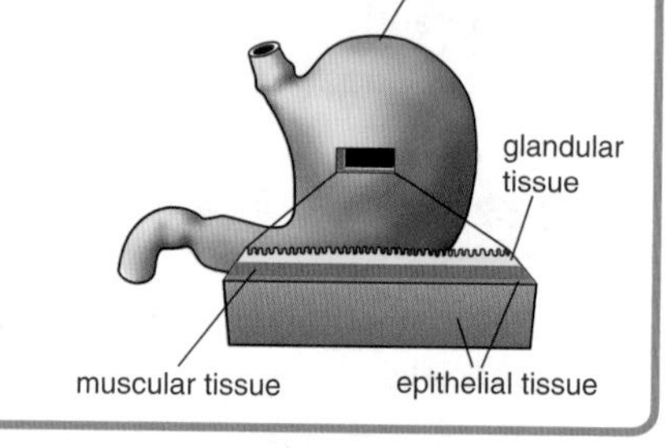

Functions of the digestive system

- A system is a group of organs that performs a particular function.
- The function of the digestive system is to break down the food you eat so the food molecules can enter the blood.
- Each of the organs shown in Figure 5 has an important role in this.

Remember!

A group of cells is called a tissue. Organs are made up of different types of tissue. The heart contains muscle tissue that can contract and relax, nervous tissue to control the heart beat, and ligaments to hold the different tissues in place.

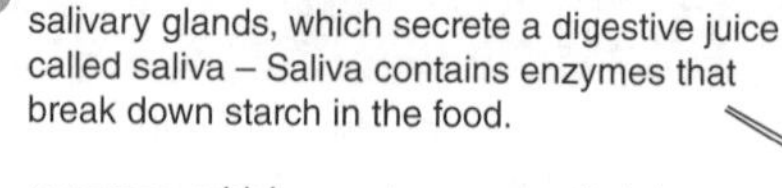

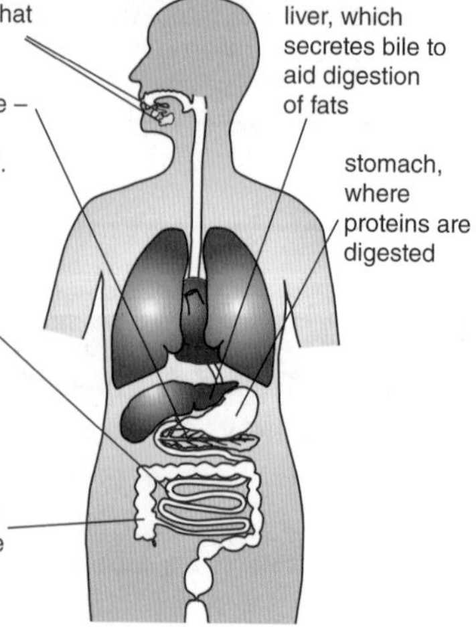

Figure 5: The functions of organs in the digestive system

Improve your grade

Describe why the stomach is classed as an organ. **AO2 (2 marks)**

Plant tissues and organs

Plant organs

- Plants contain many different organs.
  - leaves – where the plant makes its food and where photosynthesis takes place. Leaves are usually broad and flat, so they can get plenty of sunlight – the energy source that drives photosynthesis.
  - stem – this holds the flowers and leaves up in the air. It allows the leaves to get plenty of light, so that they can photosynthesise. It also allows the flowers to attract insects, so that they can be pollinated.
  - roots – these anchor the plant firmly into the soil. Roots absorb water from the soil, and the water then moves through the root and stem into all the other parts of the plant. Roots also absorb mineral ions from the soil, which the plant needs for healthy growth.

Plant tissues

- The organs in a plant are made up of tissues.
- The whole plant is covered in a layer of epidermis. This helps to protect the underlying cells, stops the leaves from losing too much water and prevents pathogens from entering the plant.
- Most of the cells in a leaf are mesophyll cells. This is where photosynthesis takes place.
- Xylem and phloem tubes run through the entire plant.
- These are tubes which make up the plant's transport system: xylem carries water from the roots to the leaves; sugars are transported around the plant in phloem.

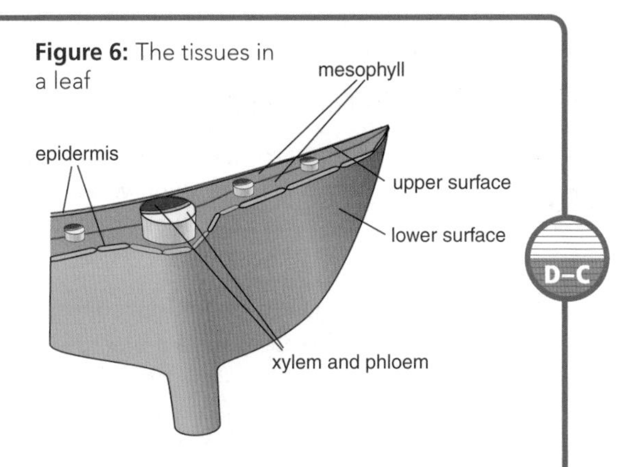

Figure 6: The tissues in a leaf

Photosynthesis

Making food

- Plants make food by photosynthesis.
- They take carbon dioxide from the air and water from the soil.
- Leaves contain a green pigment called chlorophyll. This absorbs energy to make the carbon dioxide and water react together.
- Oxygen and a sugar called glucose are produced.

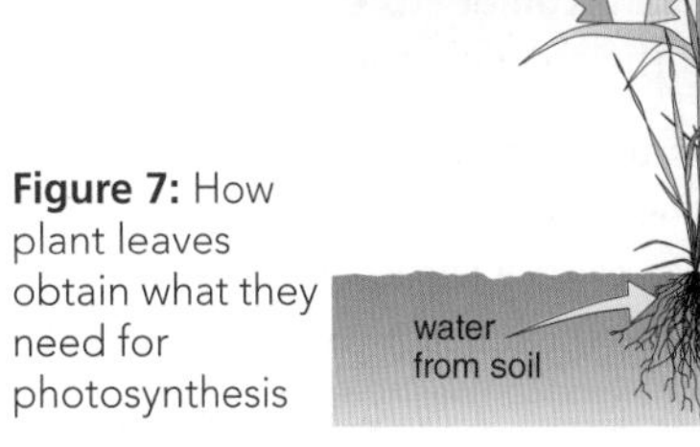

Figure 7: How plant leaves obtain what they need for photosynthesis

Oxygen and energy for living things

- The word equation for photosynthesis is:

$$\text{carbon dioxide + water} \xrightarrow{\text{energy from sunlight}} \text{glucose + oxygen}$$

- Millions of years ago there was hardly any oxygen in the air. Oxygen was first made when bacteria evolved that could photosynthesise. Gradually, over millions of years, the amount of oxygen in the air built up. Now more than 20 per cent of the air is oxygen. It has all been made by bacteria, algae and plants.

- In photosynthesis, light energy is stored in glucose molecules.
- The energy is transferred to animals when they eat the plants.
- Glucose can be converted into starch and stored for later use. Starch molecules are big and, unlike glucose, cannot diffuse out of cells.

EXAM TIP

The equation for photosynthesis shows us that water and carbon dioxide are reactants. Glucose and oxygen are products and are made from the rearrangement of the atoms in the reactants.

Improve your grade

Without photosynthesis, humans would not survive. Explain why. **AO2 (2 marks)**

Limiting factors

Speeding up photosynthesis

- The speed that plants can carry out photosynthesis and make food depends on three things.
  - Light intensity: The brighter the light, the more energy the plant is receiving and the faster it can photosynthesise.
  - Carbon dioxide concentration: Carbon dioxide is one of the raw materials for photosynthesis. If you give a plant more carbon dioxide, it will probably be able to photosynthesise faster.
  - Temperature: A warm temperature will speed up photosynthesis.

Limits to the speed of photosynthesis

- Figure 1 shows that as the light intensity increases, the rate of photosynthesis also increases.
- Light is a limiting factor for photosynthesis.
- However, there comes a point when the rate of photosynthesis does not increase any more, even when the plant is getting more light. This may be because it does not have enough carbon dioxide or the temperature is too low.
- A similarly shaped graph is produced when carbon dioxide in the air around a plant is increased.

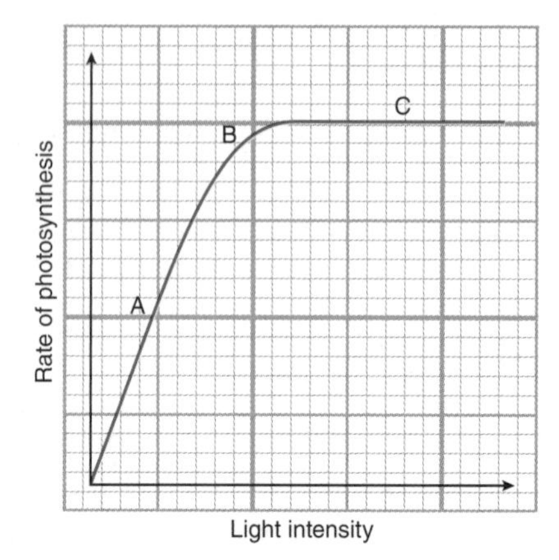

Figure 1: How light intensity affects the rate of photosynthesis

The products of photosynthesis

How plants use their food

- Plants need food for two reasons – energy and building their structure.
- When a plant cell needs energy, it breaks up glucose in a process called respiration, releasing the stored energy.
- Plants can make many different substances from glucose. These include:
  - starch, which is a carbohydrate like glucose and is good for storing energy
  - cellulose (also a carbohydrate), which is used for making cell walls
  - proteins, which are needed for making new cells and enzymes.

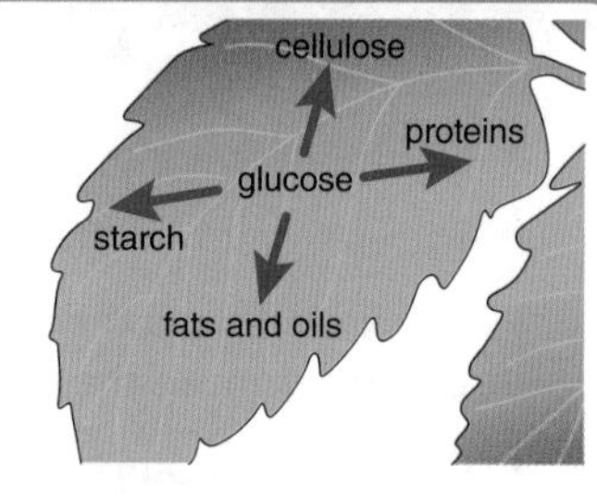

Figure 2: What plants can make out of glucose

Mineral salts

- Glucose is made from carbon, hydrogen and oxygen atoms. Plants make other substances from glucose. These are shown in Table 1.

Substance	Atoms the substance is made from			
	Carbon	Hydrogen	Oxygen	Nitrogen
carbohydrates	✓	✓	✓	
fats and oils	✓	✓	✓	
proteins	✓	✓	✓	✓

Table 1: Glucose products

- To convert glucose into proteins, plants need nitrogen which they get in the form of mineral ions. They normally use nitrate ions, NO_3^-, and absorb them from the soil through their roots.

Improve your grade

Anna uses a fertiliser that is high in nitrates on her tomato plants. Explain why she does this. **AO2 (3 marks)**

Distribution of organisms

Distribution

- The distribution of an organism means the area in which it lives.
- Organisms can only live in environments for which they are adapted.
- There are many physical factors that affect their distribution.

Factors affecting distribution

- Temperature: many species can live only in a particular temperature range.
- Availability of nutrients: very wet soils are usually short of nitrate ions, so only certain plants can live there. Animals can live only where their food is found.
- Amount of light: plants and algae must have light for photosynthesis. They cannot grow in really dark places.
- Availability of water: all organisms need water. Species that live in deserts have special adaptations that help them to obtain and conserve water. The seaweed egg wrack can only grow in places that are covered in water for most of the day or they get too hot and dry.
- Availability of oxygen and carbon dioxide: most organisms need oxygen, for respiration. They cannot live where oxygen is in short supply. Plants also need carbon dioxide, for photosynthesis.

Organisms can only live in environments for which they are adapted.

Using quadrats to sample organisms

What lives where?

- Ecosystems, such as fields, are usually too large for us to study the distribution of organisms that live in them. Instead, we study small parts called samples.
- One technique is to use a small, square area called a quadrat. You identify and count the organisms in this one area and then use the results to estimate the number in the whole field.

Using quadrats

- Quadrats can be used to measure the distribution of organisms in a habitat. It is best to place the quadrats randomly. Once you have placed your quadrat, you need to identify each species inside it.
- Then you can either:
  - count the numbers of each one
  - estimate the percentage of the area inside the quadrat that each species occupies.
- You should repeat this process many times. Work out the mean number of, or area covered by, each species.
- A transect is used to find out if the distribution of organisms changes as you move from one habitat to another.

EXAM TIP

Placing the quadrat randomly ensures the results are valid. Placing the quadrat in at least 10 different places and calculating the mean will increase reliability.

(a) counting individuals within a quadrat

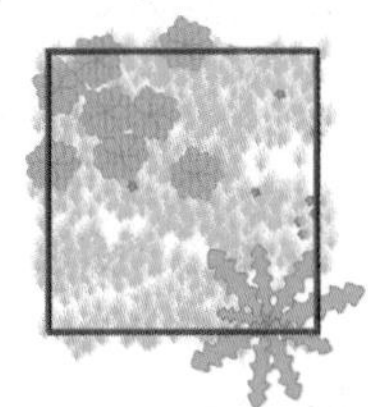

(b) estimating percentage cover in a quadrat

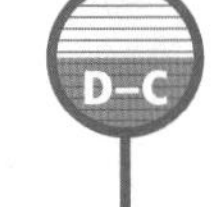

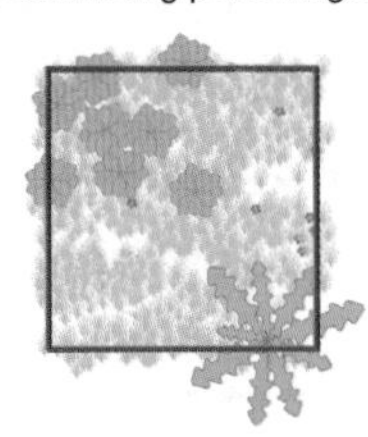

Figure 3: Two ways of collecting data from a quadrat

Improve your grade

Simon went on holiday to Mexico. He noticed that the plants growing there were very different from the plants growing at home in the UK. Explain the reasons for this difference in distribution. **AO2 (3 marks)**

Proteins

Proteins in the body

G–E

- The cell membrane, nucleus and cytoplasm all contain protein.
- Proteins in the cells in muscular tissue give the muscles their structure and strength, and bring about movement.
- Antibodies, enzymes and some hormones are examples of proteins.

Protein shapes

D–C

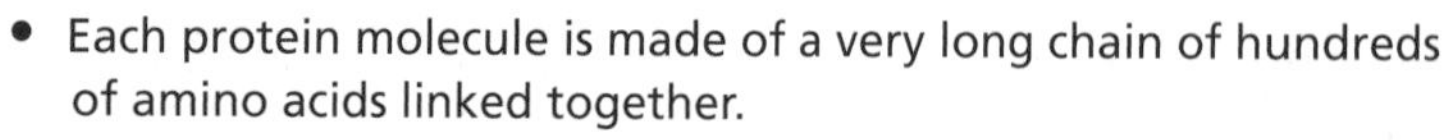

- Each protein molecule is made of a very long chain of hundreds of amino acids linked together.

Figure 1: The structure of part of a protein molecule

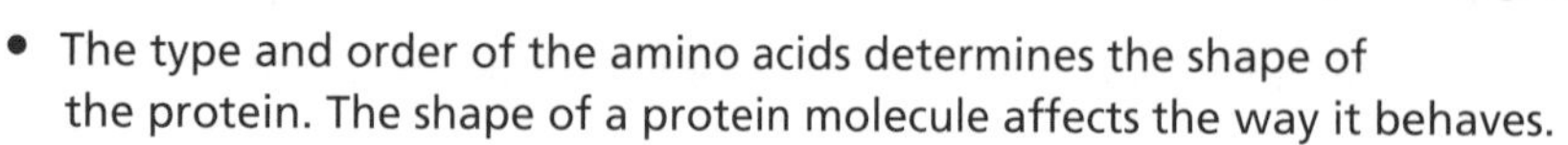

- The type and order of the amino acids determines the shape of the protein. The shape of a protein molecule affects the way it behaves.

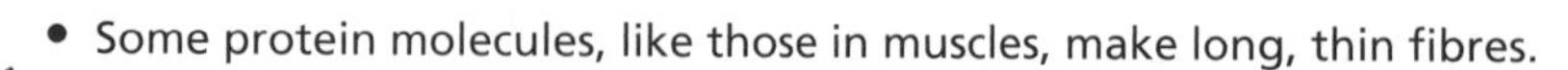

- Some protein molecules, like those in muscles, make long, thin fibres.

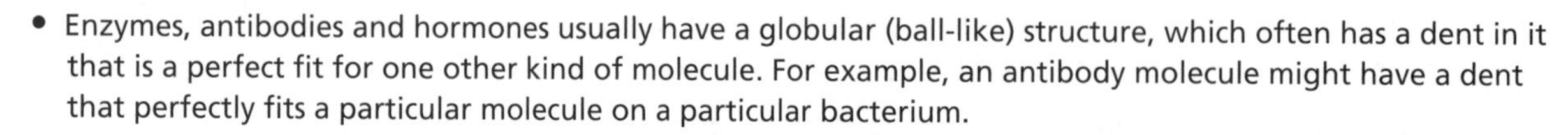

- Enzymes, antibodies and hormones usually have a globular (ball-like) structure, which often has a dent in it that is a perfect fit for one other kind of molecule. For example, an antibody molecule might have a dent that perfectly fits a particular molecule on a particular bacterium.

Enzymes

Biological catalysts

G–E

- There are hundreds of different chemical reactions going on in our bodies. Each one is controlled by an enzyme.

- Enzymes are biological catalysts. All enzymes:
  - are protein molecules
  - control one specific chemical reaction
  - make the reaction happen quickly.
- Enzymes, like a catalyst for any chemical reaction, are not used up.

Factors affecting enzyme activity

D–C

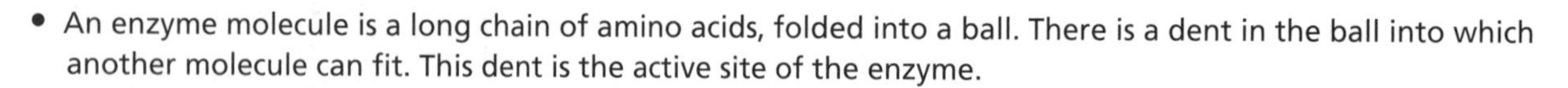

- An enzyme molecule is a long chain of amino acids, folded into a ball. There is a dent in the ball into which another molecule can fit. This dent is the active site of the enzyme.

- The molecule that fits into the enzyme is called its substrate. The enzyme makes the substrate react, changing it into a new substance.
- Most of the enzymes in the body work best at about 37 °C, which is normal body temperature. This is their optimum temperature. At temperatures above the optimum, the enzyme begins to uncurl and lose its shape. Once the active site has lost its shape, the substrate no longer fits.
- When the enzyme is permanently changed in this way, it is said to be denatured.
- Enzymes are also sensitive to pH. If the pH is a long way from the enzyme's optimum pH, then the enzyme denatures.

Improve your grade

Pepsin is an enzyme that helps break down proteins in the stomach. It has an optimum pH of 2. Use this information to explain why the stomach produces hydrochloric acid. **AO2 (2 marks)**

Enzymes and digestion

Digestion

- During digestion, food molecules are broken down into smaller ones which can get through the gut wall and into the blood.
- The blood carries them all over the body.
- Most of this breaking up of food is done by enzymes.

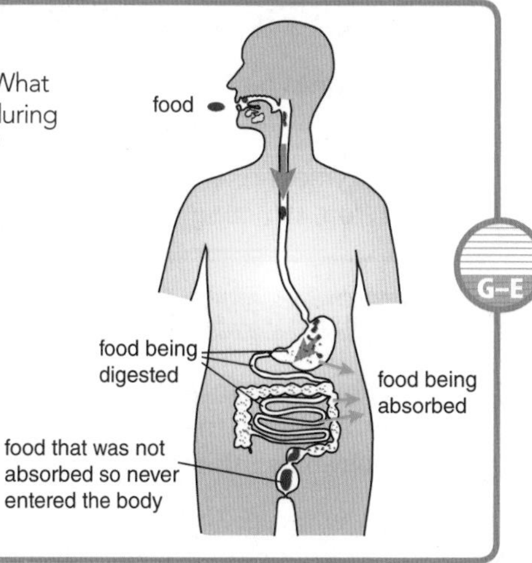

Figure 2: What happens during digestion?

G–E

Types of enzymes

- Digestive enzymes pass out of cells in glandular tissue, go into the space inside the gut and become mixed up with the food.
- They cut large food molecules into smaller bits.
- There are three main groups of digestive enzymes (see Table 1).

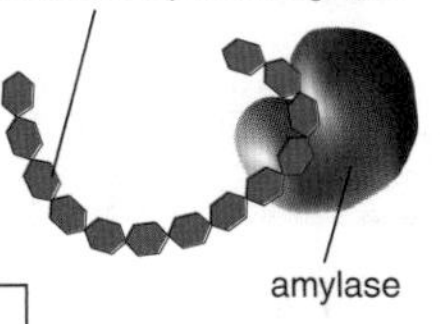

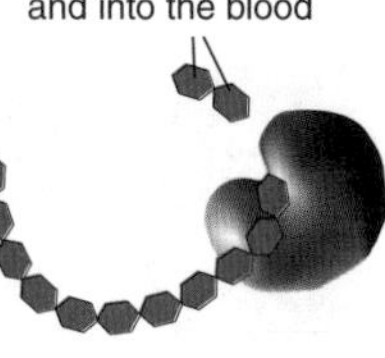

Figure 3: How amylase digests starch

D–C

	Substrate	Product	Where enzyme is produced
amylase	starch	sugars	salivary glands in the mouth and pancreas
protease	protein	amino acids	pancreas, stomach and small intestine
lipase	lipids (fats and oils)	fatty acids and glycerol	pancreas and small intestine

Table 1: The main groups of digestive enzymes

Enzymes at home

Enzymes from microorganisms

- Some microorganisms, such as yeast and bacteria, make enzymes that pass out of their cells.
- If these microorganisms are grown in large quantities, the enzymes that they produce can be collected.
- There are many different uses to which these enzymes can be put.

G–E

Biological detergents

- Biological washing powders (detergents) contain lipase and protease enzymes.
- Some stains on clothes – for example blood stains – cannot be removed using ordinary detergents. The enzymes help to break down the stains into substances that dissolve in water. The stains can then wash away.
- The enzymes in detergents often work best at about 30 °C.

EXAM TIP

You should be able to apply what you know about how temperature affects enzymes to explain why biological washing detergents should be used at temperatures no higher than around 30 °C.

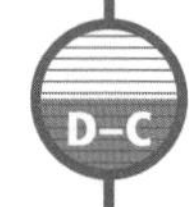

Improve your grade

Dipesh mixed some starch with amylase in a beaker and left the mixture in a water bath at 37 °C. After 30 minutes, he tested the mixture to see if there was any starch present. Predict what he will find and give a reason for your answer. **AO2 (2 marks)**

Enzymes in industry

Using enzymes for food manufacture

- Enzymes have many uses in industry. They can make reactions happen at low temperatures and pressures, which saves money.
- Enzymes are especially useful in the food manufacturing industry. They are used to make baby foods, sugar syrup, slimming foods and soft-centre chocolates.

Baby foods, sugar syrup and slimming foods

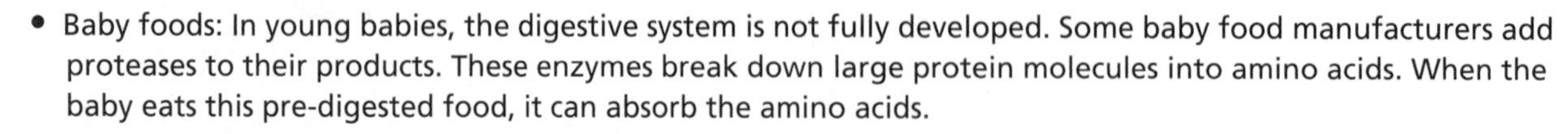

- Baby foods: In young babies, the digestive system is not fully developed. Some baby food manufacturers add proteases to their products. These enzymes break down large protein molecules into amino acids. When the baby eats this pre-digested food, it can absorb the amino acids.
- Sugar syrup: This is used in making sweets and sports drinks. Starch solution is easy to make by cooking potatoes or maize and mixing them with water. The starch can then be changed into sugar syrup by adding carbohydrase enzymes such as amylase.
- Slimming foods: These often contain a very sweet sugar called fructose instead of glucose. Fructose is made from glucose using an enzyme called isomerase.

Aerobic respiration

Getting energy

- All living cells require energy to survive.
- All of your energy comes from the food that you eat.
- The energy is locked up inside the food molecules. To release the energy, cells have to break these molecules apart. This is normally done using glucose molecules and is called respiration.

Releasing energy

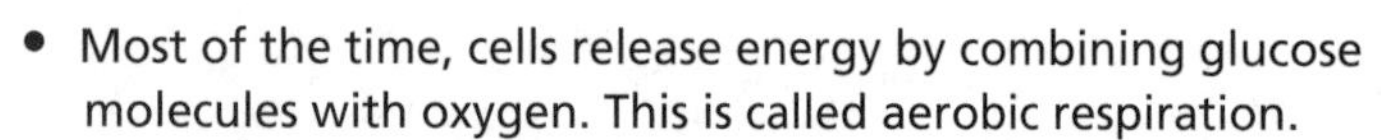

- Most of the time, cells release energy by combining glucose molecules with oxygen. This is called aerobic respiration.
- The word equation is:
  glucose + oxygen → carbon dioxide + water (+ energy)

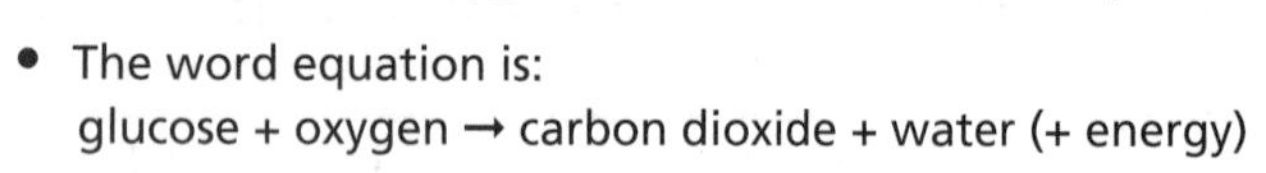

- Your body obtains oxygen from the air which enters your blood in your lungs. It is transported in the blood to all your body cells. The carbon dioxide that the cells make is carried back, in the blood, to the lungs.
- Aerobic respiration takes place in the mitochondria of cells.
- They contain all the enzymes that are needed to make the reactions of respiration happen quickly.

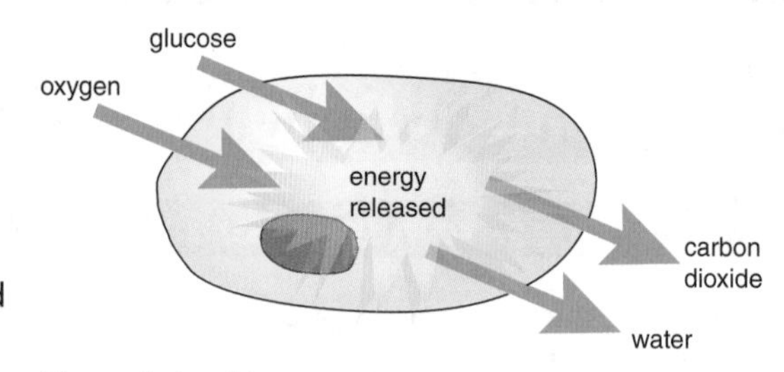

Figure 1: Aerobic respiration

Improve your grade

Explain why breathing rate increases when you exercise. **AO2 (3 marks)**

Using energy

Why do cells need energy?

- Energy is constantly being released from food molecules inside your cells, by respiration.
- Figure 2 shows what the body uses the energy for.

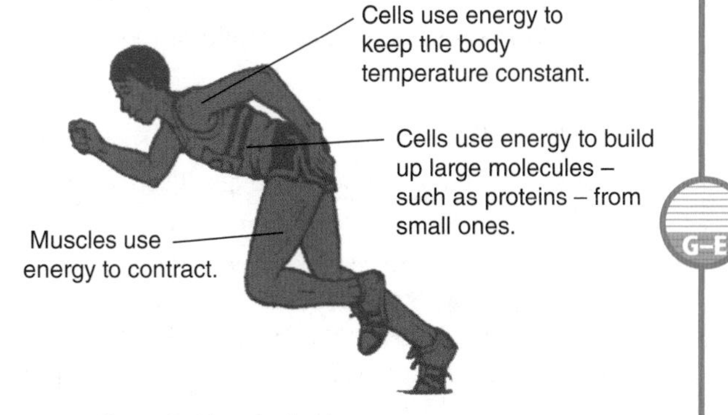

Figure 2: How the body uses energy

Using the energy

- Building large molecules: Cells use energy to make small molecules join together to make long chains.
- Muscle contraction: Muscles use energy to contract. They contain a store of glucose called glycogen, which is made up of many glucose molecules linked together. When a muscle needs energy, it breaks down the glycogen to produce glucose for use in respiration.
- Maintaining a steady body temperature: Mammals and birds keep their body temperature around 37 °C. If their environment is colder than they are, then heat is lost from their body. Respiration in cells releases energy, which increases body temperature.
- Making amino acids: Plants can make amino acids from sugar and nitrate ions. This requires energy.

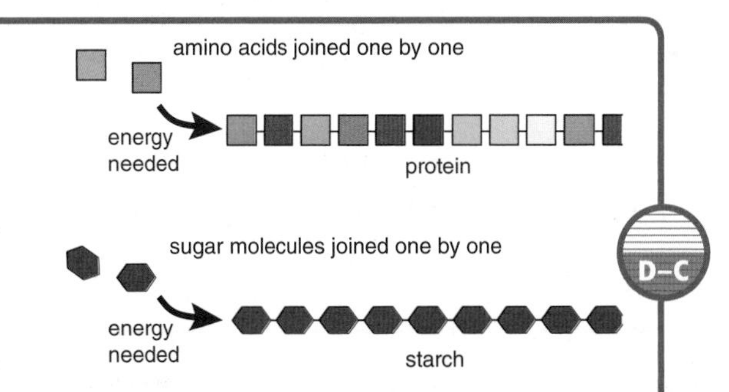

Figure 3: Linking molecules together requires energy

D–C

Anaerobic respiration

Anaerobic respiration

- Respiring muscles rely on oxygen being brought to them by the blood.
- During a sprint race, the muscles are using energy so quickly that even though the heart is working as hard as it can to pump more blood – carrying oxygen – to them, it is still not enough.
- The muscles therefore resort to releasing energy from glucose without using oxygen.
- This is called anaerobic respiration.

Lactic acid and oxygen debt

- Anaerobic respiration is incomplete breakdown of glucose to lactic acid: glucose → lactic acid (+ a little energy)
- It releases far less energy than aerobic respiration (aerobic releases 16.1 kJ per gram of glucose, anaerobic only releases 0.8 kJ).
- Lactic acid builds up in the muscles. It makes muscles feel tired and can cause cramps. They stop contracting efficiently.

Improve your grade

You are involved in a race. At first you sprint off feeling full of energy. However, halfway through, your legs start to ache and you have to stop. Explain why this happened. **AO2 (3 marks)**

Cell division – mitosis

Why do cells need to divide?

G–E

- Human bodies are made of billions of cells.
- When you grow, cells divide to make more cells, each small new cell grows to its full size and then it may divide again.
- There is a limit to how large a cell can grow and stay alive. If cells become too big, then oxygen and carbon dioxide cannot diffuse into and out of them fast enough.

Mitosis and chromosomes

D–C

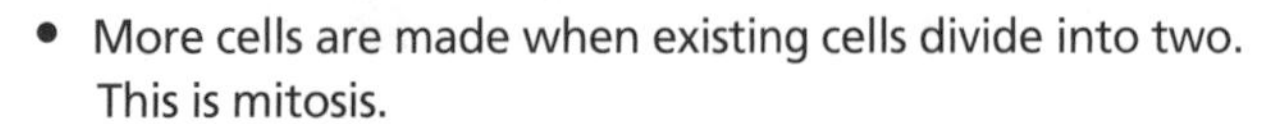

- More cells are made when existing cells divide into two. This is mitosis.

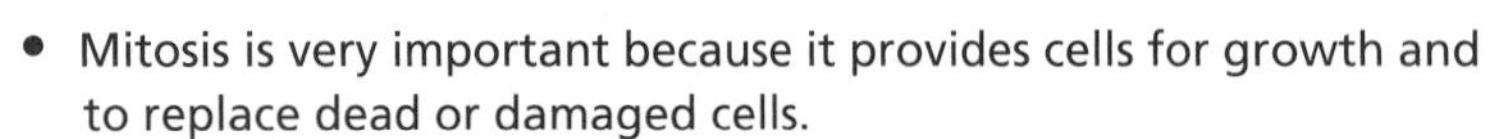

- Mitosis is very important because it provides cells for growth and to replace dead or damaged cells.
- Normal body cells have 23 pairs of chromosomes.
- Before a cell divides by mitosis, it first copies each chromosome.
- When the cell divides by mitosis, the chromosomes are shared out equally between the two new cells.

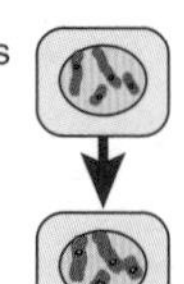

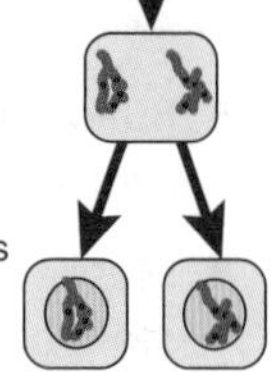

Figure 1: Before and during mitosis

Cell division – meiosis

Making gametes

G–E

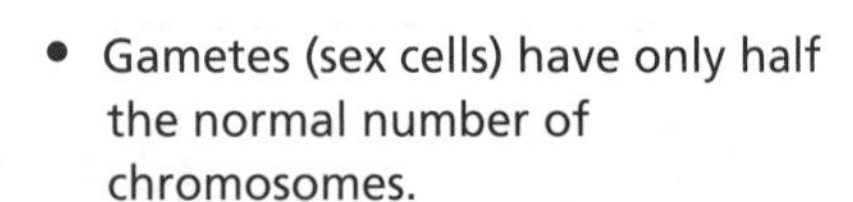

- Gametes (sex cells) have only half the normal number of chromosomes.
- When they fuse at fertilisation, the new cell ends up with the normal number again.
- In humans, gametes are made by a special kind of cell division called meiosis. Sperm cells are made in the testes. Egg cells are made in the ovaries.

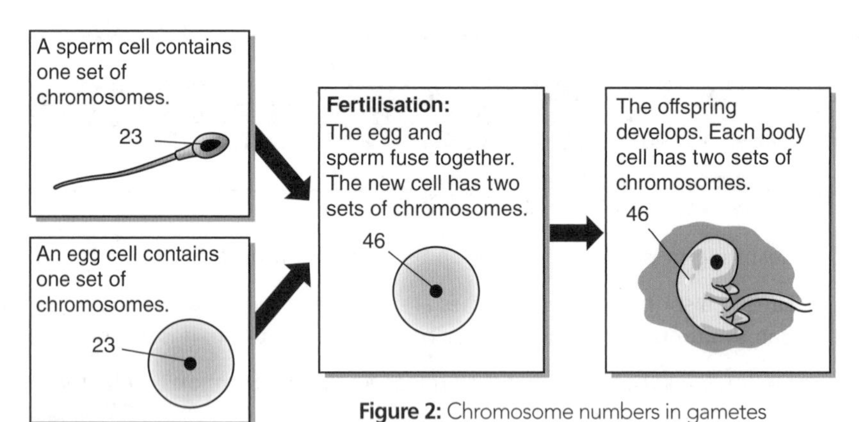

Figure 2: Chromosome numbers in gametes and other cells

EXAM TIP

It is easy to get mitosis and meiosis mixed up. Remember that meiosis is the process that produces gametes (sex cells). An easy way to remember is that meiosis contains an *e* for egg and an *s* for sperm.

Improve your grade

Explain why mitosis is an essential process in the formation of a baby from a fertilised egg. **AO2 (2 marks)**

Stem cells

What are stem cells?

- Most cells in the body are specialised. This means they are differentiated for one particular function.
- Once a cell has differentiated, it cannot change its function.
- However, some cells called stem cells have not differentiated and can turn into different types of cell.

Embryo and adult stem cells

- In animals, the fertilised egg divides over and over again to produce an embryo. For the first few days, all of these cells stay as stem cells. Each one has the potential to develop into any kind of cell in the human body.
- Most cells differentiate and form specialised cells. A few remain as stem cells and can continue to divide and specialise throughout adult life.
- In plants, many cells can differentiate at any time during their lives.
- Stem cells from embryos and adults (particularly bone marrow stem cells) could be used to replace damaged tissues in the future.

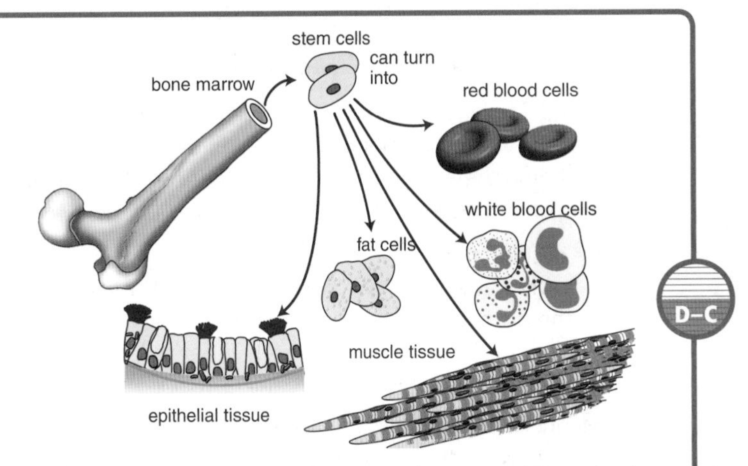

Figure 3: Bone marrow stem cells can produce many different kinds of specialised cells

D–C

Genes, alleles and DNA

Genes

- Chromosomes are made of DNA (deoxyribonucleic acid).
- Each of these chromosomes contains hundreds of genes.
- Each gene is a short length of DNA that gives instructions to the cell. Different genes give instructions for different characteristics. For example, there are genes for making the pigments that give your hair its colour.

Lactic acid and oxygen debt

- Most genes come in several different forms, called alleles. For example, a gene for hair colour might have one allele that gives black hair, another which gives brown hair, another which gives red hair and so on.
- Humans have two sets of chromosomes in each cell – one from each parent.
- There is a gene for the same characteristic at the same place on the same chromosome in each set. The two genes might be the same allele or different alleles.

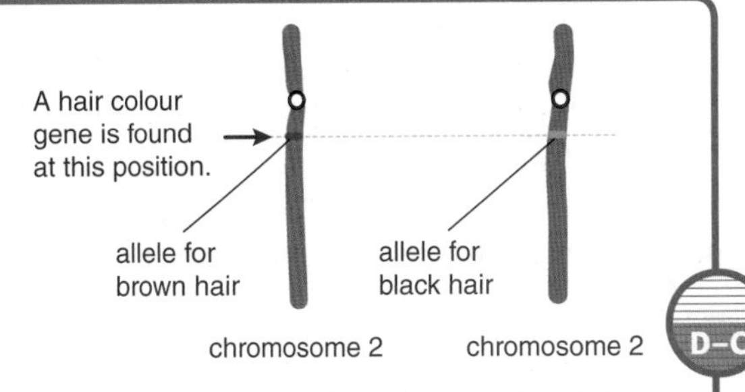

Figure 4: Genes have different forms, called alleles

D–C

- Everybody's DNA is slightly different.
- Forensic scientists often use DNA from cells in fluids, hair or bones to identify a body, or to identify a person who was at the scene of the crime. They make a DNA fingerprint by cutting up a sample of DNA into little pieces to form a series of stripes. Each person's pattern of stripes is unique.

Improve your grade

Rubina has the blood group A. Her alleles for blood group are AO. Rajesh has blood group B. His alleles are BO. Explain how their daughter, Sharmila, has a different blood group from either of them (blood group O). **AO2 (2 marks)**

Mendel

Gregor Mendel

- Gregor Johann Mendel was an Austrian monk. In 1856, he started doing experiments on inheritance in pea plants.
- His experiments involved crossing different pea plants (breeding them together).
- To do this, he took pollen containing male gametes from one flower and put it into another flower, so the male gametes could fertilise its female gametes.

Mendel's experiments

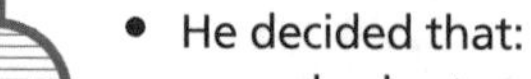

- Mendel pollinated peas with purple flowers with pollen from white flowers. All the offspring had purple flowers. Mendel then tried breeding these together. He found that the offspring were a mixture of purple and white.
- He decided that:
  - each plant must have two 'factors' for flower colour
  - the purple factors were 'stronger' than the white factors
  - each factor must be separately inherited, that is, the factors did not 'blend' together.
- Mendel presented his findings to other scientists in 1865. At that time no one knew anything about how cells divided or that chromosomes and genes existed. It was not until after Mendel died in 1900 that other scientists rediscovered his work. Today, we know that Mendel's 'factors' are alleles of genes.

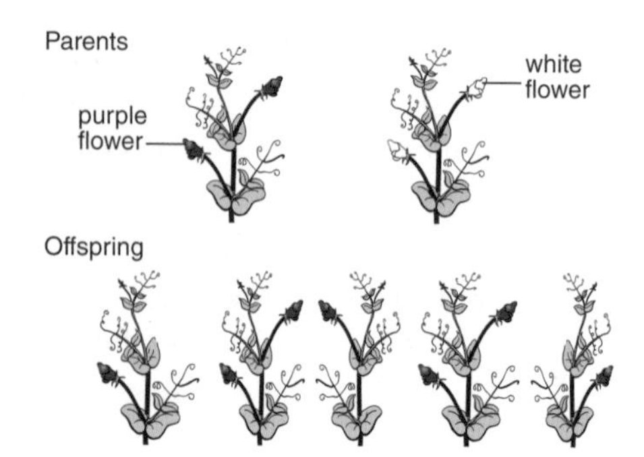

Figure 1: The result of crossing purple flowered pea plants with white flowered pea plants

How genes affect characteristics

Sex and alleles

- The nucleus of each body cell contains 23 pairs of chromosomes. One pair is called the sex chromosomes.
- There are two kinds of sex chromosomes, X and Y.
- In a female, both sex chromosomes are X. In a male, one is X and one is Y. (Females are XX, males are XY.)

Alleles come in pairs

- In rabbits, there might be a gene for fur colour: one allele of the gene may give black fur, another allele may give white fur.
- In a rabbit's cells, there are two complete sets of chromosomes. This means that there are two copies of each gene.
- If you call the allele for black fur B, and the allele for white fur b, then there are three possible combinations: BB, Bb or bb.
- Bb produces black fur. This is because the B allele is dominant. The b allele is recessive.
- When a dominant and a recessive allele are together, only the dominant one has an effect.

Figure 2: The three possible combinations of alleles for fur colour in a rabbit's cells

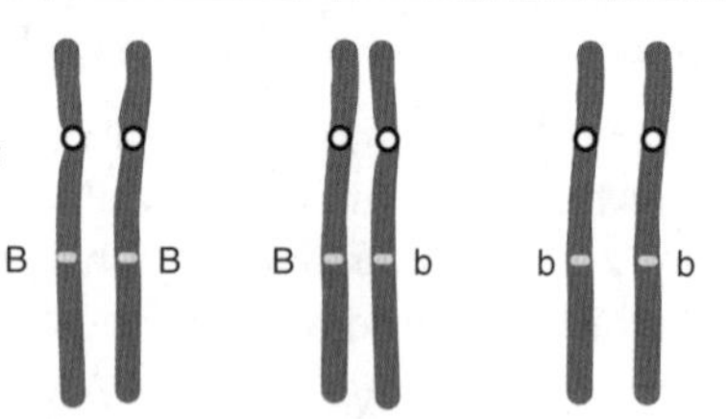

Figure 3: A rabbit with black fur could have the alleles BB or Bb. A rabbit with white fur can only have the alleles bb

Improve your grade

Katy has red hair. Both her parents have brown hair. Explain how Katy inherited red hair when her parents do not have it. **AO2 (3 marks)**

Inheriting chromosomes and genes

Genetic diagrams for gender

- A woman's eggs will all contain an X chromosome.
- A man makes two kinds of sperm cells. Half of them will contain an X chromosome and half will contain a Y chromosome.
- There is an equal chance of a sperm with an X chromosome (producing a girl), or one with a Y chromosome (producing a boy), fertilising an egg.
- This can be shown in a genetic diagram.

Parents XX female XY male

Gametes (X) (X) and (Y)

Offspring

	(X)	(Y)
(X)	XX	XY

Figure 4: A genetic diagram for gender

G–E

Genetic diagrams for alleles of genes

- Genetic diagrams are used to show how alleles of genes are inherited.
- Figure 5 shows the outcome of a male rabbit with alleles Bb for fur colour breeding with a female with bb.
- Half of the sperm cells will have allele B and the other half will have allele b.
- All of the female's eggs will contain allele b.
- The genetic diagram shows you to expect about half of the baby rabbits to have the alleles Bb and have black fur. The other half would be bb and have white fur.
- It is important to remember that a genetic diagram only shows chances, not the actual results of the cross.

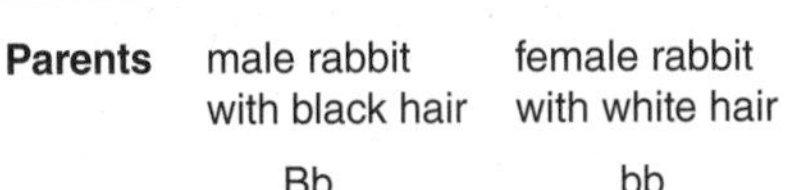

Parents male rabbit with black hair Bb female rabbit with white hair bb

Gametes (B) and (b) all (b)

Offspring

	(B)	(b)
all (b)	Bb black	bb white

Figure 5: Rabbit fur colour

Remember!

It is usually a good idea to complete a genetic diagram by summarising the approximate chances of getting each of the different genotypes and phenotypes. In the above cross, the expected genotype ratio is 1Bb : 1bb.

How genes work

DNA and codes

- Each of the chromosomes in the nucleus of a cell is made of a very long molecule of DNA twisted into a 'double helix' shape.
- Each gene is a length of DNA that codes for a particular characteristic.
- Genes store information as a type of code, called the genetic code.
- It provides instructions to the cell about which proteins it should make.

chromosomes in a nucleus

genes within a chromosome

bases

A gene is a particular length of DNA.

Figure 6: Chromosomes are made from DNA

G–E

EXAM TIP

Make sure you use the correct terms. A gene controls a certain characteristic, for example the gene for hair colour. Alleles are different versions of the same gene, for example there are alleles for brown and for blond hair.

Improve your grade

The height of pea plants is controlled by a single gene. The tall allele is dominant. A tall pea plant was bred with a short pea plant. The offspring formed was in the ratio 1 tall : 1 short. Draw a genetic diagram to show the cross. **AO2 (4 marks)**

Genetic disorders

Inherited disorders

- An inherited disorder is caused by faulty genes.
- They are inherited from a person's parents.
- Polydactyly and cystic fibrosis are two examples of inherited disorders.

How they are inherited

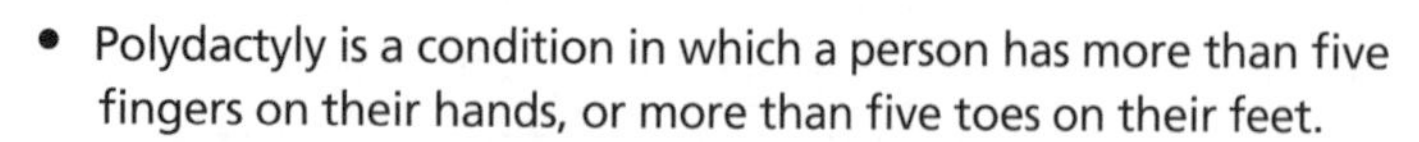

- Polydactyly is a condition in which a person has more than five fingers on their hands, or more than five toes on their feet.
- It is caused by a dominant allele, so you only need to inherit one allele in order to have this condition.
- Figure 1 shows how polydactyly was inherited in one family.
- Cystic fibrosis is a disorder of cell membranes and affects the lungs and pancreas.
- It is caused by a recessive allele.
- A person with the disorder will have two recessive alleles. Someone with one recessive allele is called a carrier.

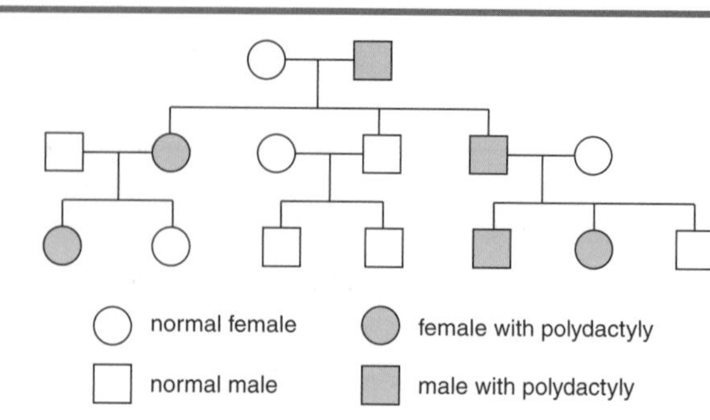

Figure 1: Example of a family tree for polydactyly

Remember!

Carriers of cystic fibrosis will not have any symptoms of the illness. They may not even know they are a carrier unless they pass this allele on and have a child with cystic fibrosis.

Fossils

Early life-forms

- Fossils are the remains or impressions made by dead organisms. They are usually preserved in layers of rock.
- The fossils in each layer are a record of the life on Earth at the time when the layer was formed.
- Fossils are evidence of how life on Earth has changed (evolved) over time.

The fossil record

- Fossils provide evidence which suggests that all species of living things that exist today have evolved from simple life-forms.
- The fossil record for most species is incomplete because most organisms do not form fossils when they die (they decay instead).

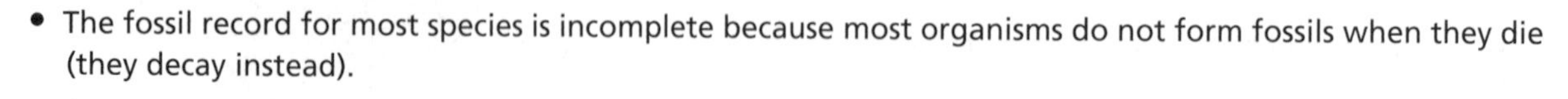

- However, there are a number of ways that the remains of plants and animals can be preserved as fossils.
  - Bones and teeth do not easily decay.
  - Some parts of organisms do not decay because conditions are not suitable for decay organisms.
  - Parts of an organism may be replaced by other materials such as hard minerals as they decay.
  - Traces of an organism, like footprints, may be preserved in rocks as prints.
- The fossil record for some organisms such as the horse, is complete and shows how it has evolved over the last 50–60 million years.

Improve your grade

Polydactyly is caused by a dominant allele. Jessica has polydactyly. Ryan does not have it. Could their children have polydactyly? Explain your answer. **AO2 (3 marks)**

Extinction

What is extinction?

- When a species completely dies out, and there are no more individuals left, the species has become extinct.
- Living things become extinct for several different reasons.
  - The environment in which they live might change.
  - All the animals in a species might be eaten by a new predator.
  - A new disease might kill them off.
  - A new, more successful, competitor might move into the habitat.

Causes of extinction

- Life on Earth began about 3.8 billion years ago. As Earth slowly changed, many different plants and animal species became extinct (died out) and were replaced by new types.
- There are several causes of extinction:
  - a change in the environment – for example, the temperature might increase and so individuals that cannot adapt or move away will die out
  - new predators – may cause the extinction of species that are its prey and are not adapted to get away
  - new diseases – if a new, fatal, disease is introduced to a habitat then the species living there may not have immunity and so will die
  - a more successful competitor – if there is a limited amount of food in a habitat and two different species are competing for it, the less well-adapted one may become extinct.

Remember!
If a species is not able to quickly adapt to new situations it may become extinct.

New species

New species

- Many biologists define a species as a group of organisms that share similar characteristics, and that can breed together to produce fertile offspring.
- Siberian and Bengal tigers live so far apart that they never come into contact with each other. They are isolated from one another. If this situation went on for long enough, the two kinds of tigers might gradually become more and more different from one another.
- Eventually, they might be so different that – if they came into contact with one another – they might not be able to breed together. They would be different species.

How new species arise

- Many new species arise through a series of steps. Here is an example.

1 Geographical isolation – a few lizards drift away from the mainland on a floating log and end up on an island. They are isolated from the rest of the species.

2 Genetic variation – in both the mainland lizards and the island lizards, there are many different alleles of genes which lead to variation.

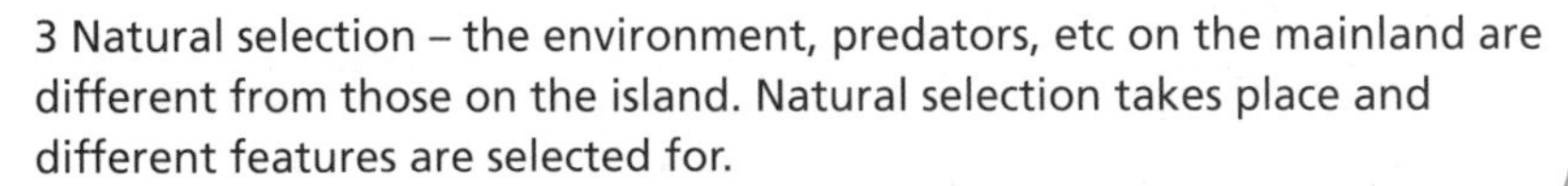

3 Natural selection – the environment, predators, etc on the mainland are different from those on the island. Natural selection takes place and different features are selected for.

4 Speciation – over time, more and more differences build up between the two populations of lizards until they are no longer the same species.

Remember!
Biologists define a species as a group of organisms that share similar characteristics, and that can breed together to produce fertile offspring.

Improve your grade

Around 300 years ago, the dodo lived on the island of Mauritius. It had no predators. People arrived on the island. They brought predators such as dogs, pigs and rats. Explain why the dodo became extinct. **AO2 (2 marks)**

Cells, tissues and organs

The parts of cells are called **organelles**. Each has an important part to play in the function of the cell.

Plant, animal and microbial cells have many similarities and some differences.

Oxygen and other dissolved substances move into and out of cells by **diffusion**.

In animals and plants, cells are grouped into **tissues**. **Organs** are made up of different tissues working together to carry out a function.

Different types of cells have different structures which enable them to carry out their specific functions. They are **specialised**.

Plants and the environment

Chlorophyll in plant cells absorbs energy from sunlight. This energy is used to convert carbon dioxide and water to glucose and oxygen in a reaction called **photosynthesis**.

The distribution of different species of organism in the environment is affected by physical factors. These include temperature, nutrients, light, water, oxygen and carbon dioxide.

Limiting factors, including light, temperature and carbon dioxide concentration, affect the rate of photosynthesis.

Glucose is used to supply energy; to make **cellulose** for cell walls; to make storage substances like **starch** and fats; and to make protein to build new cells and enzymes.

Quantitative data about the distribution of organisms can be collected using **quadrats** and **transects**.

Proteins and respiration

Protein molecules are long chains of **amino acids**. Hormones, antibodies and enzymes are all proteins.

Enzymes are biological catalysts which control all metabolic reactions, including digestion. They are also used in the home and in industry.

All living organisms release energy from glucose by respiration.

The rate at which enzymes work is affected by temperature and pH.

In **aerobic respiration**, energy is released when glucose is combined with oxygen. During exercise, when muscles are using a lot of energy, heart rate and breathing rate increase to provide muscles with extra oxygen.

In **anaerobic respiration**, a small amount of energy is released from glucose without using oxygen.

Cell division, inheritance and speciation

Cells normally divide by **mitosis**, which results in two genetically identical daughter cells. To produce gametes, cells divide by **meiosis**, which produces genetically different cells, each with half the normal number of chromosomes.

Genes are passed from one generation to the next. A gene controlling a particular characteristic may have different forms, called alleles. Some diseases, such as cystic fibrosis, can be inherited.

Genetic diagrams can be used to predict the probable characteristics of the offspring of two parents.

Clones can be created artificially by taking cuttings and carrying out **tissue culture**, **embryo transplants** and **adult cell cloning**.

New species can arise if two populations of a species become separated. Natural selection may result in them becoming so different that they can no longer interbreed.

Fossils provide evidence about some of the species that lived long ago.

Investigating atoms

What's inside an atom?

- Atoms are made up of three small particles, protons, neutrons and electrons. Table 1 shows their properties.

Table 1: Properties of the particles within atoms

Particle	Relative mass	Relative charge
proton	1	+1
neutron	1	0
electron	very small	–1

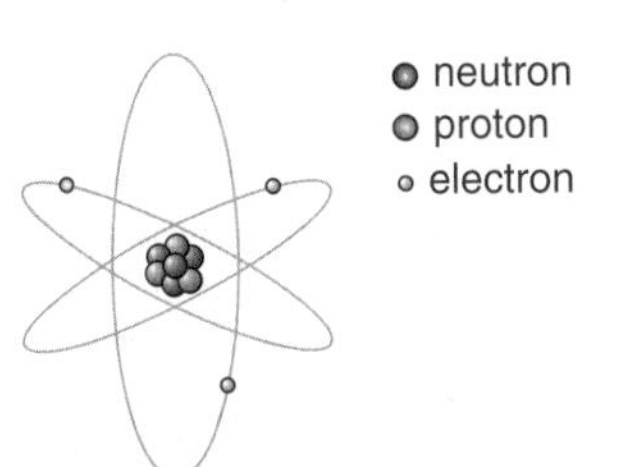

Figure 1: Protons, neutrons and electrons in an atom

- The atomic number is the number of protons. This is the same as the number of electrons in the atom.
- The mass number is the number of protons and neutrons in the atom's nucleus.

How ideas about atoms have changed

- All matter consists of atoms, which cannot be divided up.
- All atoms of the same element are identical, but different from atoms of every other element.
- The Greeks suggested the idea of atoms in the 5th century BC, John Dalton revived the ideas in the 19th century.
- In 1897 the electron was discovered, 1909 the nucleus was discovered, 1911 Rutherford suggested that electrons orbit the nucleus, 1919 protons were discovered, and 1932 neutrons were discovered.

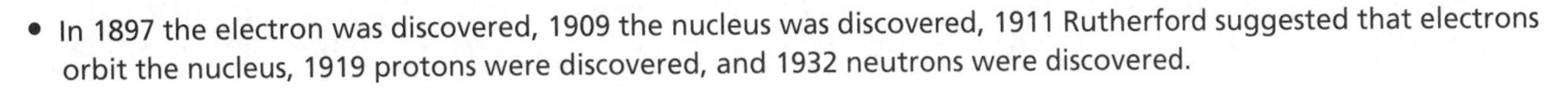

Mass number and isotopes

Mass number and isotopes

- Atoms of the same element have the same number of protons. However, they can have different numbers of neutrons. Atoms of the same element with different numbers of neutrons are called isotopes.
- Isotopes are named by giving the mass number after the element's name or symbol, for example, carbon-14.

Properties of isotopes

- Not all atoms of an element have the same number of neutrons. Atoms of the same element with different masses are called isotopes.
- The proportion of each isotope of an element is fixed, and it is the same in both the pure element and in the compounds containing that element.
- Because isotopes have the same number of electrons, they all react in exactly the same way.

How Science Works

- The relative atomic mass (A_r) of an element is the average mass of an atom. You work it out using the percentage or number of atoms of each different isotope and its mass. Potassium has two isotopes, 90 per cent of the atoms are of potassium-39; the other 10 per cent are of potassium-40. The A_r of potassium is $\frac{(90 \times 39) + (10 \times 40)}{100} = 39.1$

Improve your grade

Chlorine has two isotopes, chlorine-37, and chlorine-35. What is the difference between the two isotopes?
AO1 (2 marks)

Compounds and mixtures

Compounds and their formulae

- A chemical compound consists of two or more different elements chemically combined. It can be broken down into simpler substances only by chemical reaction and has different properties to the original elements.
- A chemical formula shows the type and number of atoms in each molecule. A glucose molecule is made from 6 carbon, 12 hydrogen and 6 oxygen atoms, so its formula is $C_6H_{12}O_6$.
- Some compounds have enormous structures. Their formulae show the ratio of the elements in the compound.

Relative formula mass

- To work out the relative formula mass (M_r) of a compound you find the atomic mass of each element multiplied by the number of atoms of that element present in the formula, then add them up.
- calcium sulfate, $CaSO_4$

  atoms and A_r Ca = 40 S = 32 O = 16

  so (40 × 1) + (32 × 1) + (16 × 4) = 136
- Atoms and molecules are far too small to be counted. To make counting atoms or molecules easy, chemists call the atomic or formula mass measured in grams a mole.

Electronic structure

Subatomic particles

- Inside the atom, the protons and neutrons are in a central nucleus with the electrons orbiting in shells around the nucleus.

Remember!
The number of outer electrons of an element can be found by looking on the periodic table and finding the group number of the element.

Electronic structures

- The electrons occupy the lowest available energy levels. They fill up each shell in turn. 2 electrons go in the first (except H which has only 1), up to 8 in the second, the third shell accepts 8 electrons and if the third shell fills up, then the remaining electrons go into the fourth shell – up to calcium.

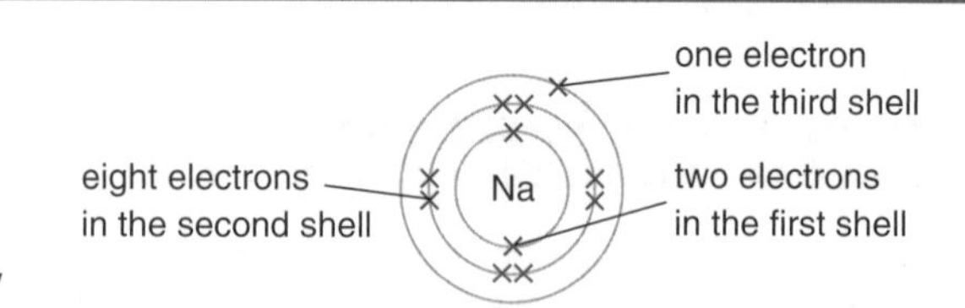

Figure 1: The electronic arrangement of a sodium atom, 2, 8, 1

- For elements beyond calcium the number of outer electrons is the same as the Group number. The number of outer electrons determines the reactivity of an element.

Improve your grade

Use the periodic table to work out the electronic structure of these elements: *magnesium, chlorine, nitrogen, aluminium, calcium.* **AO2 (5 marks)**

Ionic bonding

What is ionic bonding?

- Each atom has an equal number of protons and electrons, so it has no overall charge. An ion forms when an atom gains or loses electrons and becomes charged.

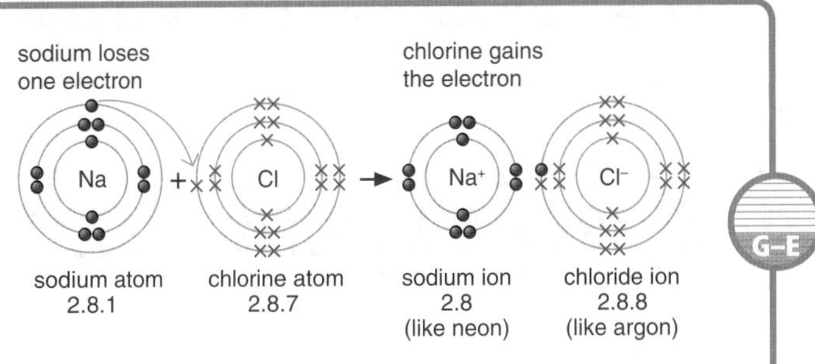

Figure 2: How sodium and chlorine use ionic bonding to form sodium chloride

How does ionic bonding happen?

- When a metal reacts with a non-metal the metal atoms lose electrons and become positive ions, the non-metal atoms gain electrons and become negative ions.
- A sodium atom (Na) loses an electron to form a sodium ion (Na^+) with a 1+ charge. It has become a positive ion. A chlorine atom (Cl) gains an electron to form a chloride ion (Cl^-) with a 1– charge. It is now a negative ion.
- The oppositely charged ions are electrostatically attracted. This is ionic bonding.

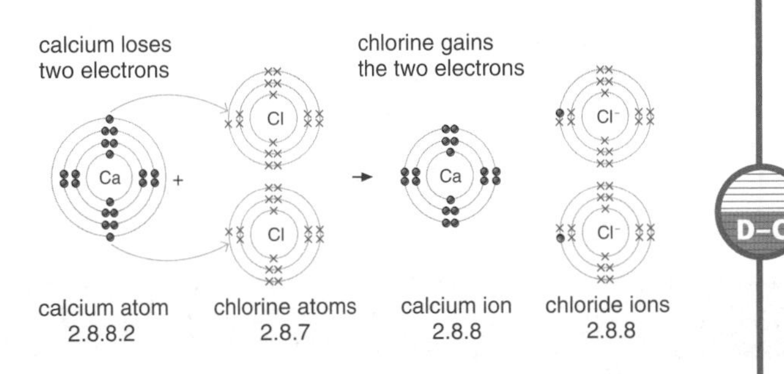

Figure 3: How calcium forms ionic bonds

Alkali metals

Properties of Group 1 elements

- Group 1, the alkali metals, react vigorously with water. These reactions produce hydrogen gas and form metal hydroxide solutions which are alkaline. This is one reason for calling them 'alkali metals'.

- They are stored under oil to prevent them reacting with air or moisture, and are shiny when first cut, but rapidly tarnish in air.
- The alkali metals react readily with non-metal elements to form ionic compounds.

Similarities and differences in reactions

- Group 1 metals react in similar ways because their atoms all have just one electron in their outer shell.
  - lithium just floats on water and fizzes gently
  - potassium bursts into flame, zooms across the water and spits
  - caesium explodes on contact with water.
- The reactivity increases down the Group.

How Science Works

- Trends or patterns, like the reactivity of Group 1 metals, are important. For example, rubidium is below potassium and above caesium in Group 1. So it will probably have a more violent reaction with water than potassium, but less than caesium.

Improve your grade

Draw diagrams to show how calcium and fluorine lose and gain electrons to make the compound calcium fluoride (CaF_2). **AO1 (3 marks)**

Halogens

Properties of Group 7 elements

- The halogens are a family of reactive non-metal elements. They make up Group 7, at the right-hand side of the periodic table.

The chemical reactivity of halogens

- This table shows information about Group 7 elements, the halogens.

Table 1: The halogens

name	symbol	description	molecule
fluorine	F	pale yellow poisonous gas	F_2
chlorine	Cl	green poisonous gas	Cl_2
bromine	Br	dark orange–red poisonous liquid that easily vaporises	Br_2
iodine	I	a dark grey solid which on heating becomes a purple gas	I_2

- When halogen elements form compounds fluorine becomes fluoride, bromine becomes bromide , and the ion always has a charge of 1–.
- Fluorine is the most reactive Group 7 element, then chlorine, bromine, and finally iodine which is least reactive.

Ionic lattices

Ionic compounds

- Ionic compounds are solid at room temperature; have high melting and boiling points; and conduct electricity when molten or dissolved in water, but not when solid.

Electronic structures

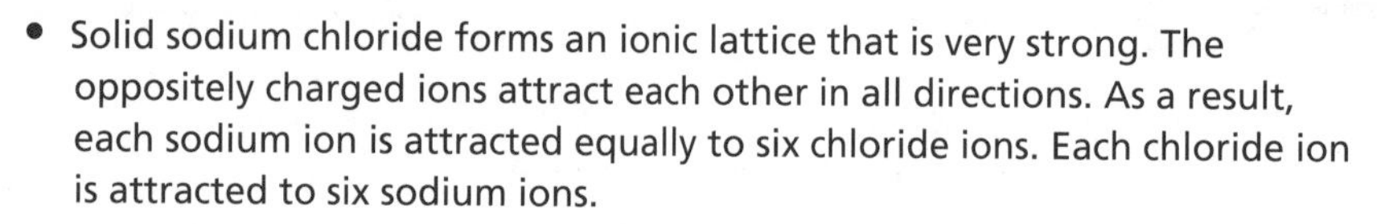

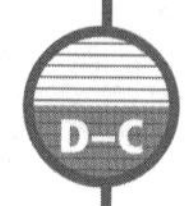

- Solid sodium chloride forms an ionic lattice that is very strong. The oppositely charged ions attract each other in all directions. As a result, each sodium ion is attracted equally to six chloride ions. Each chloride ion is attracted to six sodium ions.
- Melting and boiling points are high, as it is hard to separate the ions from each other. Only when the ions are free to move by melting or dissolving in water can electricity be conducted. Ionic lattices have melting points greater than 500 °C.
- When ionic compounds conduct electricity (electrolysis) they decompose back to their elements. The metal element is formed at the negative electrode or cathode, the non-metal element is produced at the positive electrode or anode.

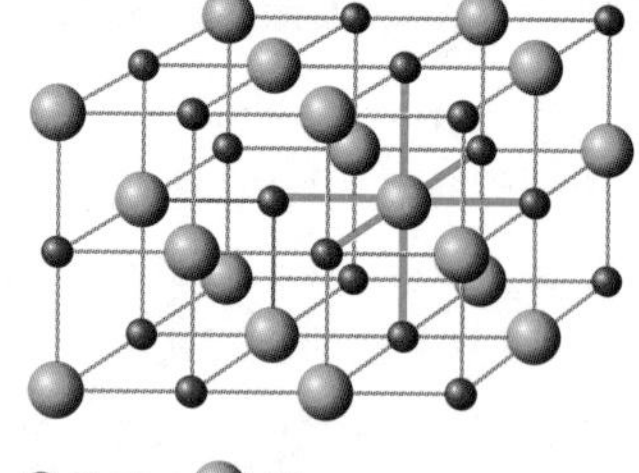

Figure 1: The sodium chloride lattice

Improve your grade

Explain in detail why sodium chloride:

a has a high melting point

b conducts electricity when in solution. **AO1 (3 marks)**

Covalent bonding

Covalent bonds

- When atoms of two non-metals combine, they share electrons to achieve noble gas electronic structures, which are stable.
- The outer electron shells of atoms overlap. This sharing forms a covalent bond which holds the two atoms together. Covalent bonds are very strong.

G–E

Sharing by numbers

- A chlorine atom has seven outer electrons, so it shares one more from another atom, see Figure 2. It can share from another chlorine atom, or any non-metal atom.
- An oxygen atom has six outer electrons, so needs two more to make eight. In water, the oxygen shares two of its electrons, one with each hydrogen to make H_2O.
- To make an oxygen molecule, oxygen atoms share two electrons with another oxygen atom to make a double bond; two shared pairs of electrons.

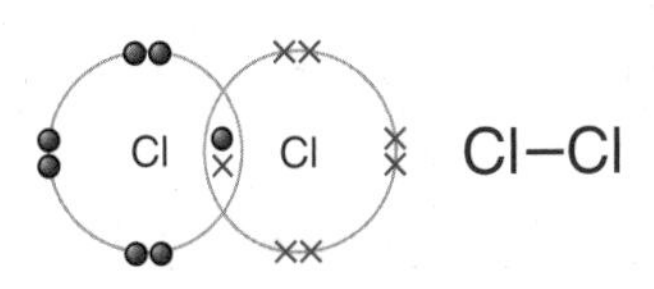

Figure 2: A chlorine molecule

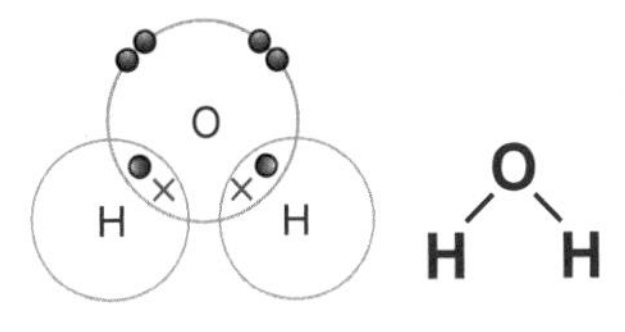

Figure 3: A water molecule

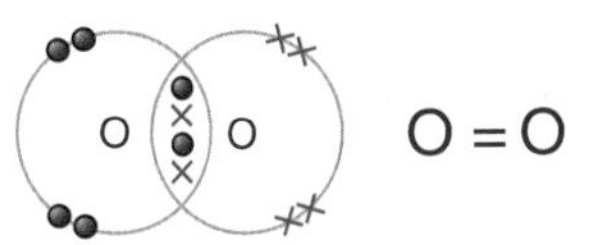

Figure 4: An oxygen molecule

D–C

EXAM TIP

When drawing electronic structures to show bonding, to save time and to make it clearer, you can miss out the inner filled electron shells.

Covalent molecules

Molecules

- Some non-metal elements exist as pairs of atoms joined together by covalent bonds, as simple molecules (H_2, N_2, O_2, Cl_2, Br_2, F_2 and I_2).
- These molecules consist of single covalent bonds, except for oxygen, which has a double covalent bond, and nitrogen, which has a triple covalent bond.
- Molecules made from just two atoms are called diatomic.

Properties

- Covalent molecules have no charged particles that are free to move. So they cannot conduct electricity.
- Whilst ionic compounds can dissolve in water, covalent compounds do not.
- Covalent bonds are very strong, but they only exist between the atoms in the molecule. There are weak attractions between the molecules, intermolecular forces. These weak forces are easily broken, so covalent molecules have low melting and boiling points.

Improve your grade

Draw diagrams to show the covalent bonds present in these compounds:
HCl, H_2, CO_2, NH_3, and CH_4. **AO1 (5 marks)**

Covalent lattices

Giant covalent structures

- Not all covalent substances exist as molecules. Some have giant structures or macromolecules.
- The whole diamond crystal or grain of sand is one enormous macromolecule, containing billions of atoms. Every atom is joined to all the others through a network of strong covalent bonds.

Bonding structure and properties

- Some covalent compounds have giant structures. These structures are called lattices.
- Examples of giant structures include diamond and graphite and silicon dioxide (sand).
- They have very strong structures so the melting and boiling points are very high.

name	structure	hardness	electrical conductor
diamond (carbon)		hardest natural substance	no
graphite (carbon)		soft, used as a lubricant	yes
silicon dioxide (SO_2)		very hard, used as sandpaper	no

Polymer chains

Chain molecules

- Polymers are very large molecules. They contain thousands of atoms joined with covalent bonds. However, they are not giant covalent structures.
- Plastics are made up of thousands of separate polymer molecules.
- Each polymer molecule is a long chain of small molecules (monomers) joined end to end.

Polymers vary

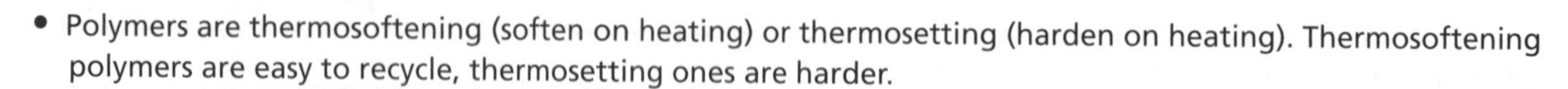

- Polymers are thermosoftening (soften on heating) or thermosetting (harden on heating). Thermosoftening polymers are easy to recycle, thermosetting ones are harder.
- A polymer's use depends on both the monomer and the process used to make it.
  - High density polyethene (HDPE) is used for plastic bottles, and water pipes.

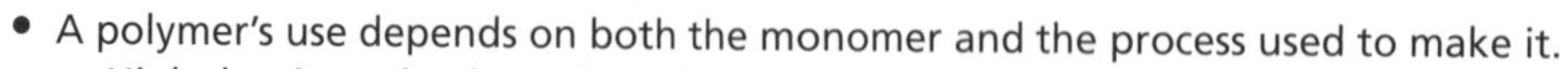

  - Low density polyethene (LDPE) is used to make film and plastic bags.

- HDPE and LDPE are made from the same monomer, but by different catalysts, temperatures and pressures, making different polyethenes with different properties.
- A thermosoftening polymer is a tangle of smooth chains. It is easy for the molecules to slide past each other and melt. A thermosetting polymer has more side chains and links to other chains. It is very hard for the chains to slide at all, so it cannot melt.

Improve your grade

Draw diagrams to show the difference between the structure of a thermosetting and a thermosoftening polymer.
AO1 (2 marks)

Metallic properties

Metals: Properties, structure and bonding

- Metals have a range of very useful properties. They are usually hard and dense, with high melting and boiling points, good conductors of heat and electricity, malleable (can be bent or hammered into shape) and ductile (can be drawn out into thin wires).
- Metal atoms are arranged in a regular lattice of rows and layers.
- Atoms in metals are held together by metallic bonding.

G–E

Alloys

- Alloys are mixtures of two or more metal elements.
- In a pure metal, all the atoms are the same size. They fit into perfectly regular layers which can slide easily. Larger or smaller atoms in alloys disrupt the regular lattice. The layers slide less easily, so alloys are harder and less malleable than pure metals.

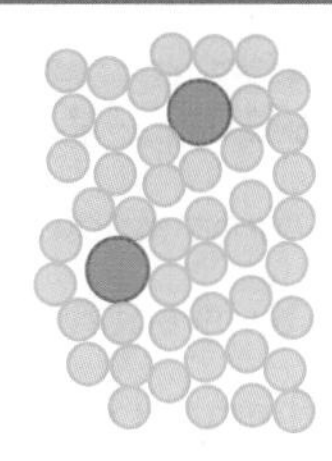

Figure 1: An alloy lattice

- Most metals in common use are alloys, because they are stronger than pure metals and have lower melting points.

- Special alloys have been developed that can remember their shape.
- These shape memory alloys such as nitinol are useful, as when warmed they return to their original shape. Uses include dental braces; and plates to hold broken bones together.
- In metals, the outer electrons of each atom are delocalised or free to move.
- As the sea of electrons can move, the metal is able to conduct electricity.
- The structure allows metals to bend, as the positive ions can move around each other.

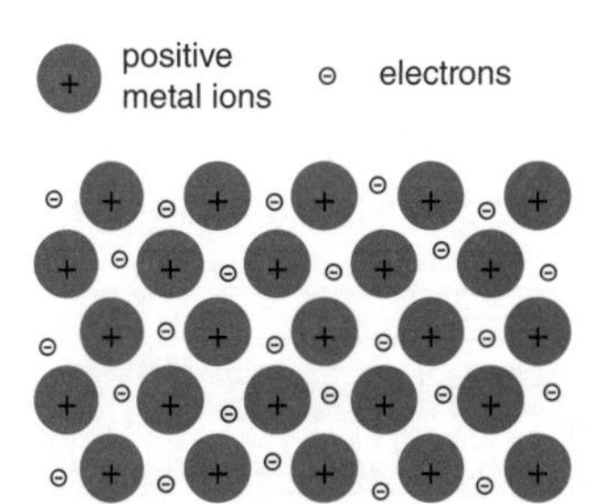

Figure 2: Metallic bonding

D–C

Modern materials

Smart materials and nanoscience

- Modern materials include 'Smart' materials, such as photochromic, thermochromic, and shape memory alloys. There are also very small materials known as nanoparticles.
  - Photochromic materials change colour according to the intensity of light.
  - Thermochromic materials change colour according to the temperature.

G–E

Applications of modern materials

- Modern materials include 'Smart' materials, such as photochromic, thermochromic, and shape memory alloys. There are also very small materials known as nanoparticles.
- Photochromic materials change colour according to the intensity of light.
- Thermochromic materials change colour according to the temperature.
- Nanoparticles are very small, being from 1–100 nm in size, with up to 300 atoms.
- Carbon has several nanoparticles, such as nanotubes; and Buckminsterfullerene (C_{60}).
- Uses of smart materials include self-cleaning and shading glass.

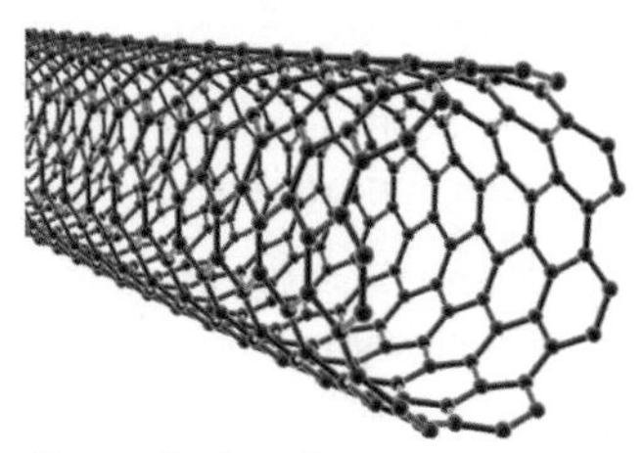

Figure 3: A carbon nanotube

D–C

Improve your grade

Explain how the metal lattice allows for metals to be stretched without breaking. **AO1 (2 marks)**

Identifying food additives

Additives and analysis

- Some food additives are man-made, and some people are allergic to certain additives judged to be safe and suffer unpleasant side-effects.
- Occasionally, illegal and/or toxic additives are found in foods. Scientists regularly perform chemical analysis on food samples to check what they contain.

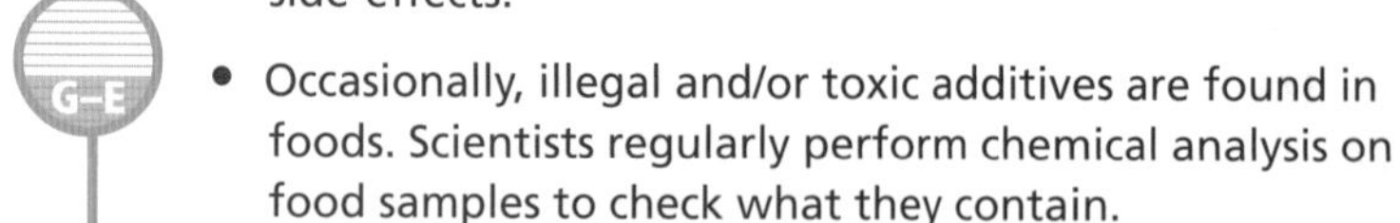

- One important method of analysis is chromatography.

Remember!
The paper used in chromatography has to dip into the solvent while leaving the sample spots above the pool of solvent – or the sample spots will dissolve in the solvent.

Chromatography

- Paper chromatography shows that the colourings in a food are those listed on the label.
- When a solvent rises up the paper, it carries the colours with it and separates them.
- Each chemical in a mixture travels up the paper at the same speed as it does when alone.
- The chromatogram in Figure 1 shows that the food contains E133 and E102, as the food has spots at the same height as the samples. It does not have any E131 and E142, but it has a third chemical that is not identified.
- Chromatography is a separating technique also used with drugs and medicines.

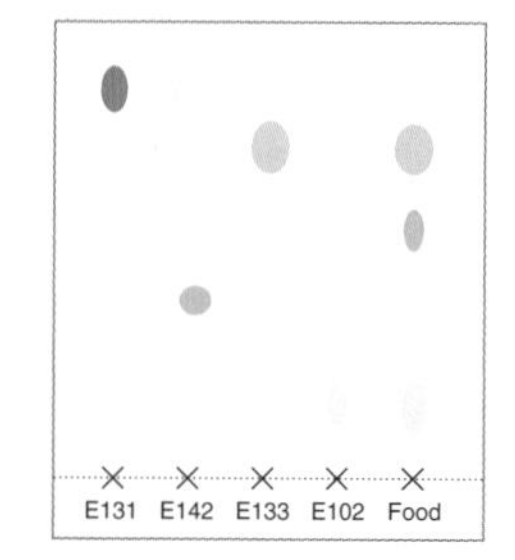

Figure 1: A chromatogram of food colours

Instrumental methods

Types of chromatography

As well as paper chromatography there is column and thin layer chromatography.

A solvent carries the mixture, and each component moves at different speeds and separates.

Retention times and mass spectrometry

- Analytical chemists use a technique called gas chromatography. They send a gas solvent with the substances to be analysed, through a tube packed with a solid material.
- The components travel through the tube at different speeds and are detected at the end. The results are shown as a graph, with each peak representing a component and the height of the peak indicating how much was present.
- The position of the peak is used to work out the retention time for each component.
- Often a mass spectrometer is attached to the gas chromatography equipment, a GC-MS method. This measures the relative atomic or molecular mass of each component.

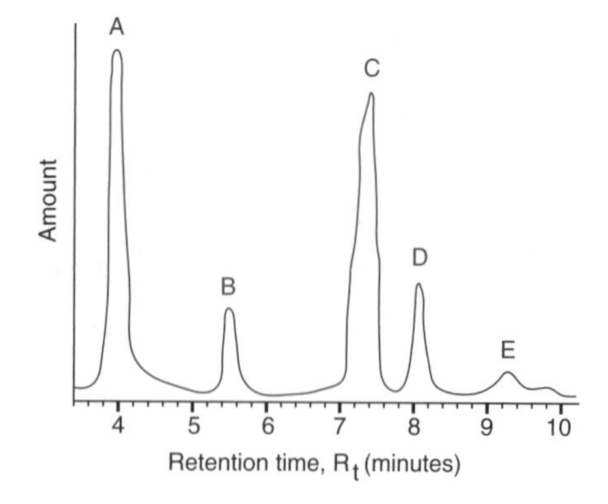

Figure 2: A gas chromatogram retention–time graph

Improve your grade

Look at Figure 2. It shows the retention–time graph for a gas chromatogram sample.

a How many different components were in the sample?

b List the components in order of the quantity of each present. Start with the most plentiful. **AO3 (2 marks)**

Making chemicals

Making chemicals is complicated

- There's more to making chemicals than just mixing them together. You need to work out how much of each reactant is needed, the best conditions to carry it out under, and how much it will all cost.
- The process must be safe, and avoid polluting the environment.

Routes and conditions

- There are several types of chemical reaction that can be used to make a chemical:
  - oxidation or reduction, losing or gaining oxygen
  - neutralisation, reacting an acid with an alkali or a base
  - precipitation, reacting two soluble compounds to make an insoluble product
  - electrolysis, using electricity to split a liquid or solution.
- Whichever type of reaction is used, the reaction conditions are important in terms of speed, costs, and safety:
  - temperature, higher temperatures make reactions go quicker
  - concentration, more concentrated solutions react faster
  - pressure, higher pressure compresses gas molecules, giving more molecules in the same volume and leading to faster reactions
  - particle size, smaller pieces react faster than larger pieces
  - catalysts, these can speed up the reaction rate, and are unchanged by the reaction.

D–C

Chemical composition

Remember!

To calculate relative formula mass you need to find the mass of each atom in the formula, multiply the mass by the number of atoms of that element in the formula, then add the mass of each element present to the others. For example: $FeSO_4$, Fe = 56, S = 32, O = 16, so (56 × 1) + (32 × 1) + (16 × 4) = 152

Percentage composition

The relative quantities of each element present in a compound can be worked out using the percentage composition method:

- find the relative formula mass (M_r), for example $FeSO_4$ = 152 (see *Remember!*)
- find the mass of the element in the compound, Fe = 56 × 1 = 56
- divide 56 by M_r and multiply by 100 to get the percentage of iron in iron sulfate = 36.8%.

How Science Works

- Farmers and gardeners like to know the NPK values of fertilisers. The NPK value tells them the percentage of each element, by mass, present in the fertiliser.

G–E

Empirical formulae

Working out a formula from experimental data

To calculate the formula of copper oxide from an experiment, you need to know: how much copper was used, how much copper oxide was produced, how much oxygen was needed.

If 1.27 g of copper was burnt, and made 1.59 g of copper oxide, what is the formula of copper oxide?

To answer this, you need to:

- calculate the mass of oxygen used as 1.59 g – 1.27g = 0.32 g
- divide the mass of copper by its
- (1.27/63.5 = 0.02) and oxygen by its A_r (0.32/16 = 0.02)
- divide each result by the smallest one, i.e. 0.02, to give the ratio of atoms in the formula, so Cu = 0.02/0.02 = 1, O = 0.02/0.02 = 1, so the formula is CuO.

Improve your grade

Calculate the formula of lithium oxide made from 5.6 g of lithium and 6.4 g of oxygen. **AO2 (2 marks)**

Quantities

Equations

- Balanced equations show the proportions of substances involved in a reaction.

Quantities from equations

- Look at this equation for the production of ammonia from nitrogen and hydrogen:
  $N_2(g) + 3H_2(g) \rightleftharpoons 2NH_3(g)$.
  The equation shows that one nitrogen molecule reacts with three hydrogen molecules to produce two molecules of ammonia gas.
- We can then work out the relative reaction masses of each reactant or product by multiplying the A_r or M_r by the number of reacting molecules for each gas.
  $N_2 = 28 \times 1 = 28$, $H_2 = 2 \times 3 = 6$, and $NH_3 = 17 \times 2 = 34$.
- 28 units of nitrogen will react with 6 units of hydrogen to make 34 units of ammonia. To work out how much ammonia can be made if 56 tonnes of nitrogen is used, you divide 56 by 28 to find the reacting amount (= 2); so 56 tonnes of nitrogen will make 34 × 2 tonnes of ammonia = 68.

EXAM TIP

Whenever there is an important or difficult calculation required that needs a balanced equation, the balanced equation will be given in the question.

How much product?

Percentage yields, economics and the environment

- How much product a reaction gives depends on the amounts of reactants used.
- The theoretical yield is calculated from the equation. It is the maximum possible yield.
- The actual yield is found by weighing the product obtained.

Calculating percentage yield

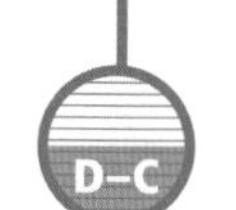

- Other factors affect the amount of product produced in a chemical reaction.
- Reactants are rarely 100% pure. The mass of actual reactant is less than the mass weighed out, so forms less product: the same reactants can form different products: if the reaction is reversible, some of the products turn back into reactants.
- To compare the effectiveness of making a chemical in different ways, we calculate the percentage yield

$$\text{percentage yield} = \frac{\text{actual yield}}{\text{theoretical yield}} \times 100$$

Remember!

When calculating percentage yields, it doesn't matter what units are used for the actual yield or the theoretical yield so long as they are both measured in the same units.

Improve your grade

Explain the difference between the theoretical yield of a reaction and the actual yield. **AO1 (2 marks)**

Reactions that go both ways

Reversible reactions used in industry

- Some chemical reactions are reversible. The reactants turn into products, but the products can turn back into the reactants.
- A double headed set of arrows like this ⇌ are used to show a reversible reaction.
- The forward reaction goes from left to right in the equation. The reverse reaction goes right to left.

Reversible reactions

- Heating solid ammonium chloride causes it to decompose into two gases, hydrogen chloride and ammonia. When the two gases cool, the ammonium chloride is reformed. This is a reversible reaction.
- Copper sulfate crystals are blue, but heat them and they turn white, as the water of crystallisation is removed. Add a little water to the white anhydrous copper sulfate and the white powder turns blue again.

$$CuSO_4.5H_2O(s) \rightleftharpoons CuSO_4(s) + 5H_2O(\ell)$$

- The double-headed arrow shows that the reaction can go both ways, left to right or forward, right to left or backward.

Rates of reaction

Rates of reaction can vary

- To measure the rate of a reaction you must measure both the quantity of a reactant or product, and the time.

Remember!
The rate of reaction is rarely constant. It starts quickly when there are lots of reactants, but as products are formed less reactants are available so the rate of reaction will slow.

Measuring rates of reaction

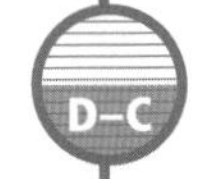

- When chemists talk about the rate of reaction they mean how much chemical reacts, or is formed, in a given time.
- rate of reaction = $\frac{\text{amount of reactant used up}}{\text{time taken}}$ or $\frac{\text{amount of product formed}}{\text{time taken}}$
- You can alter the rate of a chemical reaction by doing one or more of these: increasing the temperature, increasing the concentration of a reactant, using smaller pieces of a reactant, or adding a catalyst.
- Often a gas is produced and so it is easy to measure the quantity produced. Figure 1 shows ways to measure a gas.

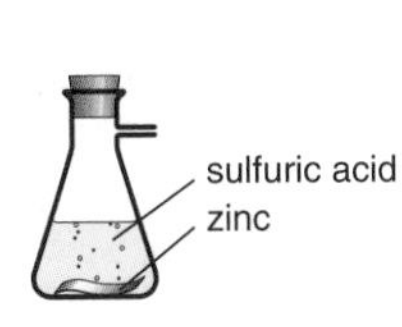

A

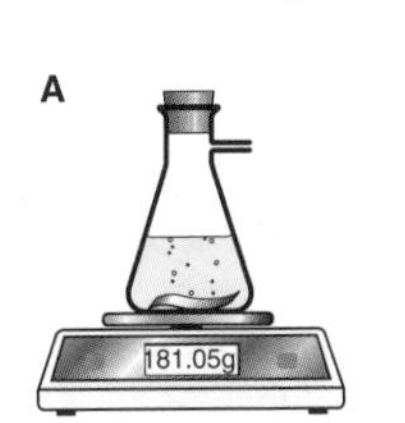

or

B

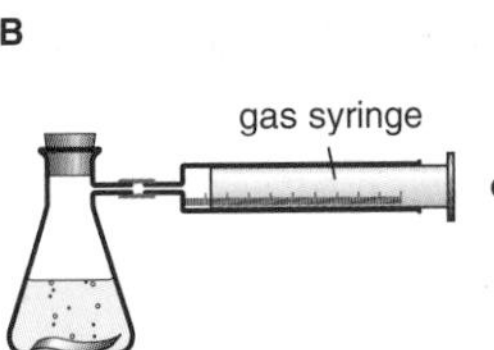

or

C

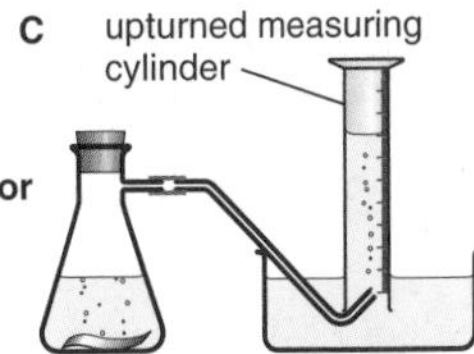

Figure 1: Ways to measure the quantity of a gas produced

Improve your grade

Figure 2 shows the volume of gas produced when some magnesium reacted with hydrochloric acid, over 4 minutes.

a What was the final volume of gas produced?

b Calculate the rate of the reaction during the first minute. **AO2 (3 marks)**

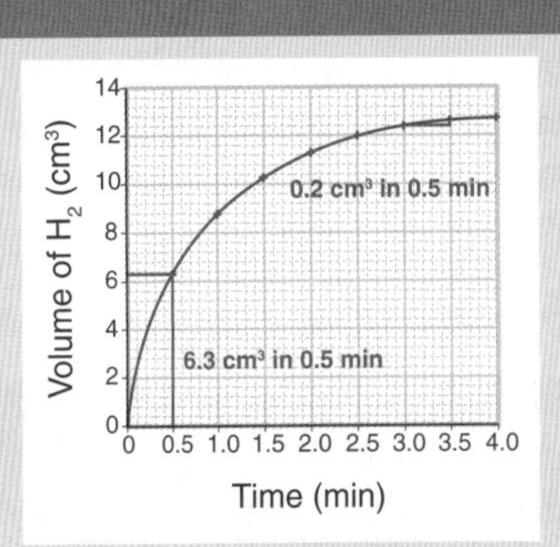

Figure 2: Measurements from a magnesium–acid reaction

Collision theory

Chemical collisions

- Reactants must collide to react. The more collisions, the higher the reaction rate will be.
- Making reactant particles collide more frequently increases the reaction rate.

Remember!
Not all collisions cause reactions.

Explaining collision theory

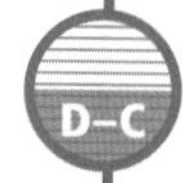

- For a chemical reaction to occur, the reactants must collide with each other. Any change that increases the number of collisions will increase the rate of a reaction. Not all collisions cause a reaction. The collision has to be hard enough to cause the reactants to react. This can be done by increasing the temperature, increasing the concentration, increasing the pressure, using smaller particles of a solid, adding a catalyst.

EXAM TIP

If asked about the effect of changing one factor, always think about the effect of the change on the **number** of particles available to react or the **energy** of the particles.

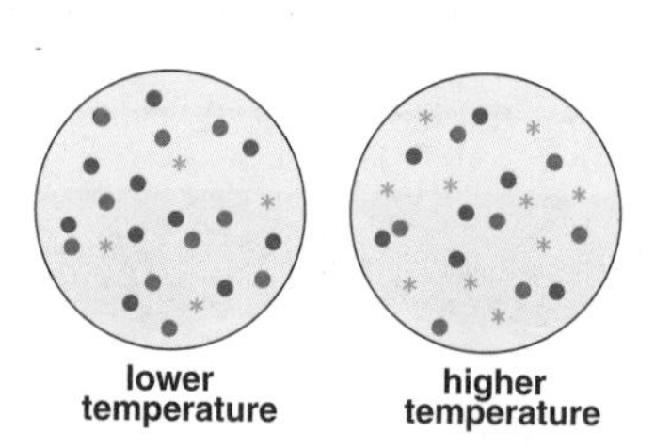

Figure 1: Effect of temperature on the number of successful collisions per second in a liquid or gas

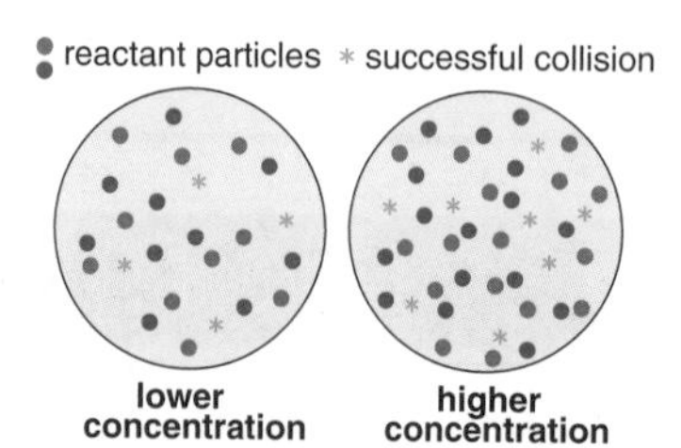

Figure 2: Effect of concentration on the number of successful collisions per second in a liquid or gas

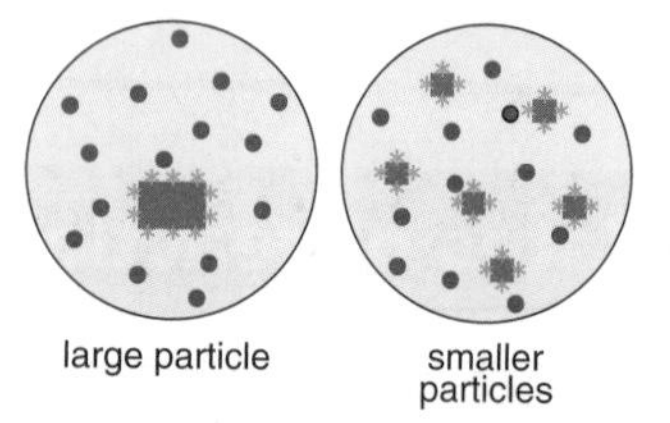

Figure 3: Effect of surface area of a solid on the number of successful collisions per second

Adding energy

Why temperature affects reaction rates

- Transferring energy to reactants by heating, increases reaction rates. It makes reactant particles move around faster. This has a double effect:
  - more collisions per second
  - more successful collisions causing reactions.

Investigating the effects of temperature

- Magnesium reacts with sulfuric acid, giving off hydrogen gas.

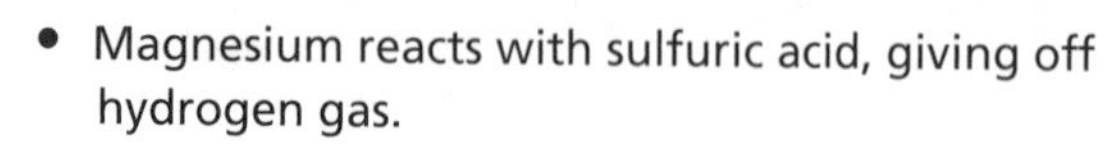

$$Mg(s) + H_2SO_4(aq) \rightarrow MgSO_4(aq) + H_2(g)$$

- Figure 4 shows results from using the same amounts of magnesium and acid at three different temperatures.
- The steeper the graph's curve, the faster the rate. All three reactions produce the same final volume of hydrogen gas, as they have the same quantities of starting reactants.

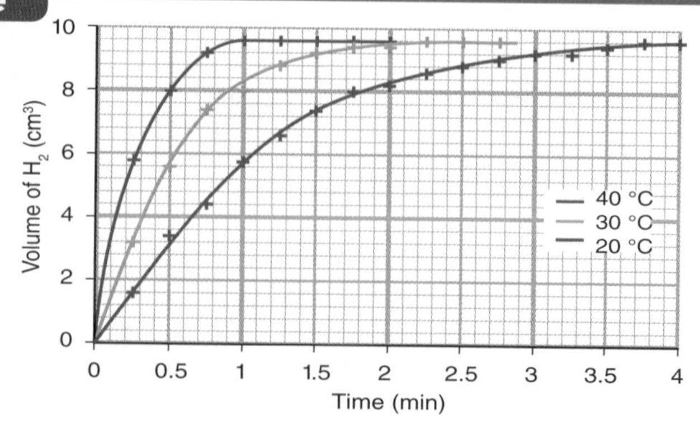

Figure 4: Graph of volume of hydrogen gas produced at different temperatures

Improve your grade

A student was investigating a chemical reaction. She decided to increase the temperature of the reaction, whilst decreasing the concentration of one reactant. Use collision theory to explain why the rate of reaction remained unchanged. **AO3 (2 marks)**

Concentration

Investigating the effects of concentration

- Higher concentration means that there are more reactant particles in the same volume. There are more collisions per second, increasing the reaction rate. Increasing the concentration does not affect the energy of collisions.

The effect of concentration

- Reactants are used up during a reaction and their concentrations decrease, so the reaction slows down.
  That is why the graphs are curves.
- To find the rate at any point during the reaction find the gradient of the graph at that point.
- Look at Figure 5. What is the rate at 0.5 mol/dm³ after 2 minutes?
  - Draw a tangent at this point, and construct a triangle as shown.
  - Find the scale lengths on each axis: x-axis = 2 min, y-axis = 3.3 cm³
  - Rate = amount of product ÷ time = 3.3/2 = 1.65 cm³ of H_2/min

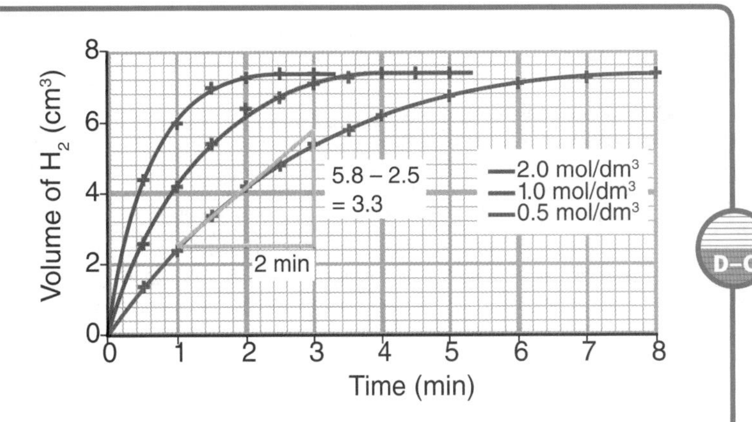

Figure 5: Rate-of-reaction graphs for magnesium reacting with sulfuric acid at different concentrations

D–C

Size matters

Size and surface area

Solids only react at their surface. Breaking a solid into smaller pieces exposes more surface area. That provides more reactant particles to be available for collisions: the reaction rate increases.

The ratio between the two surface areas shows the change in rate of reaction. For example, if you reduce particle size from 1 cm³ to 1 mm³:

- for 1 cm cube surface area:
  - each face = 10 mm × 10 mm = 100 mm²
  - total = 6 × 100 mm² = 600 mm²
- for 1 mm cube surface area:
  - each face = 1 mm × 1 mm = 1 mm²
  - total = 6 × 1 mm² = 6 mm²
- A 1 cm cube contains 1000 cubes of 1 mm side length, so:
  - surface area of 1000 cubes = 6000 mm²
  - 6000/600 = 10, so the reaction will be ten times faster.

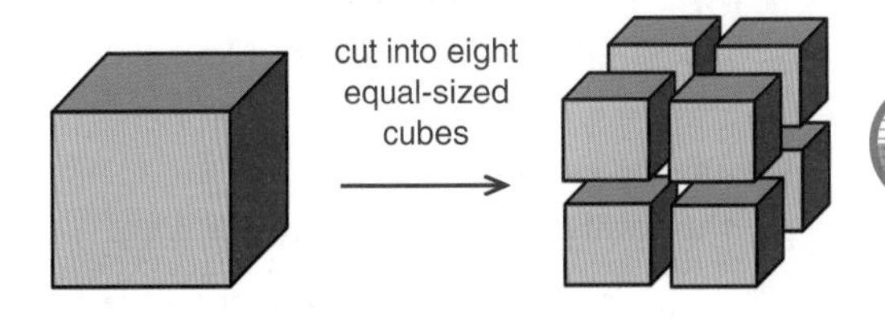

Figure 6: Smaller particles, larger surface area

D–C

Improve your grade

Use Figure 5 to find the rate of the reaction after 1 minute when the concentration is 2 mol/dm³. **AO2 (2 marks)**

Clever catalysis

Investigating catalysts

- Catalysts are chemicals that speed up reactions, but are not used up in the process. They can be used over and over again.

Catalysts are chemicals that take part in the reaction, but are unchanged at the end of the reaction.

Using catalysts

- Car exhaust fumes contain poisonous carbon monoxide and oxides of nitrogen. In a catalytic converter platinum, rhodium and palladium, catalyse reactions between the gases producing less harmful ones, for example carbon monoxide becomes carbon dioxide, nitrogen oxides are converted to nitrogen. Many catalysts are transition metals or their compounds.

Table 1: Some catalysts and their uses

catalyst	process	product
iron	Haber Process	ammonia
platinum	oxidising ammonia	nitric acid
nickel	hydrogenating vegetable oils	vegetable spreads such as margarine

Controlling important reactions

Making ammonia

- Ammonia is a really useful chemical. it is used to make fertilisers, explosives and pharmaceutical drugs.
- Ammonia is made from nitrogen from the air, and hydrogen obtained from natural gas (methane).

Economics and safety

- When planning an industrial process, chemists need to consider all the variables that affect rates of reaction. Making ammonia is a good example:
- $N_2(g) + 3H_2(g) \rightleftharpoons 2NH_3(g)$
  - the temperature used is 450 °C
  - the pressure used is 200 atmospheres (this has the effect of increasing the concentration)
  - an iron catalyst is used
  - the catalyst has a large surface area.
- The choice of conditions produces a small yield of ammonia (15–20%), but quickly and cheaply. The unreacted nitrogen and hydrogen are fed back in, so eventually all of the reactants become ammonia.
- Increasing temperature speeds up the reaction, but reduces the amount of ammonia formed, the high pressure produces more ammonia, but is very expensive, the iron catalyst speeds up the reaction, whilst being cheap.
- Many chemical reactions are exothermic, they produce heat. This can alter the rate of the reaction so the chemical plant needs to be able to keep the temperature constant.

Improve your grade

Nitrogen and hydrogen react together slowly to make ammonia. Explain why a high temperature is used, even though this reduces the percentage of ammonia made. **AO1 (2 marks)**

The ins and outs of energy

Transferring energy

- All chemical reactions transfer energy. Energy released in a chemical reaction may be transferred as heat to the surroundings, such as in combustion (burning).
- Chemical reactions in batteries produce electrical current, in glow sticks they produce light, and in explosions sound.
- Some reactions need energy, photosynthesis takes the energy it needs from the Sun.

Energy stores

- All substances store energy, but different substances store different amounts.
- If reaction products store less energy than the reactants did, the reaction is exothermic. The extra energy is transferred to the surroundings, heating them up.
- If the products store more energy (an endothermic reaction), they must have absorbed energy from somewhere. The surroundings provide the energy by cooling down.

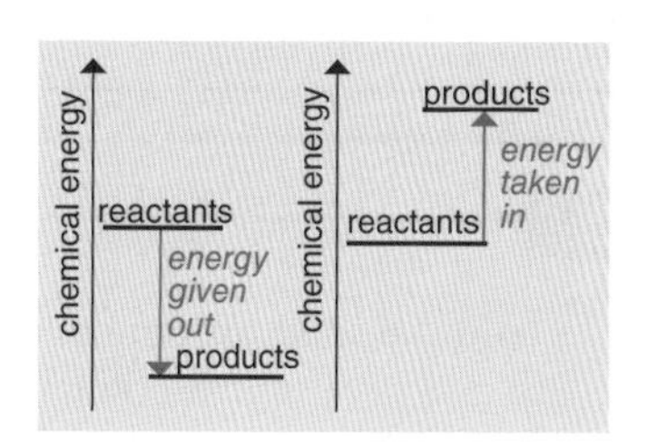

Figure 1: Exothermic and endothermic energy-level diagram

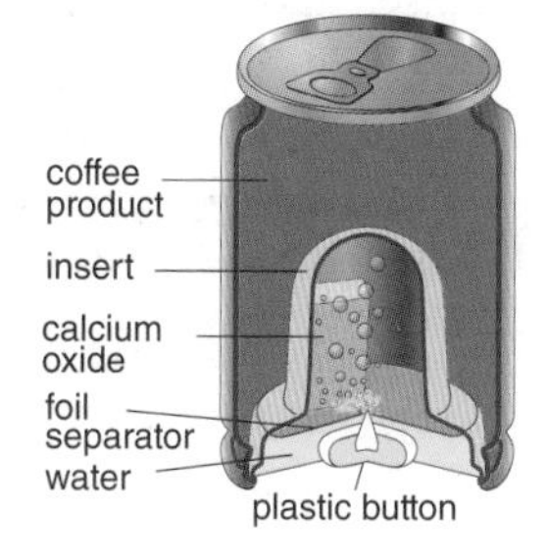

Figure 2: Self-heating coffee can

- These changes can be shown by an energy-level diagram.
- Exothermic and endothermic reactions are useful. Calcium oxide reacts exothermically with water to provide a heat source for self-heating cans. Ammonium nitrate is used to provide an endothermic reaction to make sports injury packs, to cool muscles.

Acid–base chemistry

Acids and alkalis

- Acids dissolve in water to form acidic solutions containing hydrogen ions, $H^+(aq)$.
- Alkalis dissolves in water to form alkaline solutions containing hydroxide ions, $OH^-(aq)$.
- Acidity and alkalinity are measured on the pH scale.

Acid–base reactions

- Most bases are insoluble in water. Bases that do dissolve in water are alkalis, because their solutions contain hydroxide ions in the solution. Alkalis are usually hydroxides of Group 1 and 2 metals in the periodic table.
- A base reacts with an acid to form a salt and water, for example zinc oxide and sulfuric acid.

$$ZnO(s) + H_2SO_4(aq) \rightarrow ZnSO_4(aq) + H_2O(\ell)$$

- Metal carbonates also produce carbon dioxide.

$$CaCO_3(s) + 2HCl(aq) \rightarrow CaCl_2(aq) + H_2O(\ell) + CO_2(g)$$

- Alkalis react with acids to give a salt and water only.
- Group 1 carbonates and hydrogencarbonates are bases because they react with acids to give salt and water, and carbon dioxide.
- Neutralisation reactions are exothermic and can be represented by this ionic equation:

$$H^+(aq) + OH^- (aq) \rightarrow H_2O(\ell)$$

Improve your grade

Explain why this ionic equation can be used to represent all neutralisation reactions between acids and alkalis.
$H^+(aq) + OH^-(aq) \rightarrow H_2O(l)$ **AO1 (2 marks)**

Making soluble salts

Reactions that form salts

- Soluble salts can be made from acids in one of three ways: reacting with a metal, reacting with an insoluble base, reacting with an alkali. The name of the salt that's made depends on the metal, or on the metal in the name of the alkali or base, and the acid.

Table 1: The salts made by common acids

acid		salt made
sulfuric	H_2SO_4	sulfate
hydrochloric	HCl	chloride
nitric	HNO_3	nitrate

Practical methods

- To make a salt from a solid: add the solid to some dilute acid until no more will dissolve, if the reaction is slow, then warm the mixture gently, filter the solution to remove unreacted solid, evaporate the filtered solution until it is nearly all gone, then allow it to cool and the salt crystals will appear.
- To make a salt using an alkali: place 25 cm^3 of alkali in a conical flask, and add some universal indicator, add the acid from a burette, syringe or measuring cylinder until the acid is neutralised – note the volume of acid required, then either repeat the experiment omitting the universal indicator; or add a little powdered charcoal to the coloured solution and heat, then filter to remove the charcoal, evaporate the solution until it is nearly all gone.

Remember!
The higher a metal is in the reactivity series, the easier it will react with the acid. At the top of the reactivity series the reaction will be extremely violent – so Group 1 salts are made by other methods.

Insoluble salts

Making insoluble salts

- An insoluble salt is one that will not dissolve in water. Insoluble salts are made by mixing together two solutions, each containing one part. The metals swap partners. The solid product formed by mixing two solutions is called a precipitate. Filtering removes the precipitate, which is washed before drying in an oven.

Useful precipitations

- Salts are ionic compounds. A mixture of two salts in solution contains positive ions of two metals and negative ions of two non-metal groups. If any pair combines to form an insoluble salt, they produce a solid precipitate.
- Precipitation reactions are useful to remove unwanted ions from water; calcium ions cause hardness in water and are removed using sodium carbonate solution, phosphate ions in washing powder are removed by adding aluminium ions to the waste water. Calcium carbonate is produced for toothpaste this way.

Improve your grade

Describe the method to make some solid chromium chloride from chromium metal. **AO3 (4 marks)**

Ionic liquids

Conduction needs movement

- In an ionic solid, the ions are not able to move so they cannot carry an electric current.
- When dissolved in water or heated until it melts, the ions become free to move. As the ions can move, they can now carry their electrical charge through the solution or liquid.

Conduction in liquids

- Pure water does not conduct electricity as it is made of molecules. Tap water is not pure as it contains many dissolved ions, so it does conduct electricity.
- When a liquid conducts, the electric current must enter and leave the liquid through solid conductors called electrodes. This is electrolysis.
- The ions in a solution or molten compound move around randomly. In electrolysis, positive ions are attracted to the negative cathode and negative ions are attracted to the positive anode.
- As the ions move, they carry electric charge through the liquid from one electrode to the other.
  - At the anode, the negative ions release their electrons; whilst at the cathode, the positive ions pick up electrons at the same rate as the negative ions are releasing their electrons.
  - Metal ions gain electrons at the cathode, and become atoms that coat the electrode or react with the liquid. Non-metal ions at the anode lose electrons, and may be released as a gas or react with the liquid.
- Electroplating is another use of electrolysis (see Figure 2). An object is coated with a thin layer of metal, as shown. The silver anode dissolves into the solution, maintaining the concentration of silver ions.

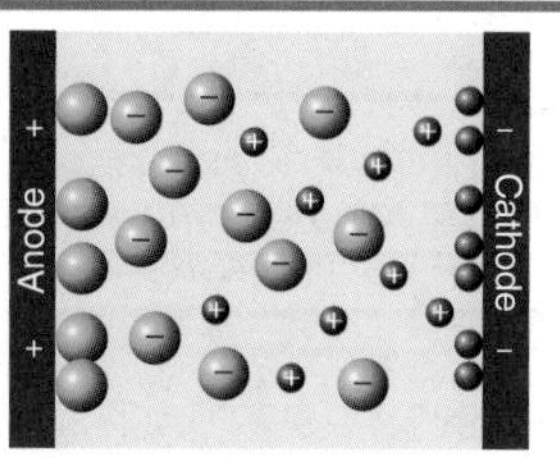

Figure 1: Ions in a solution that is conducting electricity

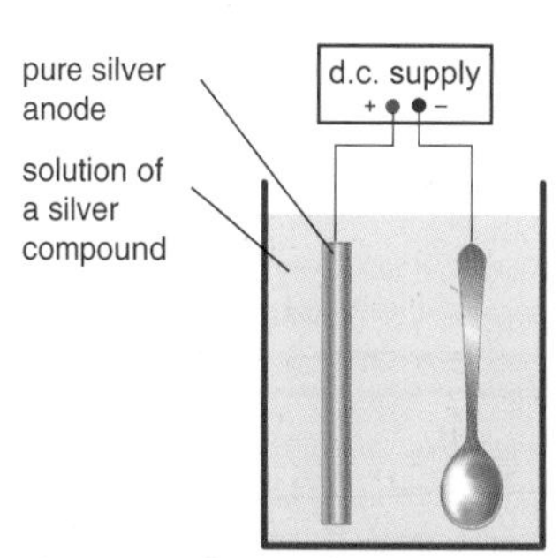

Figure 2: Electroplating a spoon

D–C

Electrolysis

Decomposing ionic compounds

- Ionic compounds decompose when they conduct electricity.
- Electrolysis of molten sodium chloride decomposes the compound, producing sodium metal at the cathode and chlorine gas at the anode.
- Electrolysis of a solution of sodium chloride (brine) is slightly different. The water makes the brine a mixture of ions (Na^+, H^+, OH^- and Cl^-). Chlorine gas is produced at the anode. At the cathode, the lower of sodium and hydrogen (the positive ions) in the reactivity series is made – which is hydrogen. The sodium ions react with the hydroxide ions to make a solution of sodium hydroxide.
- Sodium hydroxide is used for making soap; and chlorine for the production of bleach and plastics.

G–E

More applications of electrolysis

- Electrolysis of molten sodium chloride decomposes the compound producing sodium metal at the cathode and chlorine gas at the anode.
  - Electrolysis of a solution of sodium chloride (brine) is slightly different. The water makes the brine a mixture of ions (Na^+, H^+, OH^- and Cl^-). Chlorine gas is produced at the anode, at the cathode, the lower of sodium and hydrogen (the positive ions) in the reactivity series is made, which is hydrogen; and the sodium ions react with the hydroxide ions to make a solution of sodium hydroxide.
- Aluminium metal is obtained by electrolysis from aluminium oxide (Al_2O_3) dissolved in cryolite to reduce the melting point of the oxide.
- Copper is purified by electrolysis.

Improve your grade

Explain why the electrolysis of molten sodium chloride produces sodium metal and chlorine gas, but when a solution of sodium chloride is electrolysed hydrogen gas, chlorine gas and sodium hydroxide are produced. **AO2 (3 marks)**

C2 Summary

Structure and bonding

Chemical bonding involves either transferring or sharing electrons in the outer shells of atoms.

Ionic compounds are held together by strong electrostatic forces of attraction.

Atoms that lose electrons become positively charged ions. Atoms that gain electrons become negatively charged ions.

When atoms share pairs of electrons, they form covalent bonds.

Structure, properties and uses

When melted or dissolved in water, ionic compounds conduct electricity. Covalent compounds do not.

Ionic compounds have regular structures (giant ionic lattices), high melting and boiling points.

Diamond and graphite have different properties determined by their structures.

Metals consist of **giant structures** of atoms arranged in a regular pattern.

The properties of thermosoftening and thermosetting polymers depend on what they are made from and how they are made.

Nanoscience refers to structures that are 1–100 nm in size and of the order of a few hundred atoms.

Atomic structure, analysis and quantitative chemistry

Atoms of the same element can have different numbers of neutrons; these atoms are called isotopes of that element.

The relative atomic mass (A_r) and the relative formula mass (M_r) allow numbers of particles to be compared. The A_r or M_r of a substance, in grams, is one mole of that substance.

Elements and compounds can be identified using instrumental methods. Gas chromatography linked to mass spectroscopy (GC-MS) is an instrumental method.

A reversible reaction is one where the products of the reaction can react to produce the original reactants.

The amount of a product obtained is known as the yield.

Rates of reaction

Catalysts change the rate of chemical reactions but are not used up during the reaction.

The minimum amount of energy that particles must have to react is called the activation energy.

The rate of reaction $= \frac{\text{amount of reactant used}}{\text{time}}$ or $= \frac{\text{amount of product formed}}{\text{time}}$

Chemical reactions can only occur when reacting particles collide. Collision theory explains why changes to conditions affect rates.

Exothermic and endothermic reactions

An exothermic reaction transfers energy to the surroundings. An endothermic reaction takes in energy from the surroundings.

If a reversible reaction is exothermic in one direction, it is endothermic in the opposite direction.

Acids, bases and salts

Metal oxides and hydroxides are bases. Soluble hydroxides are called alkalis. Ammonia makes an alkaline solution to produce ammonium salts.

Soluble salts can be made from acids by reacting them with metals, insoluble bases, or alkalis. Salt solutions can be crystallised to produce solid salt. Insoluble salts can be made by mixing solutions of ions so that a precipitate is formed.

Hydrogen ions, $H^+(aq)$, make solutions acidic and hydroxide ions, $OH^-(aq)$, make solutions alkaline.

In **neutralisation reactions, hydrogen ions** react with hydroxide **ions** to produce **water**.

The particular salt produced depends on:
- the acid used
- the metal in the base or alkali.

Electrolysis

When an ionic substance is melted or dissolved in water, the ions are free to move about within the liquid or solution.

During electrolysis, positively charged ions move to the negative electrode and are reduced, and negatively charged ions move to the positive electrode and are oxidised.

Aluminium is manufactured by the electrolysis of a molten mixture of aluminium oxide and cryolite.

The electrolysis of sodium chloride solution produces hydrogen, sodium hydroxide and chlorine.

See how it moves

Distance–time graphs

Distance–time graphs show the distance an object moves in a period of time.

- The y-axis shows distance travelled, and the x-axis shows time.
- A line sloping up shows the object moving away from its starting point. If it slopes down, the object is moving back.

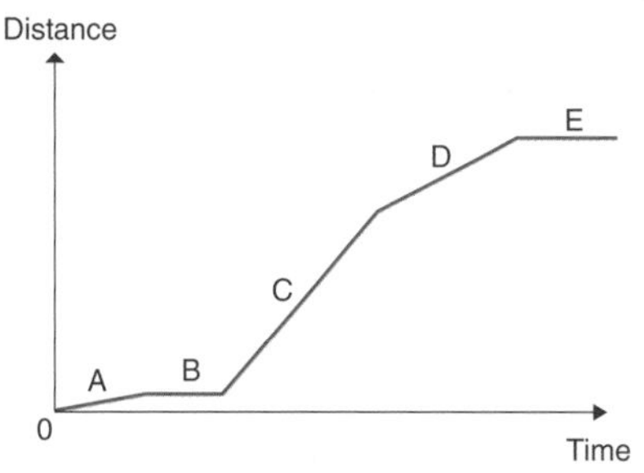

Figure 1: The object is always moving forward, but is stopped at B and E

The gradient (slope) at any point gives the speed at that time.

- A steeper gradient shows a faster speed.
- A flat gradient shows the object is stopped.

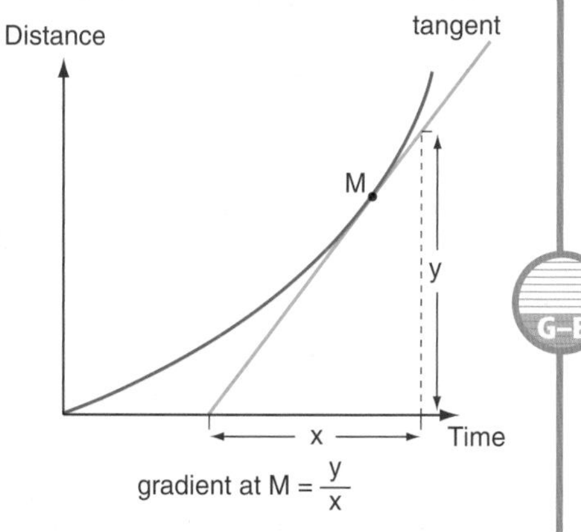

Figure 2: Use the gradient of a distance–time graph to calculate speed. If speed changes, the line curves and you must draw a tangent.

G–E

Speed and average speed

- Average speed is the speed over the whole journey. It is calculated in m/s using

$$\text{speed (m/s)} = \frac{\text{total distance travelled (m)}}{\text{time taken (s)}}.$$

EXAM TIP

The unit of speed depends on units used for distance and time, for example if distance is measured in km and time is measured in hours, speed is measured in km per hour.

How Science Works

You should be able to:

- construct distance–time graphs for objects moving in a straight line at a constant speed
- calculate the speed of objects using the slope of a distance–time graph.

D–C

Speed is not everything

Velocity and speed

Velocity is an object's speed in a given direction. Velocity–time graphs show how velocity changes with time:

- if the line slopes upwards, the object is accelerating
- if the line slopes downwards, the object is decelerating
- if the line is flat, the object is travelling at a steady speed.

Figure 3: The velocity–time graph shows a falling stone accelerating then travelling at a steady speed

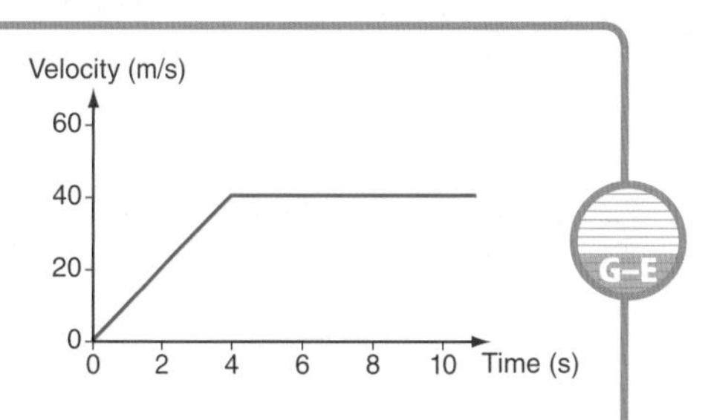

G–E

Velocity and acceleration

A change in velocity (or speed) is called acceleration.

- $\text{Acceleration (m/s}^2\text{)} = \dfrac{\text{change in velocity (m/s)}}{\text{time taken (s)}}$
- The change in velocity = final velocity – original velocity.
- A negative acceleration (deceleration) means the object slows down.

EXAM TIP

It is easy to confuse velocity–time graphs and distance–time graphs. Check the axes before writing your answer.

Remember!

The velocity–time graph shows the object is stopped when the line reaches the x-axis.

Improve your grade

Calculate the total distance travelled by the stone in Figure 3. **AO2 (3 marks)**

Forcing it

Adding up, cancelling out

- The resultant force on an object is the single force that would make an object move in exactly the same way as all the original forces acting together.
- Forces acting in the same direction add.
- Forces acting in opposite directions subtract. If they are equal in size, they cancel out and the forces are balanced.

How things move

- If there is no resultant force, forces are balanced.
  - stationary objects stay stopped
  - moving objects do not change speed.
- If there is a resultant force,
  - stationary objects start moving
  - moving objects accelerate in the direction of the force.
- If the resultant force is in the opposite direction to motion,
  - moving objects decelerate (slow down)

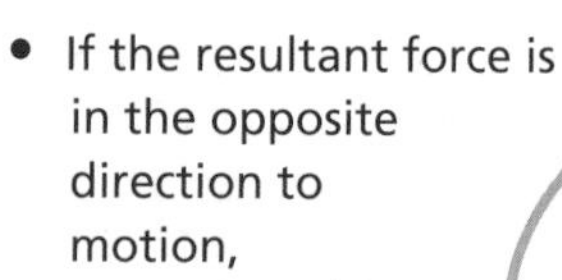

Remember!

The resultant force when a car accelerates is positive (forwards) – drag forces are smaller than the force from the engine.
There is no resultant force when the car drives at a steady speed.
When a car brakes, the resultant force is negative so the car slows down.

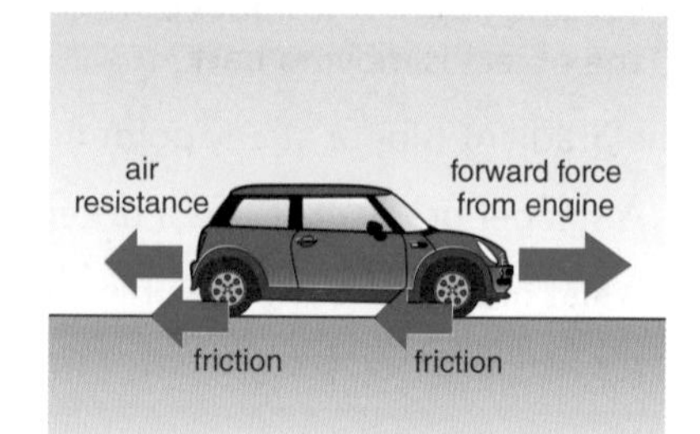

Figure 1: Unbalanced forces on an accelerating car

Force and acceleration

How quickly?

- Resultant forces change how objects move.
- A larger resultant force makes the car speed up more quickly (bigger acceleration).

Investigating acceleration

- If a resultant force is unbalanced, the object accelerates in the direction of the resultant force. Acceleration is calculated using:

$$\text{acceleration (m/s}^2\text{)} = \frac{\text{force (N)}}{\text{mass (kg)}}$$

- Acceleration increases as the force applied increases.
- Acceleration decreases if the mass of the object increases.

Improve your grade

Sam travels 5 m in 10 s, before stopping for 2 s. He then takes 6 s to return to the start. Draw a distance–time graph that shows Sam's journey. **AO2 (3 marks)**

Balanced forces

Balanced forces

For every force, there is an equal-sized force acting in the opposite direction.

- When you sit on a chair, it pushes up on you as hard as you push down on it.
- If a tray floats in water, the water pushes upwards on the tray as hard as the tray pushes downwards on the water.

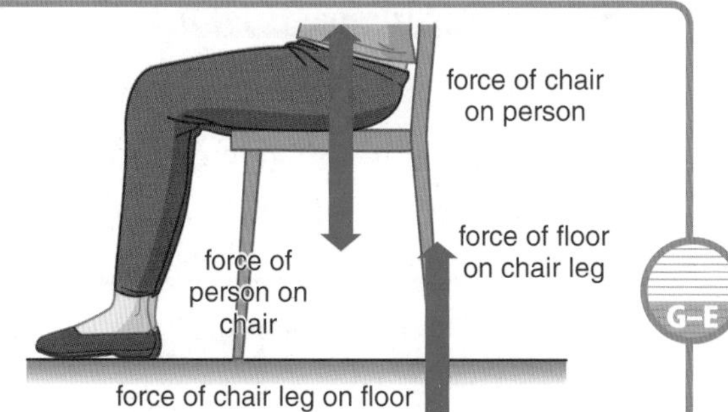

Figure 2: Upwards reaction forces balance the weight acting down

Forces on moving objects

- The weight of a skydiver pulls down on a parachute. This is matched by the upwards tension in the parachute strings.
- When a gun is fired, the bullet feels a forwards force. The bullet exerts an equal sized backwards force on the gun, which recoils. Since the gun is much heavier than the bullet, the bullet accelerates forwards much faster than the gun recoils backwards.

Stop!

Force it to stop

- When a car travels at a constant speed, force from the engine causing motion matches the friction and drag forces acting in the opposite direction.
- The faster a car travels, the more kinetic energy it has. The energy is transferred to the wheels and brake pads when braking. The driver must brake harder to stop a car travelling fast in the same distance as he would a slower car.
- Total stopping distance is thinking distance + braking distance.

Thinking and braking

- Reaction time is the time taken for a driver to react to a hazard.
- Thinking distance = speed × reaction time, and is the distance travelled before the driver reacts and brakes. It increases if the driver:
  - is distracted or tired
  - has taken alcohol or drugs (including some medicines).
- Braking distance is the distance travelled while the brakes are applied and the car is slowing down. It increases if:
  - the road is wet or icy
  - the tyres are worn down
  - the brakes are in bad condition.

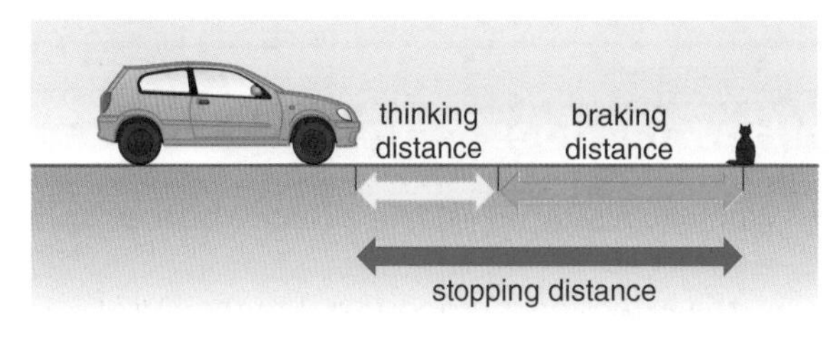

Figure 3: Stopping distance has two parts

How Science Works

- Factors that increase stopping distance also increase the risk of accidents. Drivers who fall asleep at the wheel can be prosecuted for dangerous driving.

Improve your grade

Describe how drag forces compare with forces from the engine for a car that accelerates, then reaches a steady speed then decelerates. **AO2 (4 marks)**

Terminal velocity

Faster and faster

- When an object starts moving, drag forces are smaller than the driving force.
- As it accelerates, drag forces increase: more particles are pushed out of the way each second.
- Eventually, a top speed is reached. Drag forces and driving forces are equal. There is no resultant force.

Terminal velocity

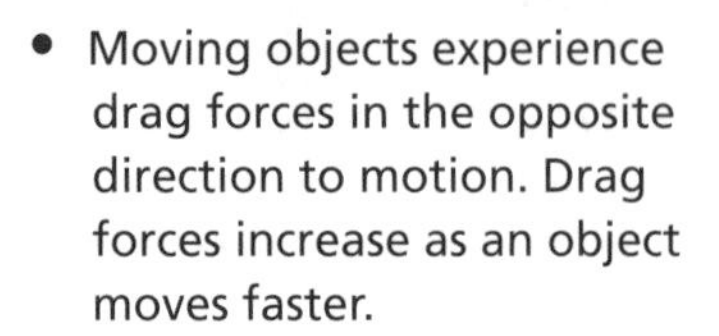

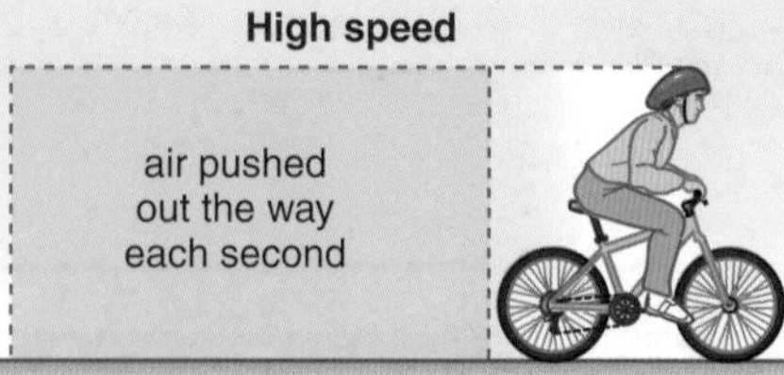

Figure 1: Air resistance and cycling speed.

- Moving objects experience drag forces in the opposite direction to motion. Drag forces increase as an object moves faster.
- The resultant force is the difference between the force causing motion and drag forces.
- A skydiver feels a constant downwards force, weight. Weight (N) = mass (kg) × gravity (N/kg).
- As the skydiver accelerates, drag forces increase until they match his weight. He reaches a top speed (terminal velocity).
- When the parachute is opened, drag forces increase suddenly. The resultant force is upwards. The skydiver decelerates.
- Drag forces decrease. The skydiver reaches a slower terminal velocity.

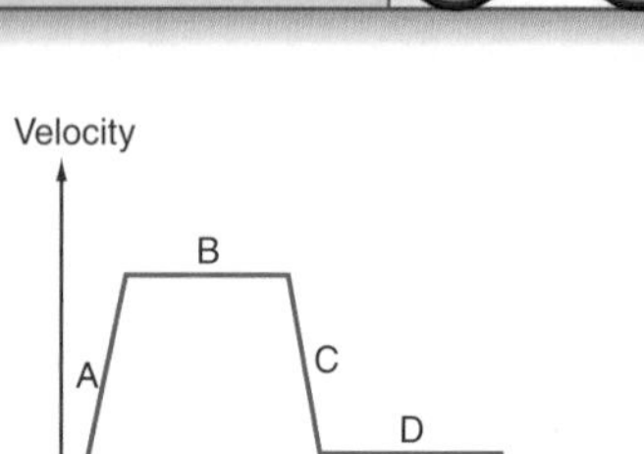

Figure 2: A velocity–time graph for a skydiver

Forces and elasticity

Changing shape

- Forces can make an object change shape, or break. Elastic objects return to their original shape when the force is removed.
- Energy is stored in an elastic object when it is stretched or squashed. It can be transferred to another object when the force is removed.

How far does it stretch?

- When an elastic object like a rubber band or spring is stretched:
  - the extension is proportional to the force applied up to a limit
  - above the limit, extension is not proportional to the force.
- This is called Hooke's law.

Remember!
Extension is the difference between the original length and the stretched length.

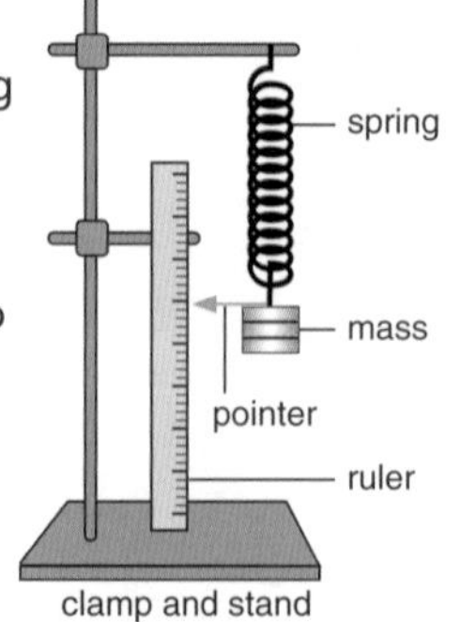

Figure 3: An experiment to test Hooke's law

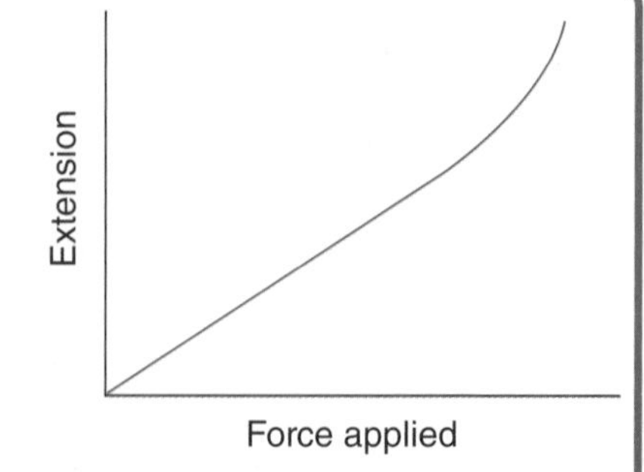

Figure 4: This graph shows Hooke's law

Improve your grade

Sketch the velocity–time graph for a marble that is released into a measuring cylinder of water. The marble reaches its terminal velocity. **AO2 (3 marks)**

Energy to move

What is kinetic energy?

- A moving object has kinetic energy. This energy has been transferred:
  - from chemical energy in food a person eats
  - from chemical energy in fuel used in an engine.
- Kinetic energy is transferred away from a loudspeaker as sound waves.

Kinetic energy transfers

- The kinetic energy from a moving object is transferred to the surroundings as a result of frictional forces. These include:
  - friction between car tyres and the road surface
  - air resistance felt by aircraft and other moving objects.
- The result is that kinetic energy is transferred to the surroundings and heats it up.
- The energy transfer can be useful.
- Regenerative brakes slow down the car using the engine. The car's kinetic energy is used to charge the car's battery as it slows down.

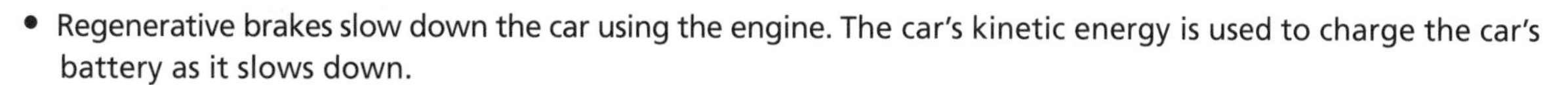

- In a car crash, car crumple zones are designed to distort, absorbing kinetic energy.

Working hard

Working

- Work done is the same as the energy transferred. Work is done, and energy is transferred whenever a force moves.
  An example of this is when energy is transferred to books when they are lifted up.

How much and how far?

- The amount of work done is calculated using:
  work done (J) = force (N) × distance moved in the direction of the force (m).
  For example, a person lifts a 60 N parcel by 1.5 m.
  Work done is 60 N × 1.5 m = 90 J.
- A smaller force is needed to drag something up a ramp compared with lifting it directly. The force along the ramp works over a longer distance.

EXAM TIP

Check diagrams so you use the correct measurements for distance and force when calculating work done.

Remember!

The force must move to do work. A person lifting a box does some work but when they stand still just holding a box they are not doing any work.

Improve your grade

a Calculate the work done lifting a piano 0.8 m into a lorry. The piano weighs 1850 N. **AO2 (3 marks)**
b Calculate the force needed to drag the piano into the lorry if a ramp 2 m long is used. **AO2 (2 marks)**

Energy in quantity

Gravitational potential energy and kinetic energy

- Gravitational potential energy is stored in an object because of its position. More energy is stored if the mass or height above ground is larger.
- Kinetic energy stores energy in an object's movement. More energy is stored if the mass or speed is larger.

Remember!
Use the vertical height above the ground when calculating gravitational potential energy.

Calculating gravitational potential energy

- You can calculate an object's gravitational potential energy using this equation:
  Gravitational potential energy (joules) = mass (kg) × gravity (N/kg) × height (m).
  For example, the gravitational potential energy gained by a 3 kg rabbit lifted 0.8 m by its owner is
  $3 \times 10 \times 0.8 = 24$ J

Remember!
Gravity on Earth is 9.8 N/kg (usually rounded to 10 N/kg).

How Science Works

- When a pendulum swings, gravitational potential energy changes into kinetic energy and back again.

Energy, work and power

Work and power

Something powerful does a lot of work in a short time. It transfers energy quickly.

Finding the power

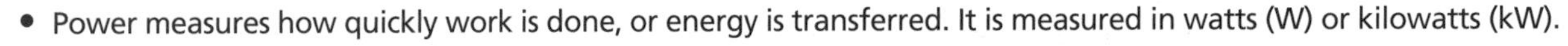

- Power measures how quickly work is done, or energy is transferred. It is measured in watts (W) or kilowatts (kW).
- Power is calculated using: power (W) = $\frac{\text{energy transferred (J)}}{\text{time (s)}}$ and $\frac{\text{force (N)} \times \text{distance (m)}}{\text{time (s)}}$

- The power of a motor that does 300 J of work in 10 seconds is
  300 J / 10 s = 30 W.

Remember!
One kilowatt is 1000 W.

Improve your grade

Calculate the power of a motor that can lift 600 N in 1 minute. **AO2 (3 marks)**

Momentum

What is momentum?

- All moving objects have momentum.
- Momentum (kg m/s) = mass (kg) × velocity (m/s)
- Momentum has a direction because velocity has magnitude and direction.

Collisions and explosions

- During elastic collisions, objects collide and bounce apart, keeping the same kinetic energy.
- During inelastic collisions, objects collide and stick together, moving as a single object.
- Momentum is conserved. If no external forces act, momentum before and after a collision or explosion is the same.
- In a collision, find the total momentum of moving objects before the collision. The total momentum before and after the collision is the same.
- If a stationary object explodes into two parts, the momentum before and after is zero. Each part after the explosion has the same momentum in opposite directions.

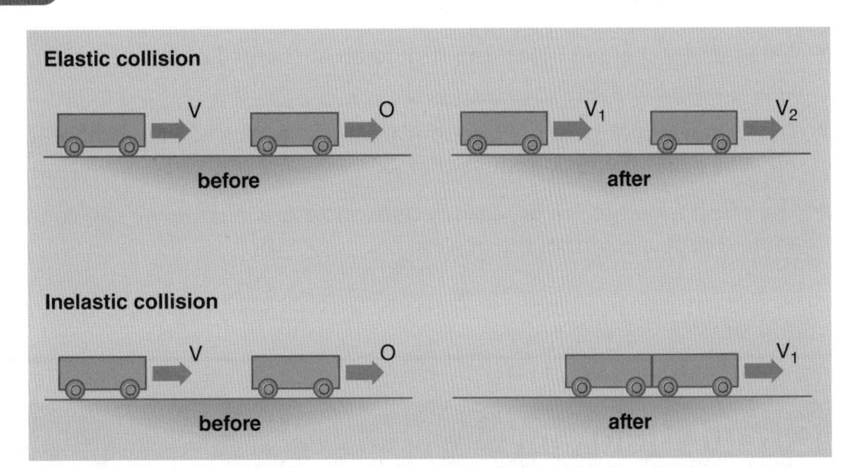

Figure 1: Elastic and inelastic collisions

Remember!

You can choose which velocity direction is positive, but always use a diagram to help you see what is happening in a question.

Static electricity

Static electricity

- Atoms contain electrons with a negative charge. When different materials are rubbed together, electrons move from one material to the other.
- Some materials gain electrons and end up with a negative charge.
- Other materials lose electrons and end up with a positive charge.

Investigating charges

- Objects with the same charge repel.
- Objects with opposite charges attract.

Remember!

Electrons are the only particles that move between different materials. Losing electrons makes a material positively charged.

How Science Works

- Materials like cling film gain or lose electrons so easily that they just need to touch other materials for an electrostatic charge to build up.

Improve your grade

Two trolleys each with a mass of 1 kg roll towards each other. One trolley travels at 2 m/s to the left and one travels at 3 m/s to the right. They collide and stick together. What is the velocity and direction of the trolleys after the collision? **AO2 (4 marks)**

Moving charges

Electric charge and movement

- Two electrically charged objects exert a force on each other.
- You cannot put a static charge on a metal object because the electrons flow easily through it. The flow of electrons is called a current.

Electrostatic induction

- A negatively charged object can stick to an uncharged object because of electrostatic induction.
- The negative charge repels electrons from the surface of the uncharged object leaving it with a positive charge.
- The effect is greater on dry days because moisture in damp air can conduct some charge away from the objects.

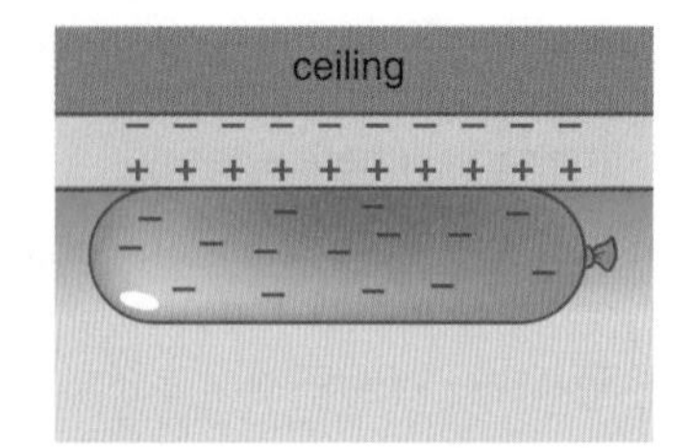

Figure 1: The balloon sticks because of electrostatic induction

Circuit diagrams

Symbols

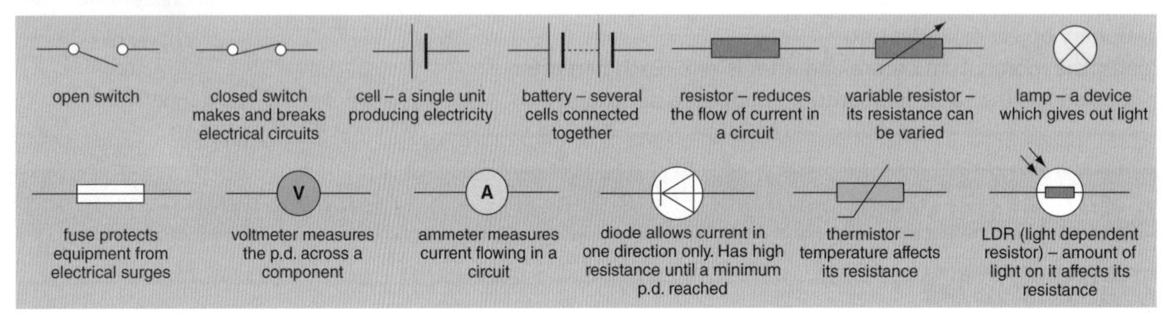

Figure 2: Circuit components

Series and parallel

- A series circuit is a loop of conductors. The current is the same throughout the circuit.
- Current is measured in amps using an ammeter connected in series.
- A parallel circuit has components connected in more than one loop.
- The current divides at the junction, and the total current flowing into the junction is the same as the total current flowing out of it.

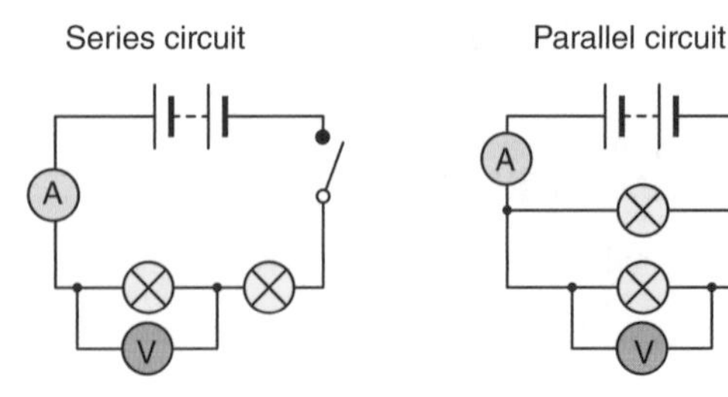

Figure 3: Comparing a series and parallel circuit

Improve your grade

Max rubbed a plastic ruler and a metal ruler with a piece of cloth.

a Which ruler became charged? **AO1 (1 mark)**

b Max held the charged ruler near a charged balloon. The balloon moved away from the ruler. Explain why. **AO2 (2 marks)**

Ohm's law

Resistance

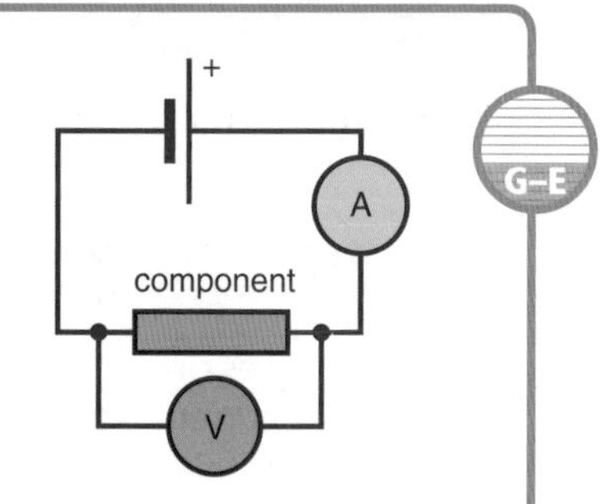

Figure 4: A circuit used to find resistance of a component

- Electric current is a flow of moving charges. The charges flow more easily through materials such as copper because copper has low resistance. If the resistance of a component or wire is high, a small current flows when a given potential difference is applied.
- Resistance is calculated in ohms (Ω) using: resistance (Ω) = voltage (V) ÷ current (A).

Factors affecting resistance

- A wire's resistance at a constant temperature depends on these factors:
  - a long wire has larger resistance than a short wire
  - a thin wire has more resistance than a thick wire
  - different materials have different resistances.

EXAM TIP

You may be asked to draw the circuit used to measure resistance.

Non-ohmic devices

Non-ohmic devices

- A filament bulb does not obey Ohm's law. As the current in a filament bulb increases, its resistance increases, because the filament wire heats up.

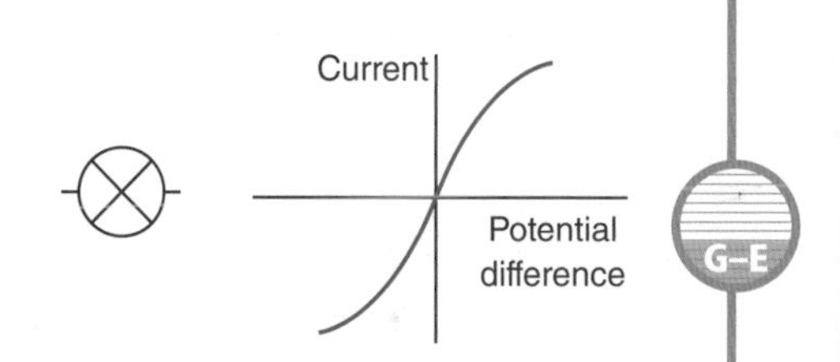

Figure 5: Current–potential difference graph for a filament bulb

Direction, light and temperature

- A diode only allows current to flow in one direction (forward). The arrow on the symbol shows the forward direction. Resistance is very high in the reverse direction.
- A light dependent resistor (LDR) has a smaller resistance when light shines on it.
  - LDRs are used in automatic light systems.
- A thermistor has a smaller resistance when it is heated.
  - Thermistors are used in fire alarms and electronic thermometers.

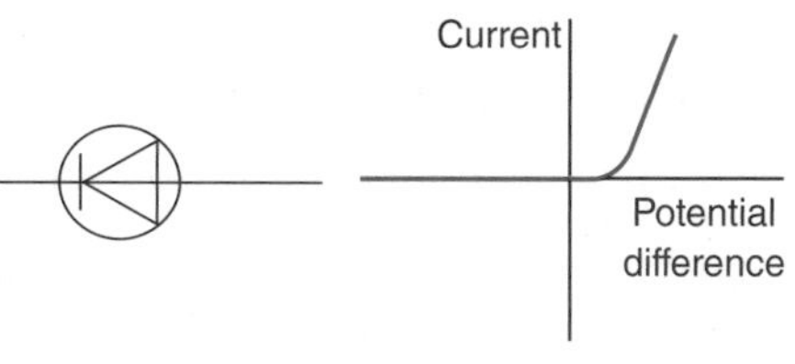

Figure 6: Current–potential difference graph for a diode

Remember!
Potential difference is also called voltage, and is measured in volts.

Improve your grade

When the potential difference across a bulb is 6V, the current through it is 0.1A. What is the resistance of the bulb? **AO2 (3 marks)**

Components in series

Series circuits

A series circuit is connected as a single loop. There is only one path that the current can flow through.

Potential difference and resistance in series

- In a series circuit the following applies:
  - the current is the same in all places
  - the total resistance of several components is each component's resistance added together. Adding more bulbs in a series circuit reduces the current. This is because the resistance increases.
  - the potential difference from the cell is shared across components. Components with a high resistance take a larger share of the potential difference.
  - V = V1 + V2 + V3

Remember!
Use voltage = current × resistance to find the voltage across a single component in a circuit.

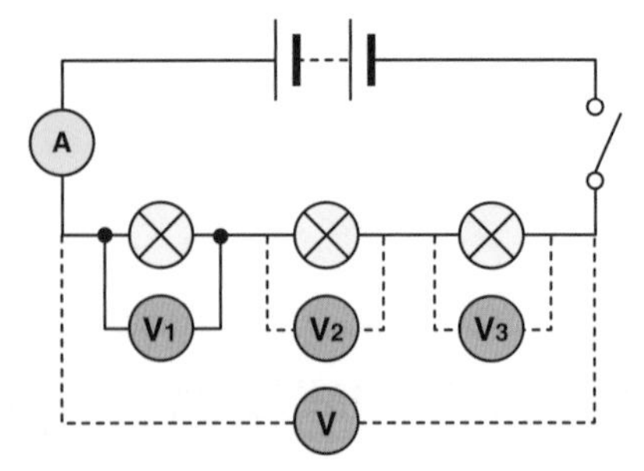

Figure 1: Measuring the potential difference and current in a series circuit

Components in parallel

Parallel circuits

- The current in a parallel circuit has more than one path it can follow.
- The current splits at junctions and joins up again. The same amount of current enters and leaves each junction.

Potential difference and resistance in parallel

- The current in each loop depends on the total resistance in that loop. A smaller current flows through loops with a high resistance compared to loops with a small resistance.
- The total current leaving and returning to the battery is the same as the current in each loop added together.
- $A_1 = A_5 = A_2 + A_3 + A_4$
- All bulbs stay as bright as if one bulb was in the circuit.
- The potential difference across each loop of a parallel circuit is the same.

EXAM TIP

You must be able to use the rules to calculate current and voltage in both series and parallel circuits.

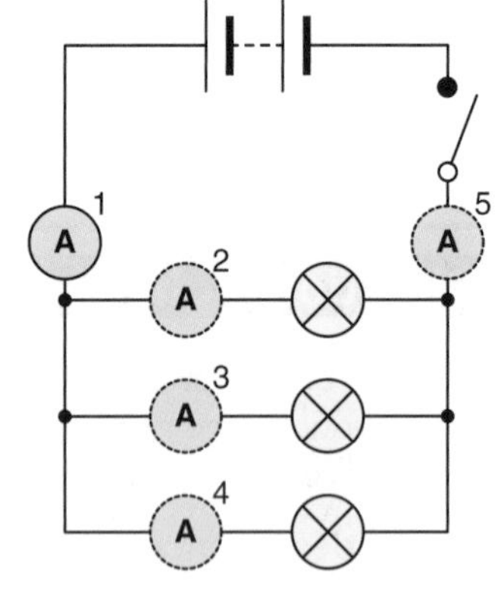

Figure 2: Current in a parallel circuit

Improve your grade

A series circuit is set up, using four 1.5 V cells in series.

a What is the total voltage supplied to the circuit?

The circuit includes a motor (resistance 10 ohms) and a bulb (resistance 5 ohms).

b What is the total resistance in the circuit?

c Calculate the current in the circuit.

d What is the potential difference across each component? **AO2 (5 marks)**

Household electricity

Direct and alternating current

- Cells and batteries produce direct current (d.c.), which flows in one direction.
- Alternating current (a.c.) repeatedly changes direction. Mains electricity changes direction 50 times per second (its frequency is 50 hertz) and is about 230 V.

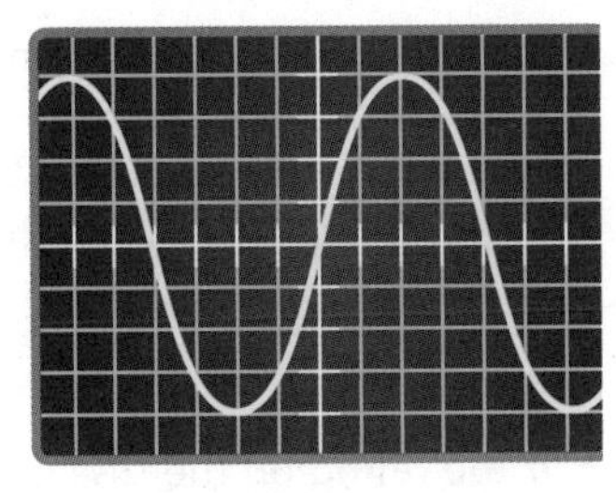

Figure 3: An oscilloscope trace of a.c. current

Using a cathode-ray oscilloscope

- An oscilloscope can display a.c. or d.c. voltage. Dials set the number of volts per vertical square, and the time per horizontal square.
- The amplitude (maximum height) of the trace shows the maximum potential difference supplied (in volts).
- The number of squares between the peaks shows time per cycle (in seconds).

Remember!
Amplitude is measured from the centre of a trace to the highest (or lowest) point.

Plugs and cables

Making the connection

- Mains electricity sockets have three holes connected to three wires:
  - two holes at the bottom connect to the live and neutral wires. These supply potential difference to equipment from the mains.
  - the third hole connects to the earth wire, which is a safety feature.
- In a three-pin plug, wires are covered in colour-coded plastic:
  - the neutral wire is blue
  - the live wire is brown
  - the earth wire is green and yellow.

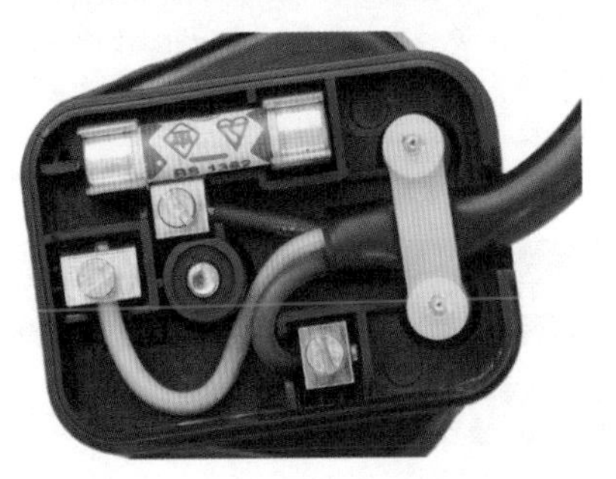

Figure 4: Inside a three pin plug

Cables and fuses

Appliances with a plastic outer case and no touchable metal parts are double insulated. They use two-core cable (live and neutral wires).

Other appliances use three-core wire (earth, live and neutral wires). If there is a fault and the equipment becomes live, the current flows through the earth wire and blows the fuse.

The fuse is connected in series with the live wire. It protects the appliance and flex from overheating. If the current is too large, the fuse melts and breaks the circuit.

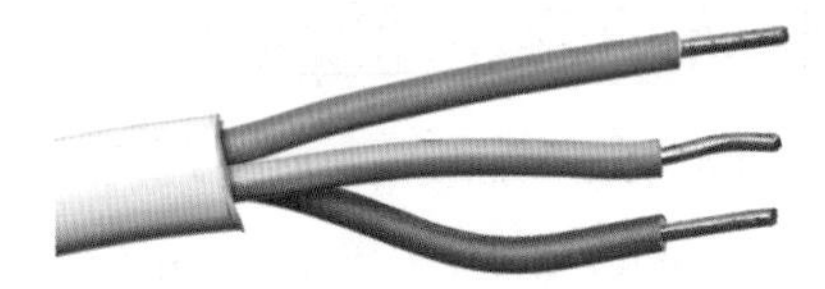

Figure 5: Three-core cable

EXAM TIP

You should be able to explain why certain materials are used for different parts of a plug and cable.

EXAM TIP

Make sure you can describe dangerous habits when using electricity and spot mistakes when wiring plugs.

Improve your grade

Explain why a double insulated appliance does not need an earth wire. **AO1 (2 marks)**

Electrical safety

Protection

- Electric appliances with a metal case must be earthed. If there is a fault and the metal case becomes live:
  - a large current flows through the earth wire connected to the metal case
  - the large current melts the fuse
  - the circuit breaks and the current stops flowing.

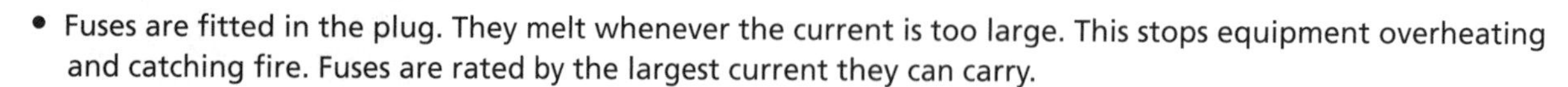

- Fuses are fitted in the plug. They melt whenever the current is too large. This stops equipment overheating and catching fire. Fuses are rated by the largest current they can carry.
- If the rating on the fuse is too low (for that plug), the fuse melts even if there is no fault.
- If the rating is too high, the fuse won't melt in time.

Residual current circuit breakers (RCCBs)

- RCCBs protect users from electrocution. The current in the live and neutral wires should always be the same. Residual current circuit breakers (RCCBs) break the circuit in less than 0.05 seconds if there is a difference in the current in these wires. The RCCB can be reset once the fault is repaired.
- Many homes now have RCCBs in them.

Figure 1: Residual current circuit breaker

How Science Works

- Mains voltage is supplied at 230 V so a 3 A fuse only works if the power of the equipment is 690 W or less. Otherwise, a 13 A fuse is needed.

Current, charge and power

Energy transfer

A current heats up any resistor that it flows through. This effect is useful in electric heaters, toasters, and light bulbs.

- An electric current transfers energy.
- The energy transferred is calculated using: energy transferred (J) = power (W) × time (s).

Current charge and work done

- The electric current is a flow of charge (electrons). Charge is measured in coulombs (C).
- Current is the rate that charge flows. It is calculated using: current (A) = $\frac{\text{charge (C)}}{\text{time (s)}}$.

EXAM TIP

When choosing the equation to use, check what you are asked for and what you are given in the question.

Remember!

You may be asked to rearrange equations. Practise doing this before the exam.

Improve your grade

Fuses rated at 3 A and 13 A are available. Which fuse should be used in each case?

a a lawn mower (power 2000 W)
b a lamp (power 60 W)
c a toaster (power 800 W). **AO2 (3 marks)**

Structure of atoms

What is inside an atom?

- Atoms are made from particles called electrons, protons and neutrons.
- The atomic number is the number of protons in the atom.
- The mass number is the number of neutrons and protons in the atom. Atoms have no overall charge as they have equal numbers of electrons and protons.

Particle	Charge	Relative mass
neutron	0	1
proton	1	1
electron	–1	negligible

Table 1: Charge and mass of subatomic particles

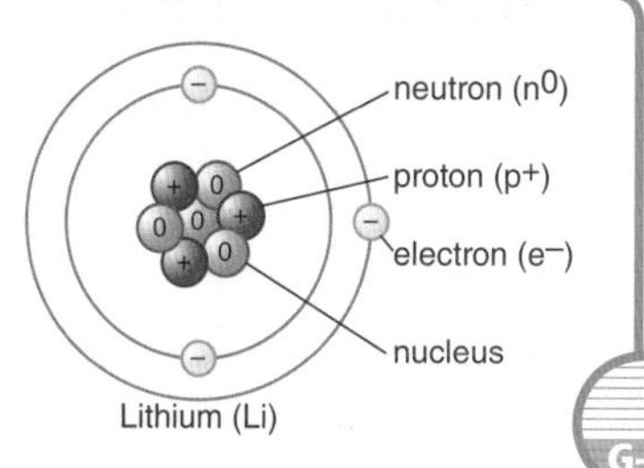

Figure 2: An atom of lithium

G–E

Electrons, protons and neutrons

- Electrons orbit the atom's nucleus. An atom that loses electrons becomes a positive ion. An atom that gains electrons becomes a negative ion.
- All atoms of the same element have the same number of protons. They can have different numbers of neutrons.
- Isotopes are atoms of the same element with different numbers of neutrons.
- Unstable atoms have different numbers of protons and neutrons. Some emit ionising radiation from the nucleus. We say the atom is decaying.

D–C

Radioactivity

Nuclear radiation

- We call alpha and beta particles, and gamma rays, ionising radiation. Radioactive atoms emit ionising radiation from the nucleus at random times. Atoms that absorb ionising radiation may change into an ion.
- Types of ionising radiation include:
  - Alpha particles which consist of two neutrons and two protons (a helium nucleus).
  - Beta particles which are electrons emitted when a neutron changes to a proton.
  - Gamma radiation which is high-energy electromagnetic radiation.

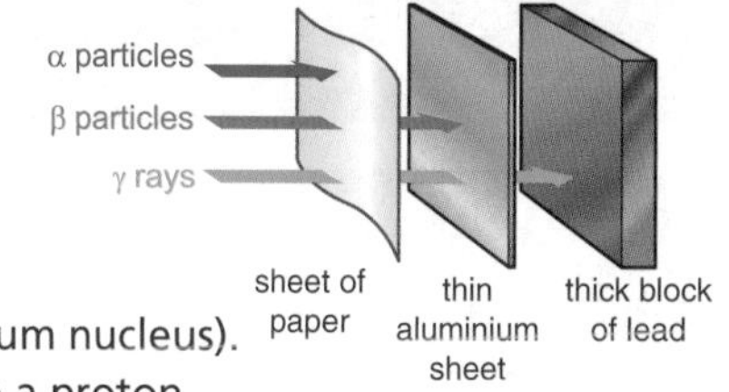

Figure 3: Penetrating power of nuclear radiation

Explaining the properties

- Alpha particles are heavy so they change atoms to ions easily. Alpha particles lose energy quickly and cannot travel far.
- Beta particles are small and do not ionise many atoms, so they can travel further.
- Alpha and beta particles change direction in electric and magnetic fields.
- Gamma rays have no charge or mass. They are not affected by electric or magnetic fields and pass easily through many materials.

Radiation	alpha (α) particle	beta (β) particle	gamma (γ) ray
Nature	positive helium nucleus $^4_2$He about 7000 times the mass of an electron	electron – negative charge	electromagnetic waves with very short wavelength and high energy
Charge	positive (+)	negative (–)	neutral
Range in air	a few centimetres	a few metres	very penetrating
Penetration	stopped by skin or a sheet of paper	stopped by 3 mm aluminium	intensity reduced, but not stopped by lead
Ionisation	strong	weak	very weak
Speed	about 5% speed of light	about 50% speed of light	speed of light

Table 2: Properties of nuclear radiation

Improve your grade

Why are the risks from handling an alpha source safely different from the risks from handling a gamma source?
AO3 (3 marks)

More about nuclear radiation

A model of the atom

- The 'plum pudding model' was an early model of the atom. It was not a successful model so a scientist called Rutherford carried out an experiment to explain the structure of an atom.
- He fired alpha particles at gold foil and observed the scattering of the particles.
- He developed a new model called the nuclear model.

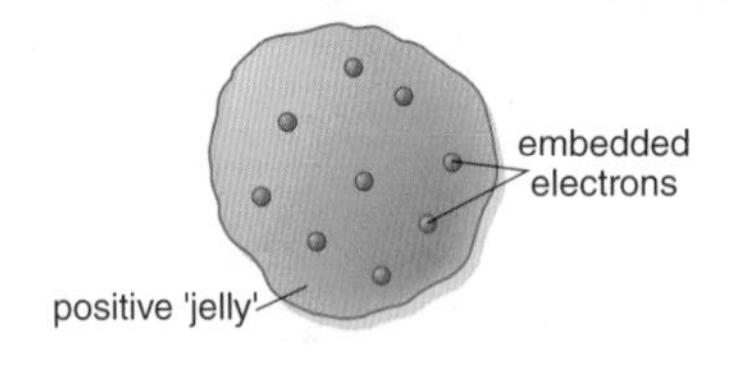

Figure 1: The plum pudding model

gold nucleii

alpha particles

Figure 2: Explaining the alpha particle tracks

Explaining the scattering experiment

- The results of Rutherford's experiment showed most alpha particles passed straight through the atom.
  - Atoms are mainly empty space.
- Positively charged alpha particles are deflected by positive charge concentrated in the nucleus.
  - The nucleus has a positive charge.
- A few massive alpha particles were reflected by the very dense nucleus.
  - Most of the atom's mass is in the nucleus.
- The pattern helped to explain the structure of atoms. Rutherford's nuclear model replaced the plum pudding model.

EXAM TIP

Make sure you can explain how the scattering experiment provided evidence for the nuclear model.

Background radiation

Radiation everywhere

- Background radiation is radiation that is all around us and comes mainly from natural sources. Some background radiation comes from man-made sources.
- Natural sources include rocks, soil and food.
- Man-made sources include X-rays and nuclear weapon fall-out.
- Sources are affected by lifestyle choices, for example, on a long flight you increase your exposure to cosmic radiation.

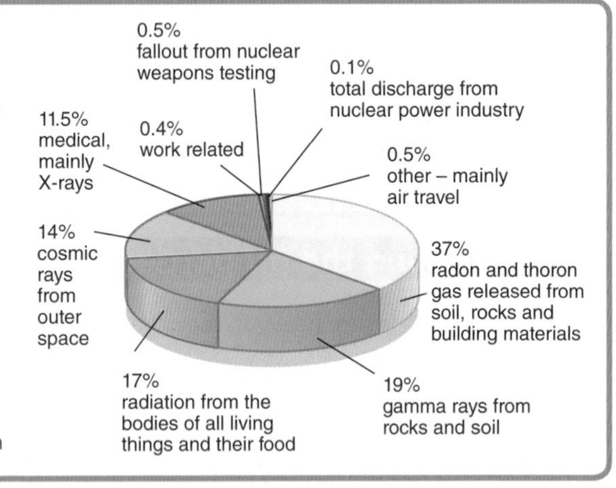

Figure 3: Sources of background radiation

Natural and lifestyle

- Some parts of the UK have higher levels of background radiation than others.
- Houses in these regions have higher levels of radon gas which can increase the risk of lung cancer.

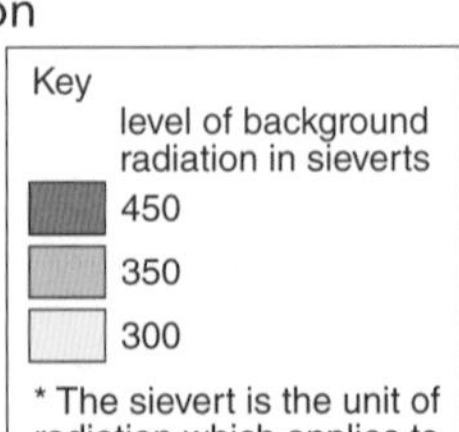

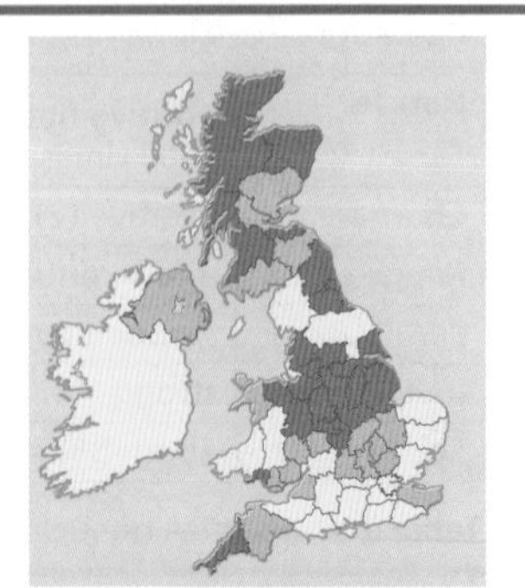

Figure 4: Levels of background radiation in the UK.

Improve your grade

Describe two differences between the plum pudding model and the nuclear model of the atom. **AO1 (4 marks)**

Half-life

Half-life

- Radioisotopes are isotopes that decay by giving out ionising radiation.
- We do not know which atoms will decay. We do know how many atoms will decay in a sample in a given time.
- Half-life is the time taken for the original radioactivity or count rate of a sample to halve.
- After a time equal to two half-lives, the activity falls to a quarter of its original value.

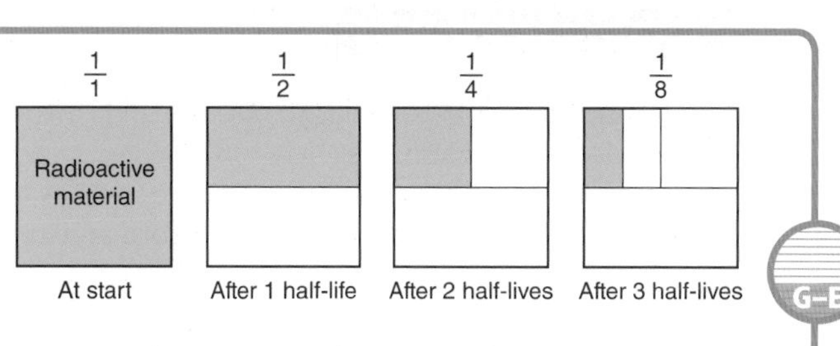

Figure 5: The amount of radioactive isotope left after each half-life

G–E

Measuring half-life

- Each radioactive material has its own half-life, which varies from millions of years to milliseconds. The half-life does not change for the given material.
- To find the half-life from a graph;
  - work out the value of half the original count rate
  - use the graph to read the time taken for the activity to reach this level.

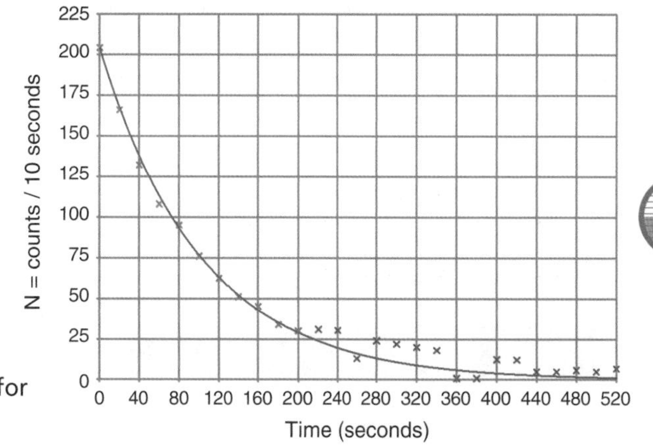

Figure 6: Radioactive decay graph for strontium-93

D–C

Remember!
Half-lives are measured as time (in hours, years, days, etc).

Using nuclear radiation

Radiation in medicine and industry

- Radioisotopes are radioactive isotopes. They have different uses.
- Gamma rays kill living cells and penetrate materials, so they are used for sterilisation and cancer treatments.
- Beta radiation penetrates thin sheets of metal and cardboard so is used to monitor the thickness of materials.

G–E

Using radioisotopes

- Medical tracers are radioactive materials injected, inhaled, or eaten by a patient. The position of the radioactivity is monitored to check blood flow, or identify blockages or tumours.
- Alpha radiation is used in smoke detectors. The particles ionise air so a current can flow. If smoke absorbs the radiation, the current cannot flow switching the alarm on.

Remember!
For use in medical tracers, a material with a short half-life is best. It reduces harm.
In monitoring equipment, material with a long half-life is best. It maintains consistent readings and the source does not need replacing.

Improve your grade

Explain whether a radioisotope with a long or a short half-life is best in each situation:
a in a smoke detector
b in a medical tracer
c in equipment used to sterilise surgical equipment. **AO2 (3 marks)**

Nuclear fission

Energy in atoms

G–E

- Nuclear fission is when a large nucleus splits into two or more parts. In nuclear power stations:
  - uranium-235 and plutonium-239 are used as fuels
  - atoms of uranium or plutonium absorb an extra neutron and become unstable
  - energy is released when the nucleus splits into two parts plus two or three neutrons.
- these neutrons can be absorbed by atoms of uranium or plutonium, and start more reactions.

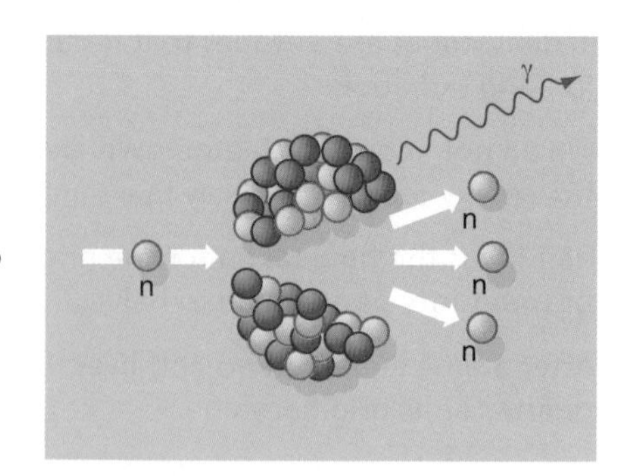

Figure 1: Splitting the atom

Critical mass and chain reaction

D–C

- If more uranium or plutonium atoms absorb the extra neutrons, this can start a chain reaction involving more atoms at each stage.
- Control rods in nuclear power stations control the size of the chain reaction by absorbing surplus neutrons.
- The critical mass of a nuclear fuel is the minimum amount needed to keep the chain reaction going.

EXAM TIP

You should be able to describe the stages taking place during nuclear fission.

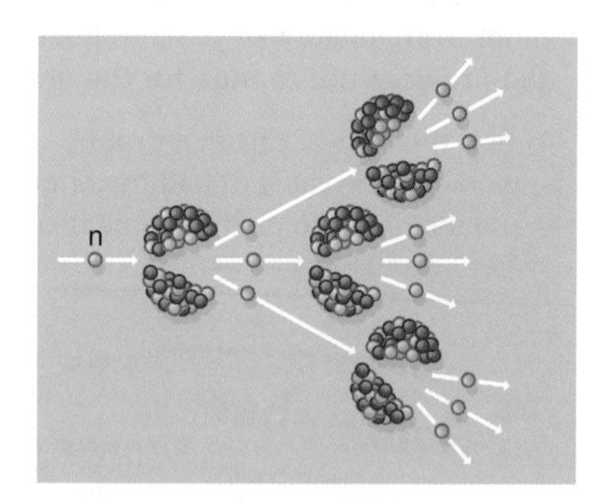

Figure 2: A chain reaction

Nuclear fusion

Fusing nuclei

- Nuclear fusion is when two small nuclei join together. This forms a nucleus of a new element and releases energy.
- Nuclear fusion is how energy is released in stars.
- Hydrogen atoms fuse to form helium atoms, releasing energy.

Using nuclear fusion

- Very high temperatures (above 15 million degrees Celsius) and very high pressures are needed for nuclear fusion. This is because positively charged protons in nuclei repel each other.
- Nuclear fusion in stars produces elements lighter than iron.
- Elements heavier than iron are produced during supernova (when massive stars explode releasing huge amounts of energy).

EXAM TIP

Do not confuse nuclear fission and nuclear fusion. Learn the correct spelling of each word.

Improve your grade

Explain why the presence of iron in the Sun is evidence that the Sun formed from the remains of older stars.
AO2 (3 marks)

Life cycle of stars

The life cycle of a star

- All stars have a life cycle. In stars about the same size as our Sun:
  - Gravity pulls dust and gas together, forming a protostar. Smaller masses (planets) may also form.
  - In the main sequence stage, hydrogen fuses to form helium which generates energy. This lasts billions of years.
  - When hydrogen fuel is used up, nuclear fusion reactions change. The star expands into a red giant.
  - When nuclear fusion reactions in the red giant change, the inner core collapses. The star becomes a white dwarf.
  - In a white dwarf, nuclear reactions form elements up to iron.
  - When its nuclear fuel runs out, the star cools and becomes a black dwarf.
- In stars much bigger than our Sun:
  - the main sequence star becomes a red supergiant.
  - when this collapses, the star explodes as a supernova.
  - a neutron star remains, or a black hole.

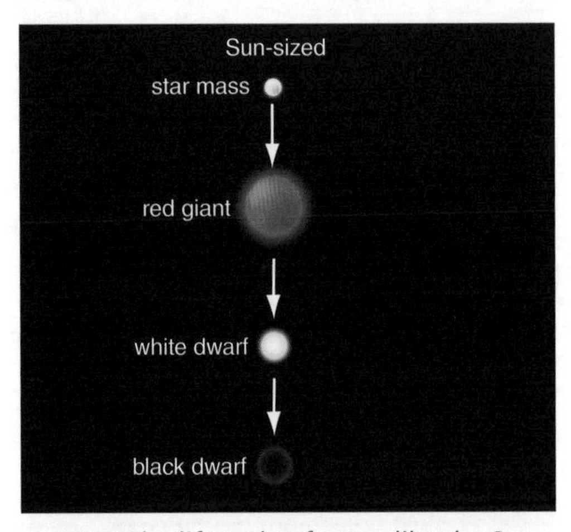

Figure 3: The life cycle of a star like the Sun

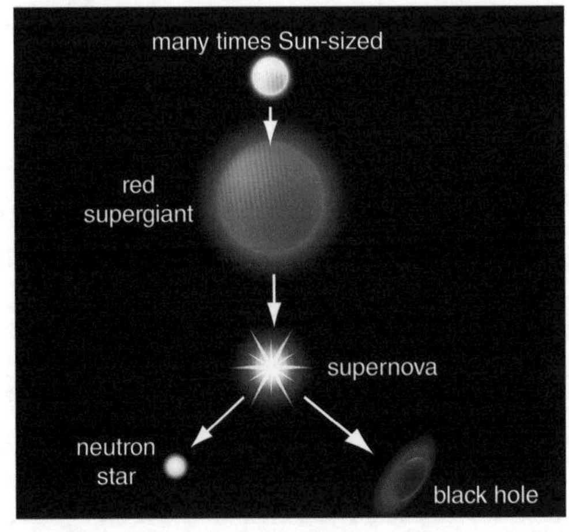

Figure 4: The life cycle of a star bigger than the Sun

G–E

Why does the main sequence last so long?

- The life cycle of a star depends on its size. A star's main sequence lasts billions of years.
  The main sequence star is stable because forces are balanced (inward gravitational forces match outward forces due to the heat).

Remember!
You should be able to explain how changes in the forces acting on a star make it move between different stages.

D–C

Improve your grade

Describe the stages in the life cycle of a star like the Sun. **AO1 (4 marks)**

P2 Summary

Forces and energy

A resultant force shows the combined effect of several forces. It makes an object accelerate (change speed or direction).

The stopping distance of a vehicle is thinking distance plus braking distance.

Moving objects reach a terminal velocity (top speed) when drag forces match driving forces.

Momentum is mass × velocity. It is conserved when objects collide or explode apart if no external forces act.

The speed of an object can be found using a distance–time graph. The acceleration and distance traveled can be found using a velocity–time graph.

Objects gain gravitational potential energy if they are lifted up. Moving objects have kinetic energy. Stretched or squashed elastic objects store elastic energy.

A force transfers energy when it does some work on an object.

Static electricity

Insulating materials can be charged by rubbing. They gain or lose electrons.

Objects carrying the same charge repel. Objects carrying the opposite charge attract.

Electrical circuits

Electric current is a flow of charge.

Resistance measures how easily a current flows through a component.
Resistance = potential difference ÷ current.

In a series circuit: the current is the same throughout, the potential difference is shared between components and the resistance of components adds.

In a parallel circuit, the current is shared between branches and the potential difference is the same for each branch.

Using mains electricity safely and the power of electrical appliances

Batteries supply direct current (d.c.); mains supply is alternating current (a.c.) at about 230 V 50 Hz.

Earthing, fuses and RCCBs protect the user and appliance.

Electric cable can be two-core or three-core. A three pin plug must be correctly wired.

Power is the potential difference supplied × current, or energy transferred ÷ time.

Energy transferred is the potential difference supplied × charge.

Atomic structure and radioactivity

An atomic nucleus contains protons and neutrons. Electrons orbit the nucleus. Radioactive materials emit ionising radiation from their nucleus.

Background radiation comes from natural sources (for example, rocks), and man-made sources (for example, medical uses).

Half-life is the time taken for the count rate of a radioactive sample to halve.

Alpha, beta and gamma radiation have different properties, uses and dangers.

Ions are atoms that have lost or gained electrons. Isotopes have the same number of protons, but different numbers of neutrons.

Nuclear fission and nuclear fusion

Nuclear fission is the splitting of an atomic nucleus.

Nuclear fusion is when two atomic nuclei join to form a larger one.

Fission of uranium-235 or plutonium-239 during a chain reaction releases energy in nuclear reactors.

All stars have a life cycle. Nuclear fusion in stars releases energy and produces all naturally occurring elements.

B1 Improve your grade

Page 6

Harry's diet is very high in saturated fats. Suggest two ways that this could affect his health. **AO2 (3 marks)**

It will make him put on weight.

This answer got 1 mark out of 3 (grade E/F). To improve the answer, the candidate should have stated another way that Harry's diet could affect his health. A diet high in saturated fat will increase the risk of developing heart disease. Also, it is scientifically correct to use the term 'mass' instead of weight in this case.

Page 7

Describe how phagocytes destroy pathogens. **AO1 (3 marks)**

They carry out phagocytosis.

This answer got 1 mark out of 3 (grade E/F). The answer needed to include a description of the process. A suitable addition to the answer might be: They surround the pathogen with their cytoplasm and pull it into the cell. They then use enzymes to destroy the pathogen.

Page 8

Suggest two ways that we can stop the spread of MRSA in hospitals. **AO1 (2 marks)**

Make sure doctors and nurses wash their hands with antibacterial hand-wash before and after treating patients. Make sure hospitals are kept clean.

This answer got full marks (grade C) as it mentioned two different ways of stopping the spread of MRSA. Another one is to encourage visitors to hospitals to use the antibacterial handwash before they enter the hospital in case they are carrying MRSA on the their hands.

Page 9

You wish to grow some harmless E. coli bacteria on some agar jelly. You use an inoculating loop to transfer the bacteria to the jelly. Explain why it is important to hold the inoculating loop in a flame first. **AO1 (2 marks)**

To make it sterile.

This answer gets 1 of the marks (grade D) as the candidate forgot to mention why making the loop sterile is important. It is to prevent the agar becoming contaminated with any other type of microorganism.

Page 10

An injury that results in breaking of the spine may result in the person being paralysed. Explain why.
AO2 (3 marks)

The nerves in the spine might be broken.

This answer only makes one suggestion and thus only achieves one of the three possible marks (grade E/F). To gain full marks, explain that the spinal cord is part of the central nervous system and contains many neurones that send messages from the brain to all parts of the body. When these impulses meet the muscles, they contract. If the nerves in the spinal cord are broken then impulses cannot cross them, so muscles will not respond.

Page 11

James is dancing in a nightclub. He starts to sweat. Explain how sweating helps to cool him down.
AO2 (2 marks)

The sweat evaporates from his skin. It needs energy to change from a liquid into a gas and so uses the heat from the skin as a source of the energy. This cools down the skin and reduces his body temperature.

This answer gets full marks and so achieves a C grade. The candidate has shown that they have a deep understanding of the science by adding the information about why the skin needs heat to evaporate.

Page 12

FSH is found in fertility drugs. Explain how taking FSH will increase a woman's fertility. **AO2 (2 marks)**

It will increase the amount of eggs that mature in her ovaries.

This answer achieves one of the two possible marks (grade D). The candidate should have continued to explain that increasing the amount of mature eggs means that more than one will be released at a time, which will increase the chance of one being fertilised.

Page 13

Explain how the hormone auxin brings about a response to light called phototropism. **AO2 (4 marks)**

Light shines on the plant from one direction. There is more auxin on the shaded side so there are more cells here. This bends the shoot towards the light.

This answer gets two marks (grade D). The candidate has misunderstood how auxins affect plant growth. They do not increase the number of cells but increase the length of them.

Page 14

Why do drugs need to be trialled before they can be prescribed by a doctor? **AO2 (2 marks)**

To check that they are safe

The candidate has achieved one mark out of a possible two (grade D). They should also mention that clinical trials need to be carried out to check that the drug actually works (and works better than any drugs already on the market). This is usually done by carrying out a double-blind trial using a placebo.

Page 15

Explain how cacti are adapted to living in the dry desert.
AO2 (3 marks)

Cacti have long roots to grow deep into the soil and spread out as well as to absorb maximum amounts of water. They can also store water in its tissues to use when they can't absorb any water from the ground. They have no leaves, which reduces water lost by evaporation. Their spikes stop animals from eating them.

This answer gets full marks (grade C). The candidate has described all of its features and explained how they enable the cactus to have a supply of water at all times, which is the main problem for a plant living in the desert. They have shown extra knowledge by explaining the function of the spikes.

Page 16

Sewage contains a lot of microorganisms. Explain why mayfly larvae cannot live in water polluted with sewage. **AO2 (2 marks)**

The microorganisms are pathogens and kill the mayfly larvae.

The candidate has not achieved any marks for this question as they have not understood how microorganisms will kill the larvae. Increasing the amount of microorganisms in the water will decrease the amount of dissolved oxygen. The oxygen-loving mayfly larvae would not be able to survive in water with low levels of dissolved oxygen.

Page 17

Explain why biomass decreases as you move along a food chain. **AO2 (3 marks)**

The amount of energy decreases as you go along a food chain. Less energy means less biomass.

This answer gets two marks (grade D). The candidate should have gone on to explain why the energy decreases. A suitable reason might be that energy is wasted as heat, and in waste such as urine.

Page 18

Write a simple plan for an investigation to prove that the decay of bread by mould requires moisture. **AO2 (2 marks)**

Take 2 slices of bread. Add a little water to one. Place both in a plastic bag and leave in a warm place.

This answer has achieved 3 marks out of 4 (grade D/C). The candidate has not explained how they would compare the decay of each one. They could just look at the growth of mould on each slice. An even better method is to measure the area of each slice that is covered with mould.

Page 19

Matthew and Emma are both very tall. Their one-year-old son, Ryan, is also expected to be tall when he is older. Why is this? **AO2 (2 marks)**

He will probably inherit a gene for tallness from both parents.

This answer gets full marks (grade C). To further show knowledge, the candidate could have explained that the gene for tallness would be carried in the egg and sperm.

Page 20

Tina has a lavender plant in her garden. She wants to produce more identical lavender plants quickly and cheaply. Outline the method she should use to do this. **AO2 (3 marks)**

She could take cuttings. She should do this by cutting off small lengths of stem and placing them in a pot of soil. Each cutting will grow into a new plant identical to the original.

This answer gets full marks (grade C), as the candidate has outlined the stages needed to grow cuttings and has realised that this will produce clones of the lavender plant. They could have added that Tina should dip the end of each cutting into hormone rooting powder. This would encourage the cuttings to grow roots and so increase the chances of them growing into new plants.

Page 21

In 1859, Charles Darwin published a book containing his ideas about natural selection and evolution. Explain why many people at the time did not believe what it said. **AO2 (3 marks)**

Most people thought that God had created all the living things on Earth in the form that they are in now.

The candidate correctly identified one of the reasons why people did not accept Darwin's theory, so achieves one mark out of three (grade E/F). They should also have explained that nobody understood how characteristics of the parent could be passed to its offspring, as genes had not yet been discovered. Also, at that time, there was not much scientific evidence to support the theory.

Page 22

The cheetah is the fastest land animal on Earth. It uses its speed to catch its prey. Use natural selection to explain how cheetahs have got faster over time. **AO2 (5 marks)**

The cheetahs that were the quickest caught prey and survived.

This answer gained two marks out of five (grade E/D). The candidate should have put much more information into their answer. When tackling questions about natural selection, apply these stages: Some cheetahs were faster than others (variation). The fastest cheetahs caught more prey than the slower ones (competition). They survived (survival) to reproduce and pass on the genes for speed onto their offspring (reproduction). If this keeps happening, then over many generations the cheetah will get faster.

C1 Improve your grade

Page 24

Draw the electronic structures of the following atoms with proton numbers 3, 9, 11, 16, and 20.
AO2 (5 marks)

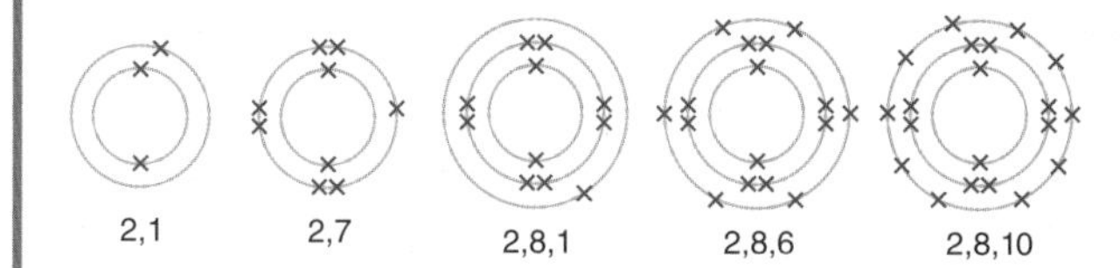

The answer gets 4 marks (grade C) as the first four are correct. In the last one, where the proton number is 20, the answer has the last 10 electrons in the third orbit rather than eight in the third and two in the fourth orbits. For full marks the structure should be 2,8,8,2.

Page 25

Explain why oxygen has a molecule with a double covalent bond and fluorine has only a single covalent bond. You should refer to the number of outer electrons in both atoms in your answer. **AO2 (4 marks)**

Fluorine has electronic structure 2,7, so needs one electron to fill the shell. So it makes a single bond. Oxygen is 2,6, and needs two electrons to fill the shell, so it makes two bonds.

This answer gets 2 marks (grade D). It correctly gives the electronic structure of the two atoms, but fails to mention sharing electrons with the other atom, or the need to achieve a Noble gas electronic structure.

Page 26

Explain the advantages and disadvantages of quarrying for a metal ore near your local town. Your answer should include two advantages and two disadvantages. **AO2 and 3 (4 marks)**

It will be good for the town as it could make lots of jobs for people. It may be bad for the town as it could increase the traffic and pollution.

This answer gets 2 marks (grade E). The candidate has failed to fully read the question, and has only given one advantage (jobs). Other possible advantages include more money to spend locally, and improved transport links. There is one disadvantage (traffic) given, the use of 'pollution' is not detailed enough, and should have included, noise, dust, and damage to the environment such as named example of harm to a specific habitat.

Page 27

Draw a table to show which of these metals is extracted by carbon and which by electrolysis. **AO2 (2 marks)**

iron, magnesium, aluminium, copper, zinc, lead, potassium, and calcium

carbon	electrolysis
iron	magnesium
zinc	aluminium
lead	potassium
	calcium
	copper

This answer gets 1 mark (grade F). The answer correctly places all the metals except for copper. Copper is below carbon and should be placed in the left hand column, the confusion is that impure copper is purified by electrolysis.

Page 28

Explain the difference between oxidation and reduction. **AO1 (2 marks)**

Oxidation is gaining oxygen and reduction is the loss of oxygen.

This answer gets 2 marks (grade C). Oxidation and reduction are the opposite of each other. In a reaction, if one substance is oxidised the other is reduced.

Page 29

Explain how copper is produced at the negative electrode when copper sulfate is electrolysed. **AO2 (2 marks)**

Copper ions gain electrons at the negative electrode, so becoming copper atoms.

This answer gets 1 mark (grade D). It recognises that copper is made of ions, but fails to state that they are positive, or have 2+ charge – this can be found on the data sheet. It also should indicate that two electrons are needed for each copper ion.

Page 30

Explain why titanium and aluminium metals are both reactive and corrosion free. **AO1 (2 marks)**

Both metals get coated with an oxide layer. This layer prevents corrosion.

This answer gets 1 mark (grade D). It clearly states that an oxide layer forms, but should say that the oxygen comes from the air. The second sentence re-states the question, without adding further explanation about how the layer prevents corrosion, or that the layer is formed quickly because of the high reactivity of the metals.

Page 31

Suggest why growing plants to use as fuels may cause more problems than it solves. **AO2 (3 marks)**

Biofuels are grown on land that can be used to grow food. There will be less food.

This answer gets 1 mark (grade D). It clearly makes the point about land use, and identifies a possible reduction in food production. But it fails to mention problems that might arise, such as: higher food prices/food becoming less affordable; starvation; it doesn't remove our reliance on carbon-based fuels; and increases the destruction of rainforests to increase land to grow fuel crops on.

Page 32

Draw out the structure of the straight chained alkanes with 5 carbon atoms and 7 carbon atoms. **AO2 (2 marks)**

```
  H H H H H
H-C-C-C-C-C-H
  H H H H H

  H H H H H H
H-C-C-C-C-C-C-H
  H H H H H H
```

This answer gets 1 mark (grade C). The second structure

should have seven carbon atoms, not six. Candidates must check their diagrams carefully.

Page 33

Draw structural diagrams to show how $C_{12}H_{26}$ can be converted to C_3H_6, and another molecule. State which of the two products is unsaturated, and why. **AO1 and 2 (3 marks)**

```
    H H H H H H H H H H H H
    | | | | | | | | | | | |
  H-C-C-C-C-C-C-C-C-C-C-C-C-H
    | | | | | | | | | | | |
    H H H H H H H H H H H H
                ↓
      H H           H H H H H H H H H
      | |           | | | | | | | | |
H—C=C-C—H   +   H—C-C-C-C-C-C-C-C-C—H
        |           | | | | | | | | |
        H           H H H H H H H H H
```

It is the C_3H_6 that is unsaturated because it has a double bond.

This answer gets 2 marks (grade C). The diagram for the conversion is good, but on the C_3H_6 diagram there is a missing hydrogen on the left. Always count that each carbon atom has 4 bonds. The answer recognises propene as the unsaturated molecule and correctly gives the reason.

Page 34

PVC (polyvinylchloride) is a commonly used polymer. Its monomer has the structure:

```
H       H
 \     /
  C = C
 /     \
Cl      H
```

Show, using three monomer molecules, the structure and bonding of a PVC polymer chain. **AO2 (2 marks)**

```
   H  H  H  Cl H  H
|  |  |  |  |  |  |  |
C- C- C- C- C- C- C- C
|  |  |  |  |  |  |  |
   H  Cl H  H  H  Cl
```

Each of the lines represents a single covalent bond or a pair of shared electrons.

This answer gets 2 marks (grade C). The diagram shows the structure and bonding correctly. The additional statement makes it clear that the candidate knows that this molecule has covalent bonding, and what covalent bonding is.

Page 35

Describe the difference between recycling polymer waste and re-using it by incineration. Give two reasons why recycling is more environmentally friendly than incineration. **AO1 (3 marks)**

Recycling polymer waste means re-using it as a polymer after some processing. Re-using it by incineration is when the polymer is burnt. The polymer is changed to heat and carbon dioxide.

This answer gets 1 mark (grade C). There is a clear description of the two processes but the second part of the question is ignored in the answer. Recycling is less damaging as it needs less energy for recycling than for making new polymers, saves on using fossil fuels, and stops polluting gases caused by incineration; it also reduces waste sent to landfill.

Page 36

Explain why biofuels are considered to be carbon neutral. **AO1 (2 marks)**

Biofuels are carbon neutral because they only put into the atmosphere the same carbon dioxide as they took out by growing.

This answer gets 2 marks (grade C). The answer scores full marks, but could be improved by adding that photosynthesis takes the carbon dioxide out of the air, rather than using the word 'growing'.

Page 37

Hand creams are oil-and-water mixtures. Explain why adding an emulsifier to the cream helps you spread the cream evenly on your hands. **AO2 (2 marks)**

The emulsifier spreads the cream evenly on your hands.

This answer gets 0 marks (grade E). There is no explanation of the need to keep the oil and water evenly mixed, so that all parts of the hands receive the correct proportions of oil and water.

Page 38

Use the theory of tectonic plates to explain how mountains are formed. **AO1 (2 marks)**

When plates collide they crumple up, making the mountains.

This answer gets 1 mark (grade E). The candidate knows about plates colliding, but fails to explain that one plate slides under the other to raise it up. It would be good to give an example such as the Himalayas, where the process is taking place.

Page 39

Sketch a timeline showing how the proportions of water vapour, carbon dioxide, oxygen, and nitrogen have changed since the Earth was formed.
AO2 (4 marks)

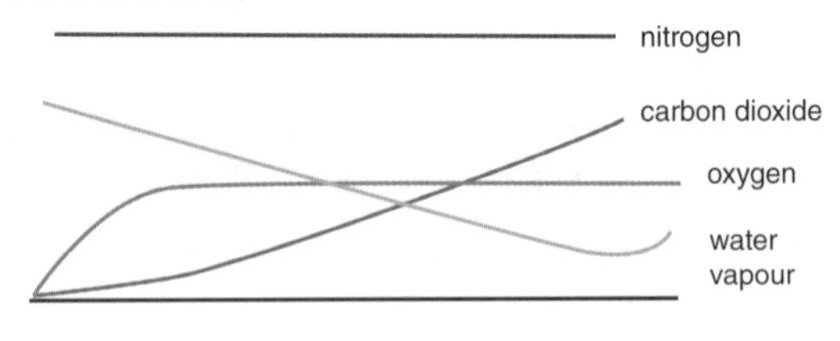

This answer gets 0 marks (grade F). The graph should be like this:

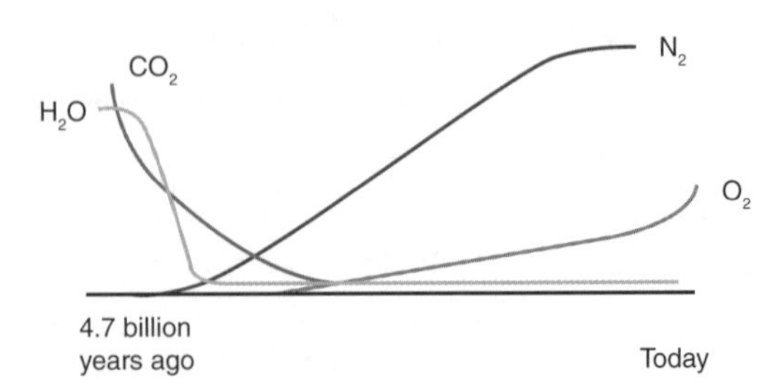

The candidate knows that carbon dioxide is increasing and has based the graph around that, and the fact that oxygen levels rose and then levelled out, but the detail is insufficient to get any marks. The timeline is not dated other than 'Today'.

Page 40

Describe how the carbon dioxide in the early atmosphere has been reduced to its current level.
AO1 (2 marks)

Photosynthesis has removed it and turned it into oxygen.

This answer gets 1 mark (grade E). There is recognition of the importance of photosynthesis, but not that the carbon dioxide has become locked away in fossil fuels, in the oceans, as animal shells or skeletons, and as limestone.

P1 Improve your grade

Page 42

Explain whether a kettle of hot water cools down quicker if its outer surface is coloured white or dark green.
AO2 (3 marks)

It cools quicker if it is dark green as more heat is lost by radiation.

This answer gets 2 out of 3 marks (grade C). The candidate correctly said that heat loss by radiation is affected by the colour of the kettle. For full marks, explain that heat is lost more quickly from dark surfaces.

Page 43

Explain why convection can take place in a liquid but not in a solid. **AO2 (3 marks)**

In a solid, particles are in fixed positions. During convection, heat is carried by particles that move.

This answer gets 2 out of 3 marks (grade C). The candidate correctly described the arrangement of particles in a solid, and could also have said that in a liquid they are free to change places. For full marks, explain that particles move from a hotter to a cooler place.

Page 44

Explain why a towel dries quicker on a windy summer day. **AO2 (3 marks)**

Water evaporates from the towel quicker.

This answer gets 1 out of 3 marks (grade D). The candidate did not describe how the wind or sun increases evaporation. For full marks, either say wind moves air away from the towel so the air just above the towel does not become saturated; or that on a summer's day, the air is warmer so particles have more energy so can break bonds more easily.

Page 45

The specific heat capacity of copper is 390 J/kg °C. Explain whether copper heats up quicker than the same mass of water when they are put in a hot place.
AO2 (3 marks)

Copper gets hotter than the water.

This answer gets 0 out of 3 marks (grade F). Copper heats up quicker but if the water and copper are left for long enough, their final temperature is the same. Remember to include 3 points for a 3 mark question, and to use the information given. For full marks, explain that copper needs less energy than water to heat up by 1 degree. If they absorb the same energy, the copper has a larger increase in temperature.

Page 46

Explain what this Sankey diagram shows in as much detail as possible. **AO2 (3 marks)**

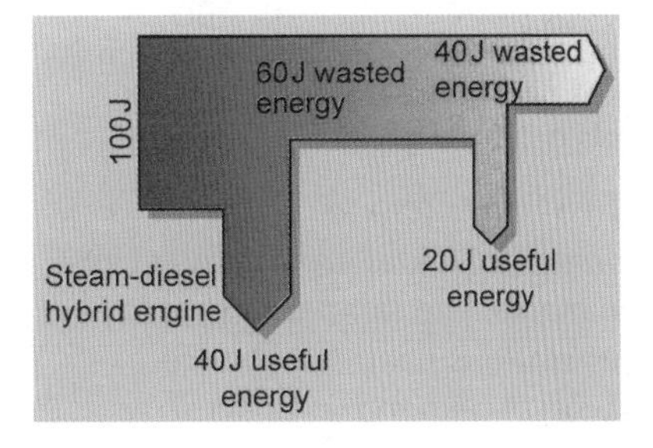

The efficiency is 20%. There are 60 J of useful energy transferred and 100 J of wasted energy.

This answer gets 0 marks (grade F). The energy is transferred in two stages, which the candidate has confused. For full marks, explain what is meant by efficiency, show calculations and discuss both stages. In the first stage, efficiency is 40%. During the second stage, efficiency is 33%. Overall efficiency is 60%.

Page 47

Two different bulbs are switched on for 10 minutes. Calculate the energy transferred by each one of:

a a 60 W filament bulb

b a 10 W energy efficient bulb. **AO2 (4 marks)**

Energy transferred = power x time. The energy transferred is 600 for bulb 'a' and 100 for bulb 'b'.

This answer gets 1 mark (grade E) as the candidate forgot to convert minutes to seconds, and did not show the unit for energy. For full marks, show working. For example, the first bulb transferred 60 x 10 x 60 = 36 000 joules.

Page 48

Describe the energy changes taking place in these parts of a coal fired power station: the burning fuel; the boiler; the turbine; the generator. **AO1 (4 marks)**

burning fuel: chemical -> heat
boiler: heat -> kinetic
turbine : kinetic (steam) -> kinetic (turbine)
generator: kinetic -> electrical

This answer gets 4 marks (grade C). The initial energy and final energy had been correctly identified in each case.

Page 49

Explain whether a hydroelectric power station or a coal fired power station is best for a city located near the coast. **AO3 (5 marks)**

Coal: Coal can be transported by water easily and it will produce enough electricity for a city.

This answer gets 3 marks (grade D). The student needs to explain why hydroelectricity is not a good choice too.

Page 50

Explain whether a coal fired power station or a hydroelectric power station is best suited to cope with surges in demand during the day. **AO3 (3 marks)**

The hydroelectric power station can be started in minutes. However, it causes flooding and can only be used in some parts of the UK. Water can be pumped into the reservoir during quiet periods ready for use later. Coal-fired power stations produce greenhouse gases.

This answer gets 2 out of 3 marks (grade C). The comments about environmental issues are irrelevant to the question. To get full marks, the candidate should explain that more power stations need to produce electricity if there is a surge in demand for electricity.

Page 51

An echo is a reflected sound wave. Explain why you can hear echoes only in certain places, and why you may hear more than one echo. **AO3 (3 marks)**

Sound waves must bounce off a surface for you to hear an echo. More than one echo is heard if the sound bounces off several objects.

This answer gets 2 out of 3 (grade C). To get full marks give examples of hard surfaces that sound can echo off, e.g. cliffs or buildings.

Page 52

Draw traces to show two sound waves. One sound is lower-pitched and twice as loud as the other sound. Label the amplitude and wavelength on each trace. **AO1 (4 marks)**

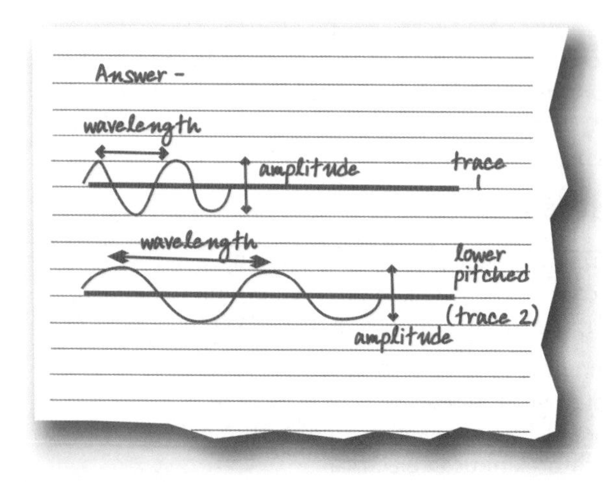

This answer gets 2 out of 4 (grade E). The amplitude is wrongly labelled, and the candidate has not shown the trace of a louder note. To get full marks, correctly label amplitude as height from mid-point to peak; and for a note twice as loud, show a wave with double the amplitude (twice as high).

Page 53

Calculate the speed of radio waves with a wavelength of 10 000 m and frequency of 30 000 Hz. **AO2 (3 marks)**

speed = 30 000 ÷ 10 000 = 3 m/s

This answer gets 1 out of 3 marks (grade D). The candidate neither wrote down the correct equation, nor substituted correctly into it. They only get marks for showing the correct unit.

Page 54

Explain which type of electromagnetic wave is the best choice for satellite communications. **AO2 (3 marks)**

Microwaves (1) because they are not absorbed by the atmosphere (1)

This answer gets 2 marks (C). The candidate should explain that microwaves can reach the satellite because they can pass through the atmosphere.

Page 55

Explain two advantages of using space-based telescopes. **AO2 (4 marks)**

They can see more as they are closer to the stars, and they can see in all directions.

This answer gets 0 out of 4 marks (grade F). The candidate does not realise that the atmosphere has a big impact on the signals reaching the Earth's surface. To get full marks, explain that the radiation is not absorbed by the atmosphere so fainter objects can be seen. Another advantage is that objects in space, which are emitting all types of electromagnetic radiation, can be detected as the telescope is outside the atmosphere.

Page 56

Explain the evidence we have that supports the Big Bang theory. **AO3 (6 marks)**

The Big Bang was a massive explosion billions of years ago: we can still hear the echo. Galaxies are still moving away from us.

This answer gets 2 out of 6 (grade F). Answers like these often allow you to include several points, but expect you to include explanations. The candidate correctly stated the Big Bang theory, but wrongly said that the 'echo' was heard – it is detected as electromagnetic radiation. To get full marks, choose points like these: the red shift shows distant galaxies are moving away from Earth; and further-away galaxies move away faster so they used to be closer together.

Background microwave radiation caused by the Big Bang is detected in all directions.

B2 Improve your grade

Page 58

The image below was taken of some cells using a light microscope. State what type of cells they are and give reasons for your answer. **AO2 (2 marks)**

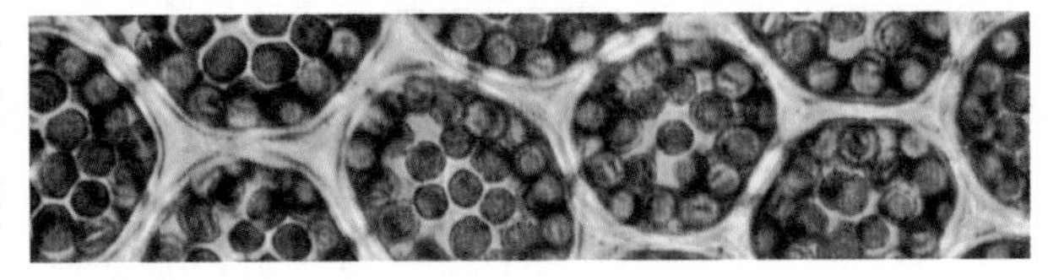

Plant cells because they are green

This answer gains 1 mark out of 2 (grade D). The candidate identified the correct type of cell was but the reason given was not sufficient. To gain full marks the answer should have talked about organelles: it could have mentioned that these cells contain chloroplasts or that they have a cell wall. Full marks would also be given if the candidate said they were algae cells.

Page 59

Explain how a sperm cell is specialised. **AO2 (4 marks)**

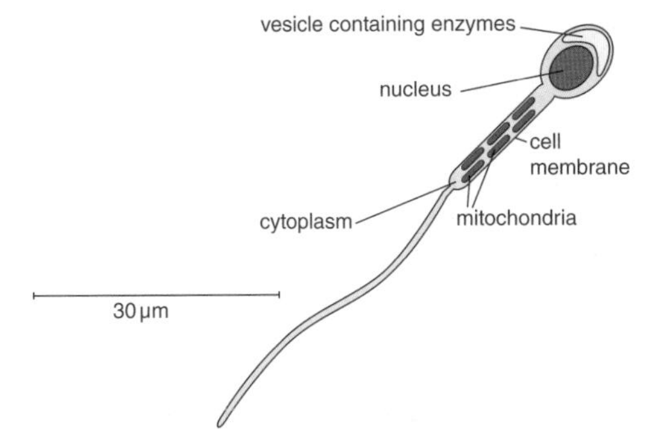

It has a tail to help it swim.

This answer gains 1 mark out of 4 (grade F). The candidate correctly identified that it has a tail for swimming but did not explain why this is required (to reach the egg for fertilisation). To gain full marks, the answer needs to include at least two of the adaptations explained. For example, the candidate could have mentioned that the sperm has a vesicle containing enzymes, which are used to digest the outside of the egg to allow the nucleus of the sperm to fuse with the nucleus of the egg during fertilisation.

Page 60

Describe why the stomach is classed as an organ. **AO2 (2 marks)**

It contains lots of cells.

This answer did not achieve any marks (grade U), as the candidate did not understand what we mean by an organ. To gain full marks the answer should explain that the stomach is an organ because it is made up of lots of different tissues, which work together to carry out a function. (It would not be necessary, although interesting, to write that the tissues are glandular tissue to secretes juices, muscular tissue to contract the stomach walls, and epithelial tissue to protect the stomach from the juices it makes.)

Page 61

Without photosynthesis, humans would not survive. Explain why. **AO2 (2 marks)**

Photosynthesis produces oxygen which we need to live.

This answer gains 1 mark out of 2 (grade D). To achieve maximum marks, the candidate should have also explained that photosynthesis produces glucose, which is a store of energy. Animals eat the plant and release the energy which is also essential for life.

Page 62

Anna uses a fertiliser that is high in nitrates on her tomato plants. Explain why she does this. **AO2 (3 marks)**

So the plants have nitrates which they can use to make protein.

This answer gains 2 marks out of a possible 3 (grade D). To achieve the missing mark, the candidate should have gone on to explain why proteins are essential for the healthy growth of the plant: they are used to build new cells. (It would not be necessary, although interesting, to write that the proteins are also used to build enzymes. Also, that without a supply of nitrate, the tomato plants would not grow to their full potential nor produce many tomatoes. Plants absorb minerals such as nitrates through their roots, so this is why fertilisers are added to the soil.)

Page 63

Simon went on holiday to Mexico. He noticed that the plants growing there were very different to the plants growing at home in the UK. Explain the reasons for this difference in distribution. **AO2 (3 marks)**

The temperature is much warmer in Mexico, so the plants that grow there are adapted to living at higher temperatures. The rainfall is also lower in Mexico, so the plants there have adaptations to help them cope with lack of water.

This answer gains full marks (grade C) as the candidate mentioned two differences in physical factors between Mexico and the UK, and explained how these affect the distribution of the plants. The plants growing in Mexico may not survive in the UK because of the low temperatures and high rainfall.

Page 64

Pepsin is an enzyme that helps break down proteins in the stomach. It has an optimum pH of 2. Use this information to explain why the stomach produces hydrochloric acid. **AO2 (2 marks)**

Pepsin has an optimum pH of 2 so it works best in acidic conditions.

This answer gained 1 mark out of a possible 2 (grade D). The answer did not really explain what is meant by optimum pH and why this is important. The optimum pH is the pH at which an enzyme works best at. The acid produced by the stomach means that pepsin can break down protein in the stomach at a fast rate.

Page 65

Dipesh mixed some starch with amylase in a beaker and left the mixture in a water bath at 37 °C. After 30 minutes, he tested the mixture to see if there was any starch present. Predict what he will find and give a reason for your answer. **AO2 (2 marks)**

There would be no starch left because the amylase breaks down starch.

This answer achieves full marks (grade C) but it would be more complete, and show the candidate's knowledge better, if it included the fact that the starch would have been broken down into sugars. Remember that enzymes are only catalysts in the breaking down of food. The starch would eventually break down by itself, it would just take a lot longer.

Page 66

Explain why breathing rate increases when you exercise. **AO2 (3 marks)**

Your muscles are contracting quickly so are using up energy at a fast rate. Breathing rate increases to get oxygen to the muscles quickly.

This answer gains 2 marks out of a possible 3 (grade D). For full marks, the candidate needs to go on to explain why the muscles need a good supply of oxygen: to carry out aerobic respiration in order to release energy from glucose. The energy is used to enable the muscles to contract.

Page 67

You are involved in a race. At first you sprint off feeling full of energy. However, halfway through, your legs start to ache and you have to stop. Explain why this happened. **AO2 (3 marks)**

My heart could not beat fast enough and my lungs could not breathe fast enough to get oxygen to my leg muscles, so they could not carry out respiration and make energy for my leg muscles to work.

This answer gains 1 mark (grade E/F). The candidate was correct in stating that the lungs and heart could not work fast enough to get the required levels of oxygen to the leg muscles but respiration would not stop – anaerobic respiration would take over. For full marks, the answer should explain this, then go on to explain that the legs would start to ache and stop working efficiently because of the build-up of lactic acid in the muscle cells.

Page 68

Explain why mitosis is an essential process in the formation of a baby from a fertilised egg. **AO2 (2 marks)**

Mitosis is needed to produce the egg and sperm cells that join to form the fertilised egg.

This answer is incorrect and will score no marks (grade U). The candidate has mixed up the terms mitosis and meiosis. For full marks the answer should state that the fertilised egg is one cell (called a zygote), which will divide by mitosis to form the many cells that make up a baby.

Page 69

Rubina has the blood group A. Her alleles for blood group are AO. Rajesh has blood group B. His alleles are BO. Explain how their daughter, Sharmila, has a different blood group from either of them (blood group O). **AO2 (2 marks)**

Sharmila has received a mixture of their alleles to give her blood group O.

This answer has achieved 1 mark out of 2 (grade D) as the candidate has not fully explained what has happened to give Sharmila blood group O. People with blood group O have two O alleles. She has received both O alleles from her parents through their gametes.

Page 70

Katy has red hair. Both her parents have brown hair. Explain how Katy inherited red hair when her parents do not have it. **AO2 (3 marks)**

Katy got the gene for red hair from her parents.

This answer gains 1 mark out of 3 (grade E/D). To achieve full marks, the candidate needs to explain why Katy's parents do not have red hair. The red-hair allele must be recessive. Her parents both have the red-hair allele but their other allele (brown hair) masks it. Katy received two red-hair alleles from her parents during fertilisation so Katy has two red-hair alleles (her genotype) and so therefore has red hair (her phenotype). Also, to show a good understanding of genetics, the candidate should have used the correct terminology. Red hair should be referred to as an allele, not a gene. The red hair allele is a type of hair-colour gene.

Page 71

The height of pea plants is controlled by a single gene. The tall allele is dominant. A tall pea plant was bred with a short pea plant. The offspring formed was in the ratio 1 tall : 1 short. Draw a genetic diagram to show the cross. **AO2 (4 marks)**

Parents	tall pea plant	short pea plant
	TT	tt
Gametes	T and T	t and t

Offspring

	T	T
t	Tt	Tt
t	Tt	Tt

This answer gains 2 marks out of 4 (grade D/E). The candidate has made a mistake in determining the allele of the tall pea plant. You can see from the cross that all of the offspring are tall, not the ratio of 1 tall : 1 short. The tall plant must have the alleles Tt.

Page 72

Polydactyly is caused by a dominant allele. Jessica has polydactyly. Ryan does not have it. Could their children have polydactyly? Explain your answer. **AO2 (3 marks)**

Yes, because they could inherit the polydactyly gene from Jessica.

This answer gains 2 marks (grade D). To gain full marks, the candidate should go on to explain that the children only need one dominant allele in order to have polydactyly and, as Jessica has the disorder, she must carry at least one dominant allele.

Page 73

Around 300 years ago, the dodo lived on the island of Mauritius. It had no predators. People arrived on the island. They brought predators such as dogs, pigs and rats. Explain why the dodo became extinct. **AO2 (2 marks)**

The dodos got eaten by the dogs, pigs and rats and they all died out.

This answer gains 1 mark out of 2 (grade D). To achieve full marks, the candidate should have added that the dodo was not adapted to escape the new predators and was not able to evolve quickly enough.

C2 Improve your grade

Page 75

Work out the relative atomic mass (A_r) of chlorine. Chlorine has two isotopes, 25 per cent is chlorine-37, 75 per cent is chlorine-35. **AO2 (2 marks)**

Chlorine -37 has 20 neutrons, and chlorine 35 has only 18 neutrons.

This answer gets 2 marks (grade C). The difference between the two isotopes is clearly given.

Page 76

Use the periodic table to work out the electronic structure of these elements: magnesium, chlorine, nitrogen, aluminium, calcium. **AO2 (5 marks)**

Mg = 2.8.2, Cl = 2.8.7, N = 2.7, Al 2.8.3, Ca = 2.8.8.2

This answer gets 4 marks (grade C). All except N are correct. The problem is that the first orbit has not been subtracted from the 7, it should be 2,5.

Page 77

Draw diagrams to show how calcium and fluorine lose and gain electrons to make the compound calcium fluoride (CaF_2). **AO1 (3 marks)**

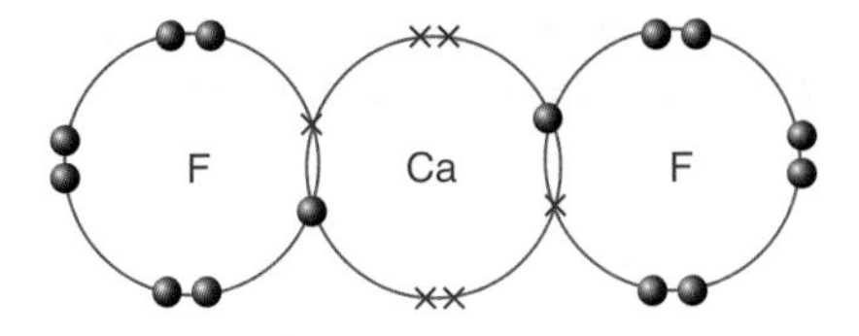

This answer gets 0 marks (grade G). The answer shows how calcium fluoride makes two covalent bonds, and not how it loses and gains electrons. Calcium has been given a wrong electronic structure, or extra electrons have been drawn in so that the outer shell is full.

Page 78

Explain in detail why sodium chloride:

a has a high melting point
b conducts electricity when in solution.
AO1 (3 marks)

a Sodium chloride has a crystal lattice so the melting point is high.

b It conducts electricity along the lattice.

This answer gets 0 marks (grade G). The answer recognises the crystal lattice, but should explain that this forms a giant structure where six ions are attracted equally to six others, making a very strong structure that needs lots of energy to break (melt). Part b) fails to mention that the ions are now free to move and can carry the electric charge/current through the solution.

Page 79

Draw diagrams to show the covalent bonds present in these compounds:

HCl, H_2, CO_2, NH_3, and CH_4. **AO1 (5 marks)**

H ×• Cl :   H •× H

O ×• C ×• O H_3 ×•× N

C ×• H ×• H

This answer gets 2 marks (grade D). Clear dot and cross diagrams are used, but only HCl and H_2 are correct. The candidate needs to learn how to draw diagrams for compounds containing more than two atoms.

Page 80

Draw diagrams to show the difference between the structure of a thermosetting and a thermosoftening polymer. **AO1 (2 marks)**

thermosetting polymer

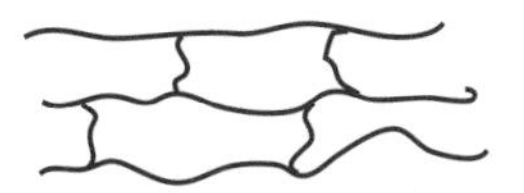

thermosoftening polymer

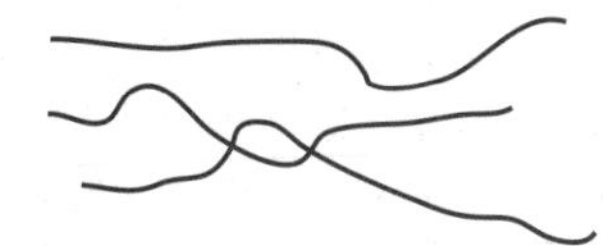

This answer gets 2 marks (grade C). Both chain structures have been clearly drawn and labelled.

Page 81

Explain how the metal lattice allows for metals to be stretched without breaking. **AO1 (2 marks)**

The atoms can be easily pulled slightly apart from each other.

This answer gets 0 marks (grade F). The atoms slide past each other distorting the original shape, but not separating out.

Page 82

Look at Figure 2. It shows the retention–time graph for a gas chromatogram sample.

a How many different components were in the sample?

b List the components in order of the quantity of each present. Start with the most plentiful. **AO3 (2 marks)**

a there are four substances

b A B C D E is the solvent.

This answer gets 0 mark (grade E). Because a) there are five peaks so there are five substances and b) the answer suggests that the candidate thinks E is the solvent. The graph has been wrongly read: the height of the peaks is the important feature. So A C D B E is the correct answer.

Page 83

Calculate the formula of lithium oxide made from 5.6 g of lithium and 6.4 g of oxygen. **AO2 (2 marks)**

Li 5.6/7 = 0.8, O = 6.4/16 = 0.4 ratio is 8:4 or 2:1, so Li_2O

This answer gets 2 marks (grade C). It clearly shows how to reach the answer.

Page 84

Explain the difference between the theoretical yield of a reaction and the actual yield. **AO1 (2 marks)**

The theoretical yield is the yield that should be obtained, and the actual yield is what you get.

This answer gets 2 marks (grade D). A good answer, explaining the difference.

Page 85

Figure 2 shows the volume of gas produced when some magnesium reacted with hydrochloric acid, over 4 minutes.

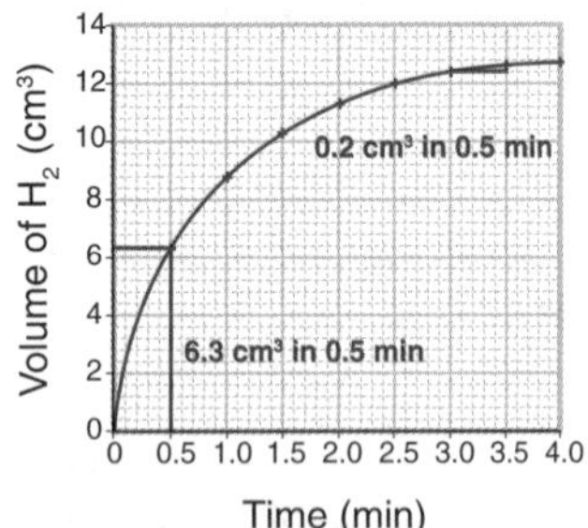

a What was the final volume of gas produced?

b Calculate the rate of the reaction during the first minute.

AO3 (3 marks)

a 12.4 cm^3

b 8.8 cm^3/1 min = 8.8 cm^3/min

This answer gets 2 marks (grade C). Part a) is wrong as the smallest square is worth 0.4 cm^3, not 0.2 cm^3; when reading graphs be sure to work out the scale. Part b) is correct, including the units.

Page 86

A student was investigating a chemical reaction. She decided to increase the temperature of the reaction, whilst decreasing the concentration of one reactant. Use collision theory to explain why the rate of reaction remained unchanged. **AO3 (2 marks)**

The increased temperature has no effect as she only changed one of the reactant's concentration not both. To change the rate, she needs to change both the reactants' concentration.

This answer gets 0 marks (grade E). Increasing the temperature does increase the rate, but reducing the concentration of one reactant will counteract the change. The increase in reacting collisions caused by the temperature rise is equalled by the reduction in reacting collisions caused by less of the reactant being present.

Page 87

Use Figure 5 to find the rate of the reaction when the concentration is 2 mol/dm^3 after 1 minute. **AO2 (2 marks)**

this is 6

This answer gets 0 marks (grade D). The answer has been simply read off the graph, and has no units. A tangent should have been drawn to enable the gradient of the curve to be found, so that the rate at 1 minute could be found. The x-axis value would be 1.8, y-axis would be 4, so 4/1.8 = 2.2 cm^3 per minute.

Page 88

Nitrogen and hydrogen react together slowly to make ammonia. Explain why a high temperature is used, even though this reduces the percentage of ammonia made. **AO1 (2 marks)**

High temperature speeds up the reaction.

This answer gets 1 mark (grade E). The answer explains why a high temperature is used, but does not answer the second part of the question. Unreacted nitrogen and hydrogen are recycled so that all of the reactants eventually become products, so percentage yield is less important than speed of reaction.

Page 89

Explain why this ionic equation can be used to represent all neutralisation reactions between acids and alkalis. **AO1 (2 marks)**

$$H^+(aq) + OH^-(aq) \rightarrow H_2O(\ell)$$

All acids produces H+ ions, and all alkalis produces OH- ions, so this equation shows what happens every time.

This answer gets 1 mark (grade D). It correctly states that acids and alkalis produce the same ions but needs to say that the other part of the acid, and part of the alkali, take no part in the reaction.

Page 90

Describe the method to make some solid chromium chloride from chromium metal. **AO3 (4 marks)**

Use hydrochloric acid, add some powdered chromium metal to it, when it stops reacting filter it, then evaporate the solution.

This answer gets 2 marks (grade C). The basic steps are

there, but you need to check the solution made is neutral with, for example, pH paper. Warming the mixture would help to ensure complete reaction. More detail is needed about the evaporation technique, including the need to partially evaporate and then to filter it to obtain some crystals to dry.

Page 91

Explain why the electrolysis of molten sodium chloride produces sodium metal and chlorine gas, but when a solution of sodium chloride is electrolysed hydrogen gas, chlorine gas and sodium hydroxide are produced. **AO2 (3 marks)**

There are only sodium and chloride ions in the molten sodium chloride so these are the two elements formed. In the solution there are four ions: sodium, chloride, hydrogen and hydroxide. The sodium reacts with the hydroxide ions to make sodium hydroxide, and hydrogen. The chlorine is made in the same way.

This answer gets 2 marks (grade C). It correctly describes what happens and why, but needs to refer to the fact that sodium is higher in the reactivity series than hydrogen, which is why the hydrogen is produced.

P2 Improve your grade

Page 93

Calculate the total distance travelled by the stone in Figure 3. **AO2 (3 marks)**

The stone travelled 80 m.

The candidate only calculated the distance travelled in the first 4 seconds. The answer is 320 m as it travelled an extra 240 m in the next 6 seconds.

Page 94

Sam travels 5 m in 10 s, before stopping for 2 s. He then takes 6 s to return to the start. Draw a distance–time graph that shows Sam's journey. **AO2 (3 marks)**

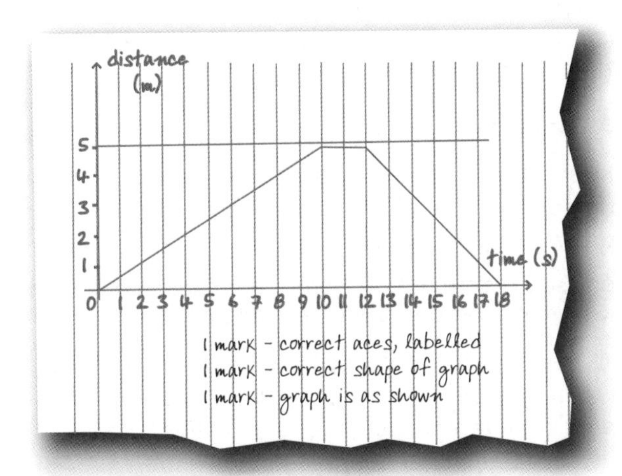

This answer gains 3 out of 3 marks (grade C).

The candidate correctly used the information from the question to plot points and then joined the points with a straight line. Distance–time graphs like these are one of the few times in physics where a graph "joins the dots".

Page 95

Describe how drag forces compare with forces from the engine for a car that accelerates, then reaches a steady speed then decelerates. **AO2 (4 marks)**

Drag forces change with speed. When the car accelerates, engine forces are larger than drag forces. At a steady speed there are no drag forces.

This answer gains 2 out of 4 marks (grade C/D).

The candidate forgot to explain what happens when the car decelerates, and was wrong to say there are no drag forces at a steady speed.

To gain full marks, they should say that at a steady speed, forces from the engine equal drag forces; and as the car decelerates, forces from the engine are smaller than drag forces.

Page 96

Sketch the velocity–time graph for a marble that is released into a measuring cylinder of water. The marble reaches its terminal velocity. **AO2 (3 marks)**

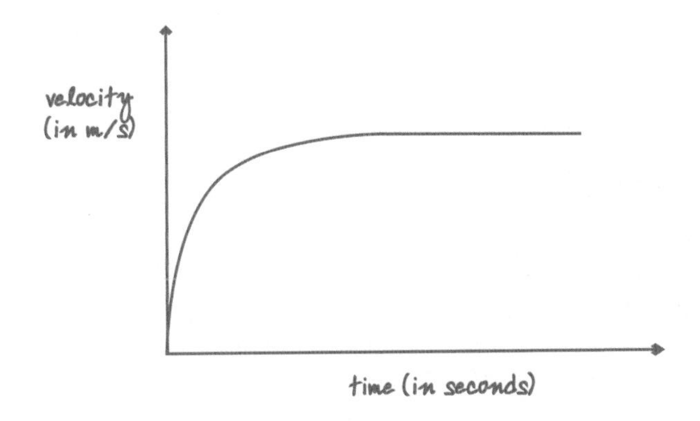

This answer gains 3 out of 3 marks (grade C).

The graph showed the rapid acceleration when the drag forces are small, which gets less as the drag forces increase. Eventually there is no acceleration when drag forces equals weight. Remember the marble's speed never decreases.

Page 97

a Calculate the work done lifting a piano 0.8 m into a lorry. The piano weighs 1850 N. **AO2 (3 marks)**

a work = force x distance = 1850 x 0.8 = 1480

This answer gains 2 out of 3 marks (grade C).

To gain full marks, always include units.

b Calculate the force needed to drag the piano into the lorry if a ramp 2 m long is used. **AO2 (2 marks)**

b force = work/distance = 1480/2 = 740 N

This answer gains 2 out of 2 marks (grade C). The candidate included the unit and showed their working in this question, which allowed them to gain full marks.

Page 98

Calculate the power of a motor that can lift 600 N in 1 minute. **AO2 (3 marks)**

600W

This answer gets 0 marks (grade F). Remember 1 minute is 60 seconds. Always show working. Power = 600/60 = 10 W

Page 99

Two trolleys each with a mass of 1 kg roll towards each other. One trolley travels at 2 m/s to the left and one travels at 3 m/s to the right. They collide and stick together. What is the velocity and direction of the trolleys after the collision? **AO2 (4 marks)**

momentum before = 1 x 2 – 1 x 3 = –1

momentum before = momentum after collision = –1 kg.m/s

speed = 1/1 = 1 m/s

This answer gains 2 out of 4 marks (grade D).

The candidate did not include a direction for the vehicle after the collision.

To gain full marks, a sketch is helpful so you don't miss details – this candidate did not realise that the masses stuck together after the collision.

Page 100

Max rubbed a plastic ruler and a metal ruler with a piece of cloth.

a Which ruler became charged? **AO1 (1 mark)**

b Max held the charged ruler near a charged balloon. The balloon moved away from the ruler.
Explain why. **AO2 (2 marks)**

a the plastic ruler

b the balloon and ruler are charged

This answer gets 2 marks (grade D). The plastic ruler gained a charge. The ruler repelled the balloon because they both have the same charge.

Page 101

When the potential difference across a bulb is 6 V, the current through it is 0.1 A. What is the resistance of the bulb? **AO2 (3 marks)**

Resistance = voltage/current

This answer gets 1 mark (grade D/E). Remember that voltage and potential difference are the same.

Resistance = 6 V/0.1 A = 60 ohms.

Page 102

A series circuit is set up, using four 1.5 V cells in series.

a What is the total voltage supplied to the circuit? (1)

b What is the total resistance in the circuit? (1)

c Calculate the current in the circuit. (1)

d What is the potential difference across each component? (2) **AO2 (5 marks)**

a 6 V (1)

b (15 ohms) (1)

c 6/15 = 0.4 A (1)

d 6 V

This answer gains 3 out of 5 marks (grade D).

The candidate has worked through the first points correctly, but forgot that in a series circuit the voltage is shared across components.

To gain full marks, include calculations like these:

bulb: 5 x 0.4 = 2 V motor: 10 x 0.4 = 4 V

Page 103

Explain why a double insulated appliance does not need an earth wire. **AO1 (2 marks)**

It has a fuse instead

This gets 0 marks (grade F). The earth wire stops a metal outer casing from becoming live. Double insulated appliances have a plastic outer casing.

Page 104

Fuses rated at 3 A and 13 A are available. Which fuse should be used in each case? **AO2 (3 marks)**

a a lawn mower (power 2000 W)

b a lamp (power 60 W)

c a toaster (power 800 W)

a 13 A

b 3 A

c 13 A

This answer gains 3 out of 3 marks (grade C).

The candidate correctly realised that the best fuse has a rating close to, but larger than, the current in the equipment. In some questions, you may be asked to show your calculations.

Page 105

Why are the risks from handling an alpha source safely different from the risks from handling a gamma source? **AO3 (3 marks)**

Alpha sources are not dangerous but gamma radiation is. Gamma radiation cannot be used safely.

This answer gets 0 marks (grade F). Alpha sources are much more dangerous inside the body because they cannot penetrate skin but are very ionising to cells. Gamma radiation is less ionising but penetrates skin and other materials easily. Both should be handled with care.

Page 106

Describe two differences between the plum pudding model and the nuclear model of the atom. **AO1 (4 marks)**

The nuclear model has a nucleus. The plum pudding model does not, it has bits of charge stuck through it.

This answer gains 1 out of 3 marks (grade D).

The candidate would get get full marks by including lots of detail such as: the nucleus is positively charged and contains most of the atom's mass. In the plum pudding model, the positive charge and the mass are spread through the atoms.

Page 107

Explain whether a radioisotope with a long or a short half-life is best in each situation:

a in a smoke detector

b in a medical tracer

c in equipment used to sterilise surgical equipment.

AO2 (3 marks)

a long

b long

c short

The answer gains 1 out of 3 marks (grade D). The candidate must remember to explain their answer. The smoke detector and sterilising equipment need sources that don't need replacing often, but the medical tracer is inside the body and will cause harm eventually.

Page 108

Explain why the presence of iron in the Sun is evidence that the Sun formed from the remains of older stars. **AO2 (3 marks)**

Iron is only formed in the later stages of a star's life cycle. The iron spreads through the Universe during a supernova.

This answer gains 2 out of 3 marks (grade C) as the candidate did not explain why iron does not form in the Sun. Iron only forms in the final stages of a massive star's life cycle.

Page 109

Describe the stages in the life cycle of a star like the Sun. **AO1 (4 marks)**

It is a main sequence star then a red supergiant and then a supernova.

This answer gains 1 out of 4 marks (grade D). The candidate has muddled up the life cycle of a star the size of the Sun, and a star much more massive than the Sun. The correct stages are: main sequence star, red giant, white dwarf, black dwarf.

Understanding the scientific process

As part of your assessment, you will need to show that you have an understanding of the scientific process – How Science Works.

This involves examining how scientific data is collected and analysed. You will need to evaluate the data by providing evidence to test ideas and develop theories. Some explanations are developed using scientific theories, models and ideas. You should be aware that there are some questions that science cannot answer and some that science cannot address.

Collecting and evaluating data

You should be able to devise a plan that will answer a scientific question or solve a scientific problem. In doing so, you will need to collect data from both primary and secondary sources. Primary data will come from your own findings – often from an experimental procedure or investigation. While working with primary data, you will need to show that you can work safely and accurately, not only on your own but also with others.

Secondary data is found by research, often using ICT – but do not forget books, journals, magazines and newspapers are also sources. The data you collect will need to be evaluated for its validity and reliability as evidence.

Presenting information

You should be able to present your information in an appropriate, scientific manner. This may involve the use of mathematical language as well as using the correct scientific terminology and conventions. You should be able to develop an argument and come to a conclusion based on recall and analysis of scientific information. It is important to use both quantitative and qualitative arguments.

Changing ideas and explanations

Many of today's scientific and technological developments have both benefits and risks. The decisions that scientists make will almost certainly raise ethical, environmental, social or economic questions. Scientific ideas and explanations change as time passes and the standards and values of society change. It is the job of scientists to validate these changing ideas.

How science ideas change

From the information you have learnt, you will know that science is a process of developing, then testing theories and models. Scientists have been carrying out this work for many centuries and it is the results of their ideas and trials that has provided us with the knowledge we have today.

However, in the process of developing this knowledge, many ideas were put forward that seem quite absurd to us today.

During the Middle Ages, *The Miasma Theory* explained how diseases were caused. Miasma was thought to be a poisonous vapour present in the air. This vapour was said to contain particles of decaying matter that created a foul smell. The name of the killer disease malaria is derived from the Italian mala, meaning 'bad' and aria, meaning 'air'.

In the nineteenth century, England was undergoing a rapid expansion of industrialisation and urbanisation. This created many foul-smelling and filthy neighbourhoods, which were focal points for disease. By improving housing, cleanliness and sanitation, levels of disease fell. This fall in the level of disease supported the miasma theory.

In 1854, John Snow confirmed a cholera outbreak in London as originating from a water pump. Within ten years, Louis Pasteur was suggesting the presence of germs in substances such as milk and meat that caused them to go off quickly. He was able to remove the germs by a process we now call pasteurisation. This process is still used to protect perishable foodstuffs today.

Reliability of information

It is important to be able to spot when data or information is presented accurately, and just because you see something online or in a newspaper does not mean that it is accurate or true.

Think about what is wrong in this example, based on a newspaper report. Look at the answer at the bottom of the page to check that your observations are correct.

MP challenges Health Minister

The Health Minister was challenged in the House by local MP, Ralph Stag, following this week's publication of figures for the number of adult males affected by oscillatory plumbosis. Mr Stag cited this year's 'sudden and dramatic increase' in the number of sufferers and demanded swift action to discover the causes of this 'worrying trend'.

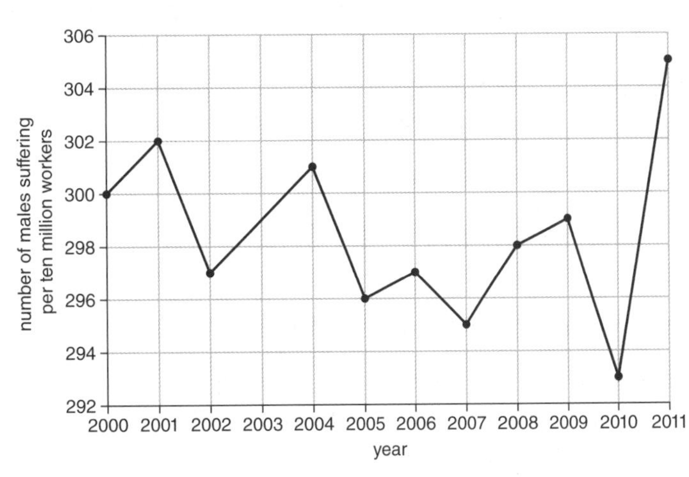

Answer

The y axis does not start at zero and so a change from 293 to 305 individuals in one year is a difference of only 12 people per ten million workers. This is unlikely to be significant and is not really a 'sudden and dramatic increase'.

How science ideas change

From the information you have learnt, you will know that science is a process of developing, then testing theories and models. Scientists have been carrying out this work for many centuries and it is the results of their ideas and trials that has provided us with the knowledge we have today.

However, in the process of developing this knowledge, many ideas were put forward that seem quite absurd to us today, such as this example from the 17th Century.

> In 1667, Johann Joachim Becher published Physical Education, in which he set out the basis of what was to become the Phlogiston Theory. The theory itself was formally stated in 1703 by Georg Ernst Stahl, a German professor of Medicine and Chemistry. He suggested that in all flammable substances there is something called 'phlogiston', a substance without colour, odour, taste, or mass that is given off during burning. 'Phlogisticated' substances are those that contain phlogiston and, on being burned, are 'dephlogisticated'. The ash of the burned material is thought to be the true material. The theory was widely supported for much of the eighteenth century. Joseph Priestley, who is credited with the discovery of oxygen, defended the theory but it was eventually disproved by Antoine Lavoisier.

Reliability of information

It is important to be able to spot when data or information is presented accurately, and just because you see something online or in a newspaper does not mean that it is accurate or true.

Think about what is wrong in this example, based on a document from a government official to a parliamentary committee. Look at the answer at the bottom of the page to check that your observations are correct.

> **RECENT DERAILMENTS IN THE STATE**
>
> *Note that this timeline is for background information only, and is not meant to be a comprehensive analysis of rail safety incidents.*
>
> Aug. 5, 2005: 9 cars of a 144-car train derail spilling highly-acidic caustic soda into the Yukon river; a preliminary government investigation states that federal safety regulations were broken by the railroad company…

Answer

Caustic soda is highly alkaline, not highly acidic.

How science ideas change

From the information you have learnt, you will know that science is a process of developing, then testing theories and models. Scientists have been carrying out this work for many centuries and it is the results of their ideas and trials that has provided us with the knowledge we have today.

However, in the process of developing this knowledge, many ideas were put forward that seem quite absurd to us today.

In 1692, the British astronomer Edmund Halley (after whom Halley's Comet was named) suggested that the Earth consisted of four concentric spheres. He was trying to explain the magnetic field that surrounds the Earth and suggested that there was a shell of about 500 miles thick, two inner concentric shells and an inner core. Halley believed that these shells were separated by atmospheres, and each shell had magnetic poles with the spheres rotating at different speeds. The theory was an attempt to explain why unusual compass readings occurred. He also believed that each of these inner spheres, which was constantly lit by a luminous atmosphere, supported life.

Reliability of information

It is important to be able to spot when data or information is presented accurately, and just because you see something online or in a newspaper does not mean that it is accurate or true.

Think about what is wrong in this example from an online shopping catalogue. Look at the answer at the bottom of the page to check that your observations are correct.

FROM BOX TO AIR IN UNDER TWO MINUTES!

Simply unroll the airship and, as the black surface attracts heat, watch it magically inflate.

Seal one end with the cord provided and fly your 8-metre, sausage-shaped kite.

- ✔ Good for all year round use.
- ✔ Folds away into box provided.
- ✔ A unique product – not for the faint hearted.
- ✔ Educational as well as fun!

Once the airship is filled with air, it is warmed by the heat of the sun.

The warm air inside the airship makes it float, like a full-sized hot-air balloon.

Answer

Black absorbs heat, it does not attract it.

Glossary / Index

The glossary contains terms useful for your revision. Page numbers are given for items that are covered in this book.

A

absorb 42, 45, 57 an object absorbs energy when the energy from infrared radiation is transferred to the particles of the object, increasing the temperature of the object

acceleration 93–94 rate at which an object speeds up, calculated from change in velocity divided by time

acid 89–90 hydrogen compound that can produce hydrogen ions, H^+

acidic solution 89 aqueous solution with pH less than 7 – the concentration of $H^+(aq)$ is higher than that of $OH^-(aq)$

activation energy 92 minimum energy needed to break bonds in reactant molecules to allow a reaction to occur

active site 64 a depression in an enzyme molecule into which its substrate fits

actual yield mass of product actually obtained from a reaction

adaptation 15, 23 the way in which an organism evolved to become better able to survive in its environment

addiction when a person becomes dependent on a drug

addition reaction reaction in which a C=C bond opens up and other atoms add on to each carbon atom

adrenaline 10 hormone that helps to prepare your body for action

aerobic respiration 66, 74 a process in which energy is released from glucose, using oxygen

aerosol tiny particles of liquid or solid dispersed in the air

agar 9 substance used to make jelly on which bacteria can be grown

aggregate stones, gravel or rock chippings used in the construction industry

air resistance 96 force resisting the movement of an object travelling through air

alcohols 34 a family of organic compounds containing an OH group, for example ethanol (C_2H_5OH)

algae 58 simple, plant-like organisms

alkali 89–90 compound that contains hydroxide ions, OH^-

alkali metal 77 very reactive metal in Group 1 of the periodic table, for example sodium

alkaline solution 89 aqueous solution with pH more than 7 – the concentration of OH^- (aq) is higher than that of H^+ (aq)

alkanes 32–33, 41 a family of hydrocarbons: C_nH_{2n+2} with single covalent bonds – found in crude oil

alkenes 33–34, 41 C_nH_{2n} with double covalent bonds (C=C), for example ethene (C_2H_4)

allele 69–72 a particular form of a gene

alloy 28, 81 a mixture of two or more metals, with useful properties different from the individual metals

alpha particle 105–107, 110 particle emitted from the nuclei of radioactive atoms and consisting of two protons and two neutrons

alternating current (a.c.) 103, 110 electric current where the direction of the flow of current constantly reverses, as in mains electricity

amino acids 40, 64, 67, 74 small molecules from which proteins are built

ampere (amp) 100 unit used to measure electrical current

amplitude 52, 57, 103 size of wave oscillations – for a mechanical wave, how far the particles vibrate around their central position

amylase 65 an enzyme that breaks down starch molecules to maltose (sugar) molecules

anaerobic respiration 67, 74 process in which energy is released form glucose, without using oxygen

angle of incidence angle between the ray hitting a mirror or lens and the normal

angle of reflection angle between a ray reflecting from a mirror and the normal

anode 78, 90 positive electrode

antibiotic resistance 8 ability of bacteria to survive in the presence of an antibiotic

antibiotic 8, 23 therapeutic drug, acting to kill bacteria, which is taken into the body

antibody 7, 9, 23, 64, 74 protein normally present in the body or produced in response to an antigen which it neutralises, thus producing an immune response

antiseptic substance that kills pathogens

antitoxin 7, 23 substance produced by white blood cells that neutralises the effects of toxins

antiviral 8 therapeutic drug acting to kill viruses

appliance device that transfers the energy supplied by electricity into something useful

aqueous solution 37 substance(s) dissolved in water

argon the most common noble gas – makes up nearly 1% of the air

atmosphere 38, 39–41 thin layer of gas surrounding a planet

atom 24–25, 28, 41, 75, 99, 105–106, 110 the basic 'building block' of an element that cannot be chemically broken down

atomic number 24, 41, 75, 105 number of protons in the nucleus of an atom

attraction 99, 110 force that pulls, or holds, objects together

auxin 13, 23 a plant hormone that affects rate of growth

average speed 93 total distance of a journey divided by total time taken

axon 10 long thread of cytoplasm in a neurone, carrying an impulse away from the cell body

B

background radiation 106, 110 low level nuclear radiation that exists everywhere, from rocks and other environmetal sources

bacteria 7–9, 21, 23, 58 single-celled microorganisms that can either be free-living organisms or parasites (they sometimes invade the body and cause disease)

bacterium (pl: bacteria) single-celled microorganisms; they do not have a nucleus

balanced diet 6 eating foods (and drinking drinks) that will provide the body with the correct nutrients in the correct proportions

balanced equation 26, 84 chemical equation where the number of atoms of each element is the same in the reactants as in the products

balanced forces 94–95 system of forces with resultant force

base 89, 92 substance that neutralises an acid to form a salt and water

bauxite 30 main ore of aluminium – impure aluminium oxide (Al_2O_3)

beta particle 105, 107, 110 type of nuclear radiation emitted as an electron by a radioactive nucleus

Big Bang theory 56, 57 theory that states that the universe originated from a point at very high temperature, and everything in the universe formed as energy and matter exploded outwards from that point and cooled down

bile 60 a liquid, produced by the liver, which flows into the small intestine where it neutralises the acidic juices from the stomach

biodegradable 35 a biodegradable material can be broken down by microorganisms

biodiesel 31, 41 fuel made from plant oils such as rapeseed

biodiversity range of different living organisms in a habitat

biofuel 31, 34, 36, 41 fuel such as wood, ethanol or biodiesel – obtained from living plants or animals

biological detergent 65 washing powder or liquid that contains enzymes

biomass 17, 23, 49 the mass of living material, including waste wood and other natural materials

biosensor sensitive material that detects very low levels of chemical or biological agents in the surroundings

biotic factor something, such as competition, that influences a living organism and is caused by other living organisms

blast furnace 28 furnace for extracting iron from iron ore

blood plasma the liquid part of blood

Bluetooth 54 low energy, short-range radio waves used to connect communications equipment such as mobile phones or laptops to other equipment nearby

boiling 32, 41, 43 change of state from liquid to gas that happens at the boiling point of a substance

braking distance 95, 110 distance a vehicle travels before stopping, once the brakes are applied

brass 28 alloy of copper and zinc

breathing movements that move air into and out of the lungs

brittle easily cracked or broken by hitting or bending

bronze alloy of copper and tin

brownfield site former industrial site that may be redeveloped for other uses

buckminsterfullerene 81 carbon molecule C_{60} – 60 carbon atoms arranged in the form of a hollow sphere

buckyballs common name for fullerenes, such as C_{60}, buckminsterfullerene

burette item of glassware used to measure the volume of liquid added during a titration

C

calcium carbonate 26, 41, 90 compound with chemical formula $CaCO_3$ – main component of limestone

carbohydrase 66 an enzyme that breaks down carbohydrates; examples include amylase and isomerase

carbon capture technology that filters the carbon and carbon dioxide gas out of waste smoke and gases from power stations and industrial chimneys

carbon cycle 19, 23, 40 the way in which carbon atoms pass between living organisms and their environment

carbon dating method of determining the age of archaeological specimens by measuring the proportion of the carbon-14 isotope present

carbon dioxide 19, 26–27, 30–31, 36, 40 one of the gases emitted from burning fossil fuels that contributes to global warming

carbon sinks carbon-containing substances, such as limestone and fossil fuels, which formed millions of years ago, removing carbon dioxide from the atmosphere

carbon-neutral fuel fuel grown from plants so that carbon dioxide is taken in as the plants are growing – this balances out the carbon dioxide released as the fuel is burned

carnivore 17 an animal that eats other animals

carrier organism that carries a recessive allele for a genetic disorder but does not itself have the disorder

cast iron 29 iron containing 3–4% carbon – used to make objects by casting

casting making an object by pouring molten metal into a mould and allowing it to cool and solidify

catalyst 64, 74, 83, 88–89, 92 substance added to a chemical reaction to improve the reaction rate, without being used up in the process

catalytic converter ('cat') 31, 88 the section of a vehicle's exhaust system that converts pollutant gases into harmless ones

catalytic cracking 33 cracking hydrocarbons by heating in the presence of a catalyst

cathode 91 negative electrode

cell body 10 the part of a nerve cell that contains the nucleus

cell membrane 58, 64 the outer covering of every cell, which controls the passage of substances into the cell

cell wall 58 a strong covering made of cellulose, found on the outside of plant cells

cellulose 62, 74 a carbohydrate; a polysaccharide used to make plant cell walls

cement 27 substance made by heating limestone with clay – when mixed with water it sets hard like stone

central nervous system (CNS) 10, 23 collectively the brain and spinal cord

chain reaction 34, 108 a series of nuclear fission reactions where neutrons released from one reaction cause another nuclear fission reaction and so on

chalcopyrite 29 common ore of copper – formula $CuFeS_2$

charge 24, 100–101, 104 particles or objects can be positively or negatively electrically charged, or neutral: similar charges repel each other, opposite charges attract

chemical analysis process of performing tests to determine what chemical substances are present in a sample, and/or the amount of each substance present

chemical bond 25, 92 attractive force between atoms that holds them together (may be covalent or ionic)

chemical equation 26 line of chemical formulae showing what reacts and what is produced during a chemical reaction

chemical formula 24 shows the elements present in a compound and the number of atoms of each, such as H_2SO_4

chemical reaction 76, 83 process in which one or more substances are changed into other substances – chemical reactions involve rearranging atoms and energy changes

chlorophyll 61, 74 green pigment inside chloroplasts in some plant cells, which absorbs energy from sunlight

cholesterol 6, 23, 37 chemical needed for the formation of cell membranes, but that increases the risk of heart disease if there is too much in the blood

chromatogram 82 pattern of spots produced by paper or thin-layer chromatography

chromatography 82 analytical technique for separating and identifying the components of a mixture, using paper or a thin layer or column of adsorbent

chromosome 19, 68–71 thread-like structure in the cell nucleus that carries genetic information

climate change changes in seasonal weather patterns that occur because the average temperature of Earth's surface is increasing owing to global warming

clone 20, 23, 74 group of genetically identical organisms

coal 48 solid fossil fuel formed from plant material – composed mainly of carbon

collision frequency 86 number of collisions per second between the particles involved in a chemical reaction

collision theory 86, 92, 99 relates reaction rates to the frequency and energy of collisions between the reacting particles

column chromatography 82 chromatography method using a solvent to carry substances down a column of adsorbent

combustion 19, 31 process where substances react with oxygen, releasing heat

communications satellite 54 artificial satellite that stays above the same point on Earth's surface as it orbits, and used to send communications signals around the world

community all the organisms, of all species, that live together in the same habitat at the same time

compact fluorescent lamp (CFL) light bulb that is efficient at transferring the energy in an electric current as light

competition 15, 23 result of more than one organism needing the same resource, which is in short supply

compost 18 partly rotted organic material, used to improve soil for growing plants

compound 24, 27 substance composed of two or more elements joined together by chemical bonds, for example, H_2O

compression region of a longitudinal wave where the vibrating particles are squashed together more than usual

concentration 87 amount of chemical present in a given volume of a solution – usually measured as g/dm^3 or mol/dm^3

concrete 26–27 mixture of cement, sand, aggregate and water

condensation 32, 43–44 change of state when a substance changes from a gas or vapour to a liquid: the substance condenses

conduction (electrical) 91 flow of electrons through a solid, or ions through a liquid

conduction (thermal) 43–44, 57 heat passing through a material by transmitting vibrations from one particle to another

conductor 28 material that transfers energy easily

consumer 17 an organism that feeds on other organisms

contact process industrial process for making sulfuric acid

continental drift 38 movement of continents relative to each other

continental plate 38 tectonic plate carrying large landmass, though not necessarily a whole continent

convection current 38, 41 when particles in a liquid or gas gain energy from a warmer region and move into a cooler region, being replaced by cooler liquid or gas

convection 43–44, 57 heat transfer in a liquid or gas – when particles in a warmer region gain energy and move into cooler regions carrying this energy with them

conventional current flow direction of flow of electric current around a circuit, from positive to negative – the opposite direction to the flow of electrons

coordination communicating between different parts of the body so that they can act together

core (of Earth) 38, 41 layer in centre of Earth, consisting of a solid inner core and molten outer core

cosmic microwave background radiation (CMBR) 56 microwave radiation coming very faintly from all directions in space

covalent bond 25, 79–80, 92 bond between atoms in which some of their outer electrons are shared

cracking 33–34, 41 oil refinery process that breaks down large hydrocarbon molecules into smaller ones

critical mass 108 the minimum mass of nuclear fuel needed to make a chain reaction happen

crumple zone 97 part of a vehicle designed to absorb energy in an accident, so reducing injuries to passengers

crust 38, 41 surface layer of Earth made of tectonic plates

crystallise form crystals from a liquid – for example, by partly evaporating a solution and leaving to cool

culture 9 a population of microorganisms, grown on a nutrient medium

current 100–102, 104, 110 flow of electricity around a circuit – carried by electrons through solids and by ions through liquids

cuttings 13, 20 small pieces of a plant that can grow into complete new plants

cystic fibrosis 72 a genetic disorder caused by a recessive allele, where lungs become clogged with mucus

cytoplasm 7, 58, 64 the jelly-like material inside a cell, in which metabolic reactions take place

D

decay (biological) 18, 23 the breakdown of organic material by microorganisms

deceleration see negative acceleration

delocalised electrons electrons not attached to any particular atom, so free to move through the structure, allowing electrical conduction – present in metals and graphite

denatured 64 the shape of an enzyme molecule has changed so much that it can longer bind with its substrate

dendrite 10 a short thread of cytoplasm on a neurone, carrying an impulse towards the cell body

dendron long thread of cytoplasm on a neurone, carrying an impulse towards the cell body

diabetes disease in which the body cannot control its blood sugar level

diatomic molecule molecule consisting of just two atoms

differentiation 69 change of a cell from general-purpose to one specialised to carry out a particular function

diffraction 51 change in the direction of a wave caused by passing through a narrow gap or round an obstacle such as a sharp corner

diffusion 59, 74 spreading of particles of a gas, or of any substance in solution, resulting in a net movement from a region where they are in a high concentration to a region where they are in a lower concentration

digestive juices 60 liquids secreted within the digestive system and containing enzymes that help to digest food molecules

digital signal 55 communications signal sent as an electromagnetic wave that is switched on and off very rapidly

diode 101 semiconductor device that allows an electric current to flow through it in only one direction

direct current (d.c.) 91, 103, 110 electric current where the direction of the flow of current stays constant, as in cells and batteries

distance–time graph 93, 110 graph showing how the distance an object travels varies with time: its gradient shows speed

distillation 32 process for separating liquids by boiling them, then condensing the vapours

distribution the transmission of electricity from a power station to homes and businesses

DNA 19, 69, 71 deoxyribonucleic acid – the chemical from which chromosomes are made: its sequence determines genetic characteristics, such as eye colour

dominant 70 a dominant allele has an effect even when another allele is present

Doppler effect 56 change in wavelength and frequency of a wave that an observer notices if the wave source is moving towards them or away from them

double covalent bond two covalent bonds between the same pair of atoms – each atom shares two of its own electrons plus two from the other atom

drug dependency 13 feeling that you cannot manage without a drug

drug 13, 23 a chemical that changes the chemical processes in the body

ductile 81 can be drawn out into thin wires

E

earth wire 103 wire connecting the case of an electrical appliance, through the earth pin on a three-pin plug, to earth

earthed 103–104 safety feature where part of an appliance is connected to earth to protect users from electrocution if there is a fault

earthquake 39 shaking and vibration at the surface of the Earth resulting from underground movement or from volcanic activity

echo 52 reflection of a sound wave

effector 10, 23 part of the body that responds to a stimulus

efficiency 46, 57 a measure of how effectively an appliance transfers the energy that flows into the appliance into useful effects

egg cell 12, 19, 68, 71 female gamete

elastic collision 99 collision where colliding particles or objects bounce apart after collision

elastic potential energy 110 energy stored in an object because it is stretched, compressed or deformed, and released when the object returns to its original shape

elastic 96 material that returns to its original shape when the force deforming it is removed

electrical power 47–50, 57, 103–104 a measure of the amount of energy supplied each second

electricity generator device for generating electricity

electrode 29, 91 solid electrical conductor through which the current passes into and out of the liquid during electrolysis – and at which the electrolysis reactions take place

electrolysis 29, 83, 91–92 decomposing an ionic compound by passing a d.c. electric current through it while molten or in solution

electrolyte solution or molten substance that conducts electricity

electromagnetic (EM) radiation 53–54, 105 energy transferred as electromagnetic waves

electromagnetic spectrum 53 electromagnetic waves ordered according to wavelength and frequency – ranging from radio waves to gamma rays

electromagnetic waves 51, 53–54, 57 a group of waves that transfer energy – they can travel through a vacuum and travel at the speed of light

electron 24–25, 41, 75, 99, 105, 110 small particle within an atom that orbits the nucleus (it has a negative charge)

electronic configuration 24 the arrangement of electrons in shells, or energy levels, in an atom

electronic structure (or configuration) 24 arrangement of electrons in shells, or energy levels, in an atom

electrostatic induction 100 electric charge induced on an object made of an electrical insulator, by another electrically charged object nearby

element 24 substance made out of only one type of atom

embryo transplant 20, 23, 74 taking an embryo that has been produced from one female's egg and placing it into another female

embryo 12, 20, 23 a very young organism, which began as a zygote and will become a fetus

emit 42 an object emits energy when energy is transferred away from the object as infrared radiation, decreasing the temperature of the object

empirical formula ratio of elements in a compound, as determined by analysis – for example, CH_2O for glucose (molecular formula $C_6H_{12}O_6$)

emulsifier 37, 41 a substance that prevents an emulsion from separating back into oil and water

emulsion 37, 41 a thick, creamy liquid made by thoroughly mixing an oil with water (or an aqueous solution)

endothermic reaction 89, 92 chemical reaction which takes in heat, or energy from other sources

energy input 45–47, 57 the energy transferred to a device or appliance from elsewhere

energy levels 24 electrons in shells around the nucleus – the further from the nucleus, the higher the electron's energy level

energy output 46–47, 51 the energy transferred away from a device or appliance – it can be either useful or wasted

energy 42, 67, 89, 97 the ability to 'do work'

environment 16 an organism's surroundings

enzyme 11, 18, 60, 64–66, 70 biological catalyst that increases the speed of a chemical reaction but is not used up in the process

epidemic 7 many people having the same infectious disease

epidermis 61 a tissue covering the outer surface of a plant's leaf, stem or root

epithelial tissue 60 tissue forming a covering over a part of an animal's body

equal and opposite forces 94–95 balanced forces equal in size but acting in opposite directions

essential oils oils found in flowers, giving them their scent – they vaporise more easily than natural oils from seeds, nuts and fruit

ethanol 31, 34, 36, 41 an alcohol that can be made from sugar and used as a fuel

evaporation 44 change of state where a substance changes from liquid to gas at a temperature below its boiling point

evolution 21–23 a change in a species over time

exothermic reaction 88, 92 chemical reaction which gives out heat

extinct 73 no longer existing

extremophile an organism that can live in conditions where a particular factor, such as temperature or pH, is outside the range that most other organisms can tolerate

F

fermentation 34, 41 process in which yeast converts sugar into ethanol (alcohol)

fertilization 12 fusion of the nuclei of a male and a female gamete

fertility drug 12 hormone given to women to cause the ovaries to produce eggs

fibre optic cable 55 glass fibre that is used to transfer communications signals as light or infrared radiation

filament bulb 101 lightbulb giving out light by current flowing through a fine wire and heating the wire until it glows white hot

finite resource material of which there is only a limited amount – once used it cannot be replaced

flammable catches fire and burns easily

flexible (material) can be bent without the material breaking

food chain 17–18, 23 flow diagram showing how energy is passed from one organism to another

force meter device measuring the size of a force, by measuring how much the force stretches a spring

formula (for a chemical compound) 24, 26, 76, 83 group of chemical symbols and numbers, showing which elements, and how many atoms of each, a compound is made up of

forward reaction 85 reaction from left to right in an equation for a reversible reaction

fossil fuel 19, 30–31, 40, 49, 50 fuel such as coal, oil or natural gas, formed millions of years ago from dead plants and animals

fossil 72, 74 preserved remains of a long-dead organism

fractional distillation 32, 34, 41 process that separates the hydrocarbons in crude oil according to size of molecules

fractionating column 32 tall tower in which fractional distillation is carried out at an oil refinery

fractions 32 the different substances collected during fractional distillation of crude oil

'free' electrons electrons that move readily from one atom to another to transmit an electric current through a conductor

freezing 43 change of state in which a substance changes from a liquid to a solid

frequency 52, 57 the number of waves passing a set point per second

friction 95, 97 force acting at points of contact between objects moving over each other, to resist the movement

FSH 12 hormone, produced by the pituitary gland, that causes eggs to mature in the ovaries

fuel cell device that generates electricity directly from a fuel, such as hydrogen, without burning it

fuel a material that is burned for the purpose of generating heat

fullerenes cage-like carbon molecules containing many carbon atoms, for example, C_{60}, buckminsterfullerene

fungus (pl. fungi) 18 living organisms whose cells have cell walls, but that cannot photosynthesise

fuse 103–104, 110 a fine wire that melts if too much current flows through it, breaking the circuit and so switching off the current

G

gamete 12, 19, 20, 68, 74 sex cell – a cell, such as an egg or sperm, containing the haploid number of chromosomes

gamma rays 54, 105, 107, 110 ionising electromagnetic radiation – radioactive and dangerous to human health

gas chromatography 82, 92 method that uses a gas to carry the substances through a long, thin tube of adsorbent

gasohol mixture of gasoline (petrol) and alcohol (ethanol) used as a vehicle fuel

gene 6, 19, 23, 69–72, 74 section of DNA that codes for a particular characteristic

genetic diagram 71, 74 a format used to describe and explain the probable results of a genetic cross

genetic engineering 21, 23 changing the genes in an organism, for example by inserting genes from another organism

genetically modified 21, 23 organism that has had genes from a different organism inserted into it

genotype the pair of alleles that an organism possesses for a particular gene

geographical isolation 73 the separation of two populations of a species by a geographical barrier, such as a mountain chain

geothermal power station 49 power station generating electricity using the heat in underground rocks to heat water

giant covalent structure 80 solid structure made up of a regular arrangement of covalently bonded atoms – may be an element or a compound

giant ionic structure solid structure made up of a regular arrangement of ions in rows and layers

gland 10 organ that secretes a useful substance

glandular tissue 60 tissue made up of cells that are specialised to secrete a particular substance

global dimming 31 gradual decrease in the average amount of sunlight reaching Earth's surface

Global Positioning System (GPS) navigation system using signals from communications satellites to find an exact position on the surface of Earth

global warming 16, 31, 49 gradual increase in the average temperature of Earth's surface

glucose 61–62, 67, 74 a simple sugar, made by plants in photosynthesis, and broken down in respiration to release energy inside all living cells

glycogen 67 carbohydrate used for energy storage in animal cells

gravitational potential energy 98, 110 energy that an object has because of its position, for example, increasing the height of an object above the ground increases its gravitational potential energy

gravitropism 13, 23 a growth response to gravity

gravity 98 the attractive force acting between all objects with mass – on Earth the attractive force due to gravity pulls objects downwards

green fuel fuel that does less damage to the environment than fossil fuels

greenhouse gas 57 a gas such as carbon dioxide that reduces the amount of heat escaping from Earth into space, thereby contributing to global warming

group 25 within the periodic table the vertical columns are called groups

H

Haber process industrial process for making ammonia

haemoglobin chemical in red blood cells which carries oxygen

haemophilia disease where blood lacks the ability to clot

halide ion ion of a halogen – halide ions have a 1^- charge

halogens 78 reactive non-metals in Group 7 of the periodic table

hard water water supply containing dissolved calcium or magnesium salts – these react with soap, making it hard to form a lather

HDL a type of cholesterol that does not appear to cause heart disease and may help to protect against it

heart disease blockage of blood vessels that bring blood to the heart

herbivore 17 an animal that eats plants

heterozygous possessing two different alleles of a gene

homozygous possessing two identical alleles of a gene

Hooke's Law 96 for an elastic object, the extension is proportional to the force applied, provided the limit of proportionality is not exceeded

hormones 10, 64, 74 chemicals that act on target organs in the body (hormones are made by the body in special glands)

hot spot area of Earth's crust heated by rising currents of magma – Hawaii is above a mid-Pacific hot spot

hydrocarbon 32–33, 41 compound containing only carbon and hydrogen

hydroelectric power station 49–50 power station generating electricity using the energy from water flowing downhill

hydrogen ion H^+ ion – hydrogen ions in solution, $H^+(aq)$, make the solution acidic

hydrophilic water-loving (attracted to water, but not to oil) – used to describe parts of a molecule

hydrophobic water-fearing (attracted to oil, but not to water) – opposite of hydrophilic

hydroxide ion OH^- ion – hydroxide ions in solution, $OH^-(aq)$, make the solution alkaline

hydroxide ion consisting of an oxygen and a hydrogen atom (written as OH^-)

hypothesis an idea that explains a set of facts or observations – a basis for possible experiments

I

image 52 an image is formed by light rays from an object that travel through a lens or are reflected by a mirror

immiscible 37 liquids that do not mix, but form separate layers, are immiscible

immune system a body system that acts as a defence against pathogens, such as viruses and bacteria

immunity 7, 9 you have immunity if your immune system recognises a pathogen and fights it

incident ray 52 the ray of light hitting a mirror or lens

inelastic collision 99 collision where the colliding particles or objects stick together after collision

infrared radiation 42, 44, 53–54, 57 energy transferred as heat – a type of electromagnetic radiation

inhibitor substance used to decrease a reaction rate – also called a 'negative catalyst'

initial reaction rate reaction rate at the start of the reaction

inoculating loop 9 metal loop that is used to transfer microorganisms

insoluble 90 not soluble in water (forms a precipitate)

insoluble salt 90 salt which is not soluble in water, so forms a precipitate

instantaneous reaction rate reaction rate at a particular instant during the reaction

insulator 45 material that transfers energy only very slowly – thermal insulators transfer heat slowly, electrical insulators do not allow an electric current to flow through them

insulin 10, 21 hormone made by the pancreas that reduces the level of glucose in the blood

intermolecular forces 79 forces between molecules

ion 25, 105, 110 atom (or group of atoms) with a positive or negative charge, caused by losing or gaining electrons

ionic bonding 25, 77 chemical bond formed by attractions between ions of opposite charges

ionic compound 77–78, 90–92 compound composed of positive and negative ions held together in a regular lattice by ionic bonding, for example, sodium chloride

ionic equation shows only the ions that actually react – anything that does not change during the reaction is omitted

ionise 105, 107 to cause electrons to split away from their atoms (some forms of EM radiation are harmful to living cells because they cause ionisation)

isomerase 66 enzyme that changes glucose to fructose

isotopes 75, 92, 105, 107 forms of element where their atoms have the same number of protons but different numbers of neutrons

IVF 12 *in vitro* fertilisation – the fertilisation of an egg by a sperm in a glass container

J

joule 42, 47 unit used to measure energy

K

kilowatt-hour 48 the energy transferred in 1 hour by an appliance with a power rating of 1 kW (sometimes called a 'unit' of electricity)

kinetic energy 57, 95, 97–98, 110 energy an object has because of its movement – it is greater for objects with greater mass or higher speed

kinetic theory 57 model used to explain how energy is transferred by particles in a substance

L

lactic acid 67, 69 a waste product of anaerobic respiration in muscle cells

laterally inverted image 52 left and right are reversed, when seen in a mirror

lattice 78, 80, 92 regular arrangement of ions or atoms in a solid – may be covalent or ionic

lava 39 magma that has erupted onto the surface of Earth

Law of Conservation of Energy 46 energy can be transferred but cannot be created or destroyed

Law of Conservation of Momentum total momentum before a collision is equal to total momentum after the collision, if no outside forces are acting

LDL a type of cholesterol that increases the risk of heart disease

LDR (light dependent resistor) 101 resistor with a resistance that decreases when light is shone on it

leaching using a chemical solution to dissolve a substance out of a rock

LED (light emitting diode) diode that gives off light when a current flows through it

LH 12 hormone produced by the pituitary gland, which causes an egg to be released from an ovary

lichen 16 small organism that consists of both a fungus and an alga

limestone 26–27, 40–41 type of rock consisting mainly of calcium carbonate

limewater calcium hydroxide solution

limit of proportionality (for Hooke's Law) the point for an elastic material when the extension stops being proportional to the force: materials break or are permanently damaged when stretched beyond this point

limiting factor anything that is in short supply and therefore stops a process from happening faster

lipase 65 enzyme that breaks down fat molecules to fatty acids and glycerol molecules

lithosphere 38 the rocky, outer section of the Earth, consisting of the crust and upper part of the mantle

longitudinal wave 51, 57 a wave in which the direction that the particles are vibrating is the same as the direction in which the energy is being transferred by the wave

low-grade ore ore containing only a small percentage of metal

lymphocyte 7 type of white blood cell

M

macromolecule 80 very large molecule made up of hundreds of thousands, or millions, of atoms, for example, a polymer or crystal with a giant covalent structure

magma 39 molten rock found below Earth's surface

main sequence star a star in which nuclear fusion reactions combine small atomic nuclei into elements with larger nuclei

mains supply domestic electricity supply – in the UK, mains supply is 230 V at 50 Hz

malleable 81 can be hammered into shape without breaking

malnourished 6 not having a balanced diet

mantle 38, 41 semi-liquid layer of the Earth beneath the crust

mass number 24, 75, 105 total number of protons and neutrons in the nucleus of an atom – always a whole number

mass spectrometer 82 instrument for identifying chemicals by measuring their relative formula mass very accurately

mass a measure of the amount of 'stuff' in an object

mechanical wave 51 wave in which energy is transferred by particles or objects moving, such as a wave on a string or a water wave

meiosis 68, 74 type of cell division producing four genetically different daughter cells, each with half the normal number of chromosomes

melting 43 change of state of a substance from liquid to solid

menstrual cycle 12 monthly hormonal cycle that starts at puberty in human females

menstruation 12 monthly breakdown of the lining of the uterus leading to bleeding from the vagina

metabolic rate 6, 23 rate at which chemical reactions take place in the body

metallic bonding 81 type of bonding in metals – a regular lattice of metal ions is held together by delocalised electrons

methane 26, 32 the simplest hydrocarbon, CH_4 – main component of natural gas

microorganism 7, 18, 58, 65 very small organism (living thing) that can be viewed only through a microscope – also known as a microbe

microwaves 53–54 non-ionising radiation – used in telecommunications and in microwave ovens

mid-ocean ridge underwater mountain range formed by magma escaping from the seabed where continental plates are drifting apart

mitochondrion (pl: mitochondria) 58, 60, 66 organelle in which the reactions of aerobic respiration take place

mitosis 68, 74 type of cell division producing two genetically identical daughter cells

mixtures 24 two or more substances mixed together – they can usually be separated by physical methods such as filtration

MMR 9 vaccine for measles, mumps and rubella

mole 76 unit for counting atoms and molecules – one mole of any substance contains the same number of particles

molecular ion ion formed when an electron is knocked off a molecule in a mass spectrometer

molecular structure arrangement of atoms from which a molecule is made

molecule two or more atoms held together by covalent chemical bonds

molten made liquid by keeping the temperature above the substance's melting point

momentum 99, 110 mass of a moving object multiplied by its velocity – a vector quantity having both size and direction

monomers 34 small molecules that become chemically bonded to each other to form a polymer chain

mortar 26–27 mixture of cement, sand and water

motor neurone 10, 23 nerve cell carrying information from the central nervous system to muscles

MRSA 8 a form of the bacterium *Staphylococcus aureus* that is resistant to many antibiotics

muscular tissue 60 a tissue that is specialised for contraction, causing movement

mutation 8, 22 a change in the DNA in a cell

myelin sheath 10 insulating layer around a nerve fibre

N

nanometer unit used to measure very small length (1 nm = 0.000 000 001 m, or one-billionth of a metre)

nanoparticles 81 very small particles (1–100 nanometres in size)

nanoscience the study of nanoparticles

nanotube 81 carbon molecule in the form of a cylinder

National Grid 50, 57 network that distributes electricity from power stations across the country

native (relating to metals such as gold) occurs in rocks as the element – not combined in compounds

natural gas gaseous fossil fuel formed from animals and plants that lived 100 million years ago – composed mainly of methane

natural oil oil produced by plants or fish

natural selection 8, 21, 22, 73, 74 the increased chance of survival of individual organisms that have phenotypes that adapt them successfully to their environment

negative acceleration 93–94 rate at which an object slows down, or decelerates, calculated from change in velocity divided by time

nerve 10 group of nerve fibres

neurone 10 nerve cell

neutral solution aqueous solution with pH 7 – the concentrations of $H^+(aq)$ and $OH^-(aq)$ ions are equal

neutralization 83, 89, 92 reaction between an acid and a base to make a salt and water (H^+ ions react with OH^- or $O2^-$ ions)

neutron 24, 41, 75, 105, 110 small particle that does not have a charge – found in the nucleus of an atom

newton standard unit of force – one newton is about the same as the weight of a small apple

Newton's Third Law when two objects are in contact with each other, they exert equal and opposite forces on one another

noble gas structure stable arrangement of electrons achieved by gaining, losing or sharing electrons to obtain an outer shell of eight (two in the case of hydrogen)

noble gas 25 unreactive gas in Group 0 of the periodic table

non-ohmic device 101 device that does not obey Ohm's law, so the current through it is not directly proportional to the potential difference across it

non-renewable something that cannot be replaced when it has been used, such as fossil fuels and metal ores

normal 51 line at right angles to a boundary, such as the line drawn for mirrors or glass blocks to help draw ray diagrams

nuclear fission 48 108, 110 nuclear reaction in which large atomic nuclei split into smaller ones, giving off large amounts of energy

nuclear fusion 108–109, 110 nuclear reaction in which small atomic nuclei join together into larger ones, giving off large amounts of energy

nuclear power station 48, 50, 108 power station generating electricity from the energy stored inside atoms – energy is released by the controlled splitting apart of large atoms (nuclear fission)

nuclear radiation 105–106 radiation given out by nuclear reactions – three types: alpha particles, which are helium nuclei, beta particles, which are electrons and gamma rays, which are electromagnetic radiation

nucleus (cells) 58, 70 a structure found in most animal and plant cells, which contains the chromosomes made of DNA, and which controls the activities of the cell

nucleus 24, 41, 76, 105–106, 108, 110 central part of an atom that contains protons and neutrons

nutrient medium 9, 23 liquid or jelly in which microorganisms can be grown

nutrient 6 substance in food that we need to eat to stay healthy, such as protein

O

oceanic plate tectonic plate under the ocean floor – it does not carry a continent

oestrogen 10 female hormone secreted by the ovary and involved in the menstrual cycle

OH group an oxygen atom bonded to a hydrogen atom and found in all alcohols

Ohm's Law 101 for a device with a constant value of resistance, the current through it is always directly proportional to the potential difference across it

oil (crude) 32–33, 41 liquid fossil fuel formed from animals and plants that lived 100 million years ago

oil (from a plant) 36, 41 liquid fat obtained from seeds, nuts or fruit

opencast mining 30 mining by digging out ore at the surface, rather than underground

optical fibre or cable 54 glass fibre that is used to transfer communications signals as light or infrared radiation

oral contraceptive 12 pills that prevent a woman releasing eggs

ore 27–30, 41 rock from which a metal is extracted, for example iron ore

organ 60, 74 structure within an organism's body, made up of different types of tissues, that carries out a particular function

organelle 58, 74 structure within a cell

organic compound a compound containing carbon and hydrogen, and possibly oxygen, nitrogen or other elements – living organisms are made up of organic compounds

oscillate vibration to and fro of particles in a wave

oscilloscope 103 device with screen to show how amplitude and frequency of an input wave varies – also called a cathode ray oscilloscope

ovary 12, 68 organ in a female in which eggs are made

ovulation 12 release of an egg from the ovary

oxidation 83 process that increases the amount of oxygen in a compound – opposite of reduction

oxygen debt 67, 69 the extra oxygen that has to be taken into the body after anaerobic respiration has taken place

P

pandemic 7 when a disease spreads rapidly across many countries – perhaps the whole world

Pangea 38 huge landmass with all the continents joined together before they broke up and drifted apart

parallel circuit 100, 102, 110 electrical circuit with more than one possible path for the current to flow around

pathogen 7, 23 harmful microorganism that causes disease

payback time 45 time taken for a type of domestic insulation to 'pay for itself' – to save as much in energy bills as it cost to install

pepsin protease enzyme found in the stomach

percentage yield 84 theoretical yield actually obtained (percentage yield = actual yield ÷ theoretical yield × 100)

period 25 horizontal row in the periodic table

periodic table 25 a table of all the chemical elements based on their atomic number

petrochemical substance made from petroleum

petroleum liquid fossil fuel formed from animals and plants that lived 100 million years ago

pH scale 23, 64 scale from 0 to 14 which shows how acidic or alkaline a substance is

phagocytes 7, 23 white blood cells that surround pathogens and digest them with enzymes

phenotype appearance or characteristics or an organism, affected by its genes and its environment

phloem 61 tissue made up of long tubes that transports sugars from the leaves to all other parts of a plant

photochromic 81 photochromic materials change colour in response to changes in light level

photosynthesis 13, 15, 17, 19, 61–63, 74 process carried out by green plants where sunlight, carbon dioxide and water are used to produce glucose and oxygen

phototropism 13, 23 a growth response to light

photovoltaic cell 49 device that converts the Sun's energy into electricity

physical factor something that influences a living organism that is caused by non-living aspects of their environment, such as temperature or light intensity

phytomining using growing plants to absorb metal compounds from soil, burning the plants, and recovering metal from the ash

phytoremediation cleaning up contaminated soil by using growing plants to absorb harmful metal compounds

pipette used to measure out an exact volume of liquid

placebo 'dummy' treatment given to some patients, in a drug trial, that does not contain the drug being tested

plane mirror 52 mirror with a flat surface

planet large ball of gas or rock travelling around a star – for example Earth and other planets orbit our Sun

plaque 6 build-up of cholesterol in a blood vessel (which may block it)

plastics 35, 80 compounds produced by polymerisation, capable of being moulded into various shapes or drawn into filaments and used as textile fibres

plate boundaries 39 edges of tectonic plates, where they meet or are moving apart

plum pudding model 106 model that said the atom was like a positively charged 'jelly' with negatively charged electrons dotted through it – later shown to be incorrect

pollution 16, 31, 49, 57 presence of substances that contaminate or damage the environment

poly(ethene) 34–35, 41 plastic polymer made from ethene gas (also called polythene)

polydactyly having more than five fingers or toes on a hand or foot

polymer 34–35, 41 large molecule made up of a chain of monomers

polymerization chemical process that combines monomers to form a polymer: this is how polythene is formed

power 47–48, 50, 98, 110 amount of energy that something transfers each second and is measured in watts (or joules per second)

power rating a measure of how fast an electrical appliance transfers energy supplied as an electrical current

power station 48, 50 place where electricity is generated to feed into the National Grid

precipitate 90 solid product formed by reacting two solutions

precipitation 83, 90 reaction between two solutions to form a solid product (a precipitate)

producer 17 organism that makes its own food from inorganic substances

products 26 chemicals produced at the end of a chemical reaction

progesterone 12 hormone, produced by the ovary, that prepares the uterus for pregnancy

protease 65 enzyme that breaks down protein molecule to amino acid molecules

protein 64 molecule made up of amino acids (found in food of animal origin and also in plants)

proton 24–25, 41, 75, 105, 110 small positive particle found in the nucleus of an atom

protostar 109 dense cloud of dust and gas that can form a new star, if it contains enough matter

pyramid of biomass 17, 23 a diagram in which boxes, drawn to scale, represent the biomass at each step in a food chain

Q

quadrat 63, 74 a square area within which type and numbers of living organisms can be counted or estimated

quarry 26 place where stone is dug out of the ground

R

radio wave 53–4 non-ionising radiation used to transmit radio and TV

radioactive 105, 107, 110 materials giving off nuclear radiation

radioisotope 107 a radioactive isotope of an element

rarefaction areas of a longitudinal wave in which the vibrating particles are spread out more than usual

rate of energy transfer a measure of how quickly something moves energy from one place to another

ray diagram 52 diagrams showing how light rays travel

RCCB (residual current circuit breaker) 104 measures the current flowing into and out of an appliance, and switches the current off if they are not equal

reactants 26, 92 chemicals that are reacting together in a chemical reaction

reaction conditions physical conditions under which a reaction is performed, for example, temperature and pressure

reaction rate (average) total amount of reaction ÷ total time

reaction rate (initial) reaction rate at the start of the reaction

reaction rate 85, 92 the speed at which a chemical reaction takes place – measured as the amount of reaction per unit time

reaction time 95 time between when a driver sees a hazard and when they begin to respond to it – increased by tiredness, drugs, or distractions

reactivity series 27, 77 list of metals in order of their reactivity with oxygen, water and acids

receptor 10, 23 nerve cell that detects a stimulus

recessive 70 a recessive allele only has an effect when a dominant allele is not present

red shift 56, 57 when lines in a spectrum are redder than expected – if an object has a red-shift it is moving away from the observer

reduction 27–28, 83 process that reduces the amount of oxygen in a compound, or removes all the oxygen from it – opposite of oxidation

reflected ray 51 ray of light 'bouncing off' from a mirror or reflecting surface

reflection 57 change of direction of a wave when it 'bounces off' from a surface

reflex action 11 a fast, automatic response to a stimulus

reflex arc 11 pathway taken by nerve impulse from receptor, through nervous system, to effector

refraction 51 change of direction when a wave hits the boundary between two media at an angle, for example when a light ray passes from air into a glass block

regenerative braking 97 type of braking which transfers some of the kinetic energy wasted by the slowing car into electricity which is used to operate the car brakes

relative atomic mass 97 average mass of all the atoms in an element, taking into account the presence of different isotopes – often rounded to the nearest whole number

relative formula mass 75, 83 total mass of all atoms in a formula = each relative atomic mass × number of atoms present

relative molecular mass same as relative formula mass, but limited to elements or compounds that have separate molecules

renewable resource 49 energy resource that is constantly available or can be replaced as it is used

repeatability consistent results are obtained when a person uses the same procedure a number of times

reproducibility how likely it is that measurements, made again under similar conditions, give the same results

resistance (electrical) 101–102, 110 measure of how hard or how easy it is for an electric current to flow through a component

resistant strain (of bacteria) 23 a population of bacteria that is not killed by an antibiotic

respiration 66–67, 74 process occurring in all living cells, in which energy is released from glucose

resultant force 94, 110 the single force that would have the same effect on an object as all the forces that are acting on the object

retention factor (Rf) used to help identify individual spots in a chromatogram (Rf = distance moved by the spot ÷ distance moved by the solvent)

retention time (Rt) 82 time taken for a component to travel through the tube of adsorbent during gas chromatography

reverse reaction 85 reaction from right to left in the equation for a reversible reaction

reversible reaction 85, 92 a reaction that can also occur in the opposite direction – that is, the products can react to form the original reactants again

ribosomes 58 tiny structures within a cell, where protein synthesis takes place

rod cell receptor cell in the eye that detects light

rutile an ore of titanium – impure titanium oxide (TiO_2)

S

salt compound composed of metal ions and non-metal ions – formed by acid–base neutralisation

sample 63 take measurements, or make counts, in a small area rather than over the entire area in question

Sankey diagram 46 diagram showing how the energy supplied to something is transferred into 'useful' or 'wasted' energy

saturated fat 33, 37 solid fat, most often of animal origin, containing no C=C double bonds

saturated hydrocarbon hydrocarbon containing only single covalent bonds

scalar quantity quantity that only has size, but not direction, for example, energy is a scalar quantity

secretion 60 production and release of a useful substance

sensory neurone 10, 23 nerve cell carrying information from receptors to the central nervous system

series circuit 100, 102, 110 electrical circuit with only one possible path for the current to flow around

sex chromosomes 70–71 the X and Y chromosomes

shape memory alloy 81 alloy that 'remembers' its original shape and returns to it when heated

shells 24–5, 41 electrons are arranged in shells (or orbits) around the nucleus of an atom – also known as 'energy levels'

slag 27–28 waste material produced during smelting of a metal – it contains unwanted impurities from the ore

smart material 81 material which changes in response to changes in its surroundings, such as light levels or temperature

smelting 29–30 extracting metal from an ore by reduction with carbon – heating the ore and carbon in a furnace

soften (water) treat water so as to remove the calcium ions that cause hardness

solar cell device that converts the Sun's energy into electricity

solar panel 57 panel that uses the Sun's energy to heat water

solar power station 49 power station generating electricity using energy transferred by the Sun's radiation

soluble salt 90, 92 salt which dissolves in water

solute substance which dissolves in a liquid to form a solution

solvent 82 liquid in which solutes dissolve to form a solution

speciation 73 formation of a new species

species 73–74 group of organisms that share similar characteristics and that can breed together to produce fertile offspring

specific heat capacity 45, 57 a measure of the amount of energy needed to raise the temperature of 1 kg of a substance by 1 °C

speed of light 57 speed at which electromagnetic radiation travels through a vacuum – 300 000 000 metres per second

speed 93, 110 how quickly an object is moving, usually measured in metres per second (m/s)

sperm cell 19–20, 59, 68, 71 male gamete

sperm male sex cell of an animal

stainless steel 29 steel alloy containing chromium and nickel to resist corrosion

starch 61–62, 65, 74 a carbohydrate; a polysaccharide that is used for storing energy in plant cells, but not in animal cells

state symbol symbol used in equations to show whether something is solid, liquid, gas or in solution in water

states of matter 43 substances can exist in three states of matter (solid, liquid or gas) – changes from one state to another are called changes of state

static electricity 99 an electric charge on an insulating material, caused by electrons flowing onto or away from the object

statin drug that reduces cholesterol level in the blood

steam cracking cracking hydrocarbons by mixing with steam and heating

steam distillation process of blowing steam through a mixture to vaporise volatile substances – used to extract essential oils from flowers

steel 28, 29 alloy of iron and steel, with other metals added depending on its intended use

stem cell 69 a cell that has not yet differentiated – it can divide to form cells that form various kinds of specialised cell

step-down transformer 50 transformer that changes alternating current to a lower voltage

step-up transformer 48, 50 transformer that changes alternating current to a higher voltage

sterile technique 9 handling apparatus and material to prevent microorganisms from entering them

sterile 9, 23 containing no living organisms

stimulus 10 a change in the environment that is detected by a receptor

stopping distance 95, 110 total distance it takes a vehicle to stop – the sum of thinking distance and braking distance

sub-atomic particle 24, 75 particle that make up an atom, such as proton, neutron or electron

subcutaneous just under the skin

subduction zone area of ocean floor in which an oceanic plate is sinking beneath a continental plate

sublime turn directly from solid into a gas without melting

substrate 64 molecule on which an enzyme acts – the enzyme catalyses the reaction that changes the substrate into a product

successful collisions 86 collisions with enough energy to break bonds in the reactant particles, and thus cause a reaction

sugar 66 sweet-tasting compound of carbon, hydrogen and oxygen such as glucose or sucrose

sulfur dioxide 30–31 poisonous, acidic gas formed when sulfur or a sulfur compound is burned

surface area (of a solid reactant) 87 measure of the area of an object that is in direct contact with its surroundings

sweat 11, 23 liquid secreted onto the skin surface that has a cooling effect as it evaporates

symbol (for an element) 24 one or two letters used to represent a chemical element, for example C for carbon or Na for sodium

synapse 11 gap between two neurones

synthetic artificial or made by people

syrup concentrated solution of sugar

T

target organ 10 the part of the body affected by a hormone

tarnish go dull and discoloured by reacting with oxygen, moisture or other gases in the air

tectonic plate 38, 41 section of Earth's crust that floats on the mantle and slowly moves across the surface

telecommunications 54 communications over long distances using various types of electromagnetic radiation

terminal velocity 96, 110 maximum velocity an object can travel at – at terminal velocity, forward and backward forces are the same

testes 10, 68 organs in a male in which sperms are made

tetrahedral structure structure in which atoms have four covalent bonds to other atoms positioned at the four corners of a tetrahedron, for example, diamond

thalidomide 14 a drug that was originally prescribed to pregnant women but was found to cause deformities in fetuses

theoretical yield 84 mass of product that a given mass of reactant should produce according to calculations from the equation – the actual yield is always less than this

thermal decomposition 27 chemical reaction in which a substance is broken down into simpler chemicals by heating it

thermistor 101 resistor made from semiconductor material: its resistance decreases as temperature increases

thermochromic 81 thermochromic materials change colour in response to changes in temperature

thermosetting polymer 80, 92 plastic polymer that sets hard when heated and moulded for the first time – it will not soften or melt when heated again

thermosoftening polymer 80, 92 plastic polymer that softens and melts when heated and reheated

thin layer chromatography (TLC) 82 chromatography using a plate coated with a thin layer of powdered adsorbent

thinking distance 95 distance a vehicle travels while a signal travels from the driver's eye to brain and then to foot on the brake pedal: thinking distance increases with vehicle speed

three core cable 103, 110 electrical cable containing three wires, live, neutral and earth

three-pin plug 103, 110 type of plug used for connecting to the mains supply in the UK: it has three pins, live, neutral and earth

tidal power station 49 power station generating electricity using the energy transferred by moving tides

tissue 60, 74 group of cells that work together and carry out a similar task, such as lung tissue

titration procedure to determine the volume of one solution needed to react with a known volume of another solution

tonne 1 tonne = 1000 kg (1 million grams)

toxin 7 poisonous substance (pathogens make toxins that make us feel ill)

tracer 107 radioactive element used to track the movement of materials, such as water through a pipe or blood through organs of the body

transect 63, 74 line along which organisms are sampled – transects are often used to investigate how the distribution of organisms changes when one type of habitat merges into another

transfer (energy) energy transfers occur when energy moves from one place to another, or when there is a change in the way in which it is observed

transformer 50 device by which alternating current of one voltage is changed to another voltage

transition metals 28, 88 group of metal elements in the middle block of the periodic table – includes many common metals

transmitter chemical chemical that transfers a nerve impulse across a synapse

transverse wave 51, 53 a wave in which the vibration of particles is at right angles to the direction in which the wave transfers energy

triple covalent bond three covalent bonds between the same two atoms – each atom shares three of its own electrons plus three from the other atom

tropism 13 response of a plant to a stimulus, by growing towards or away from it

tsunami 39 huge waves caused by earthquakes – can be very destructive

turbine 49 device for generating electricity – the turbine has coils of wire that rotate in a magnetic field to generate electricity

U

ultraviolet radiation 54 electromagnetic radiation that can damage human skin

unsaturated fats 33, 37, 41 liquid fats, containing C=C double bonds – usually from plants or fish

unsaturated hydrocarbon hydrocarbon containing one or more C=C double bonds

upright image 52 image that is the same way up as the object

upthrust 95 upward force on an object in water – for a floating object, the upthrust is equal to the weight of the object

U-value 45, 57 a measure of how easily energy is transferred through a material as heat

V

vaccine 9, 23 killed microorganisms, or living but weakened microorganisms, that are given to produce immunity to a particular disease

vacuole 7, 58 liquid-filled space inside a cell – many plant cells contain vacuoles full of cell sap

vacuum 51, 57 a space in which there are no particles of any kind

validity how well a measurement really measures what it is supposed to be measuring

Van der Graaff generator device for investigating static electricity: a large static electricity charge builds up on a metal dome insulated from earth

vaporise change from liquid to gas (vapour)

variation differences between individuals belonging to the same species

vector quantity a quantity that has both size and direction: velocity is a vector quantity, having size in a particular direction

velocity 53, 93, 96, 99, 110 measure of how fast an object is moving in a particular direction

velocity–time graph 93, 96 graph showing how the velocity of an object varies with time: its gradient shows acceleration

vent 39 crack or weak spot in the Earth's crust, through which magma reaches the surface

virtual image 52 image that can be seen but cannot be projected onto a screen (a mirror forms a virtual image behind the mirror)

virus 7 very small structure made of a protein coat surrounding DNA (or RNA); viruses can only reproduce inside a living cell

VO_2max the maximum volume of oxygen the body can use per minute

volcano 39 landform (often a mountain) where molten rock erupts onto the surface of the planet

voltage 50, 101 a measure of the energy carried by an electric current (the old name for potential difference)

W

wasted energy 17, 23 energy that is transferred by a device or appliance in ways that are not wanted, or useful

water of crystallization 85 water molecules present in crystals of some metal salts – shown separately in the formula, for example, hydrated copper sulfate, $CuSO_4.5H_2O$

watt 47 unit of energy transfer – one watt is a rate of energy transfer of one joule per second

wave equation 53 the speed of a wave is always equal to its frequency multiplied by its wavelength

wave power 49, 57 electricity generation using the energy transferred by water waves as the water surface moves up and down

wavelength 52, 56 distance between two wave peaks

weight 95 the downward force on a mass due to gravity, measured in newtons (N)

wind turbine 49–50, 57 device generating electricity by using the energy in moving air to turn a turbine and a generator

work 97–98 amount of energy transferred to an object by a force moving the object through a distance: work done = force × distance moved in the direction of the force

X

X-rays 54, 106 ionising electromagnetic radiation – used in X-ray photography to generate pictures of bones

xylem 61 tissue made up of long, empty, dead cells that transports water from the roots to the leaves of a plant

Y

yeast 34 single-celled fungus used in making bread and beer

yield 92 mass of product made from a chemical reaction

Z

zygote 20, 58 a diploid cell formed by the fusion of the nuclei of two gametes

Group 1	2											Group 3	4	5	6	7	0
		$^{1}_{1}$H hydrogen															$^{4}_{2}$He helium
$^{7}_{3}$Li lithium	$^{9}_{4}$Be beryllium											$^{11}_{5}$B boron	$^{12}_{6}$C carbon	$^{14}_{7}$N nitrogen	$^{16}_{8}$O oxygen	$^{19}_{9}$F fluorine	$^{20}_{10}$Ne neon
$^{23}_{11}$Na sodium	$^{24}_{12}$Mg magnesium											$^{27}_{13}$Al aluminium	$^{28}_{14}$Si silicon	$^{31}_{15}$P phosphorus	$^{32}_{16}$S sulfur	$^{35}_{17}$Cl chlorine	$^{40}_{18}$Ar argon
$^{39}_{19}$K potassium	$^{40}_{20}$Ca calcium	$^{45}_{21}$Sc scandium	$^{48}_{22}$Ti titanium	$^{51}_{23}$V vanadium	$^{52}_{24}$Cr chromium	$^{55}_{25}$Mn manganese	$^{56}_{26}$Fe iron	$^{59}_{27}$Co cobalt	$^{59}_{28}$Ni nickel	$^{64}_{29}$Cu copper	$^{65}_{30}$Zn zinc	$^{70}_{31}$Ga gallium	$^{73}_{32}$Ge germanium	$^{75}_{33}$As arsenic	$^{79}_{34}$Se selenium	$^{80}_{35}$Br bromine	$^{84}_{36}$Kr krypton
$^{85}_{37}$Rb rubidium	$^{88}_{38}$Sr strontium	$^{89}_{39}$Y yttrium	$^{91}_{40}$Zr zirconium	$^{93}_{41}$Nb niobium	$^{96}_{42}$Mo molybdenum	$^{99}_{43}$Tc technetium	$^{101}_{44}$Ru ruthenium	$^{103}_{45}$Rh rhodium	$^{106}_{46}$Pd palladium	$^{108}_{47}$Ag silver	$^{112}_{48}$Cd cadmium	$^{115}_{49}$In indium	$^{119}_{50}$Sn tin	$^{122}_{51}$Sb antimony	$^{128}_{52}$Te tellurium	$^{127}_{53}$I iodine	$^{131}_{54}$Xe xenon
$^{133}_{55}$Cs caesium	$^{137}_{56}$Ba barium	$^{139}_{57}$La lanthanum	$^{178}_{72}$Hf hafnium	$^{181}_{73}$Ta tantalum	$^{184}_{74}$W tungsten	$^{186}_{75}$Re rhenium	$^{190}_{76}$Os osmium	$^{192}_{77}$Ir iridium	$^{195}_{78}$Pt platinum	$^{197}_{79}$Au gold	$^{201}_{80}$Hg mercury	$^{204}_{81}$Tl thallium	$^{207}_{82}$Pb lead	$^{209}_{83}$Bi bismuth	$^{210}_{84}$Po polonium	$^{210}_{85}$At astatine	$^{222}_{86}$Rn radon
$^{223}_{87}$Fr francium	$^{226}_{88}$Ra radium	$^{227}_{89}$Ac actinium															

Collins

Workbook

NEW GCSE SCIENCE

Science and Additional Science

for AQA A Foundation

Authors: Nicky Thomas
Rob Wensley
Gemma Young

Revision guide +
Exam practice workbook

The key to successful revision is finding the method that suits you best. There is no right or wrong way to do it.

Before you begin, it is important to plan your revision carefully. If you have allocated enough time in advance, you can walk into the exam with confidence, knowing that you are fully prepared.

Start well before the date of the exam, not the day before!

It is worth preparing a revision timetable and trying to stick to it. Use it during the lead up to the exams and between each exam. Make sure you plan some time off too.

Different people revise in different ways and you will soon discover what works best for you.

Some general points to think about when revising

- Find a quiet and comfortable space at home where you won't be disturbed. You will find you achieve more if the room is ventilated and has plenty of light.
- Take regular breaks. Some evidence suggests that revision is most effective when tackled in 30 to 40 minute slots. If you get bogged down at any point, take a break and go back to it later when you are feeling fresh. Try not to revise when you're feeling tired. If you do feel tired, take a break.
- Use your school notes, textbook and this Revision guide.
- Spend some time working through past papers to familiarise yourself with the exam format.
- Produce your own summaries of each module and then look at the summaries in this Revision guide at the end of each module.
- Draw mind maps covering the key information on each topic or module.
- Review the Grade booster checklists on pages 252–256.
- Set up revision cards containing condensed versions of your notes.
- Prioritise your revision of topics. You may want to leave more time to revise the topics you find most difficult.

Workbook

The Workbook allows you to work at your own pace on some typical exam-style questions. You will find that the actual GCSE questions are more likely to test knowledge and understanding across topics. However, the aim of the Revision guide and Workbook is to guide you through each topic so that you can identify your areas of strength and weakness.

The Workbook also contains example questions that require longer answers (**Extended response questions**). You will find one question that is similar to these in each section of your written exam papers. The quality of your written communication will be assessed when you answer these questions in the exam, so practise writing longer answers, using sentences. The **Answers** to all the questions in the Workbook are detachable for flexible practice and can be found on pages 257–278.

At the end of the Workbook there is a series of **Revision checklists** that you can use to tick off the topics when you are confident about them and understand certain key ideas.

Remember

There is a difference between learning and revising.

When you revise, you are looking again at something you have already learned. Revising is a process that helps you to remember this information more clearly.

Learning is about finding out and understanding new information.

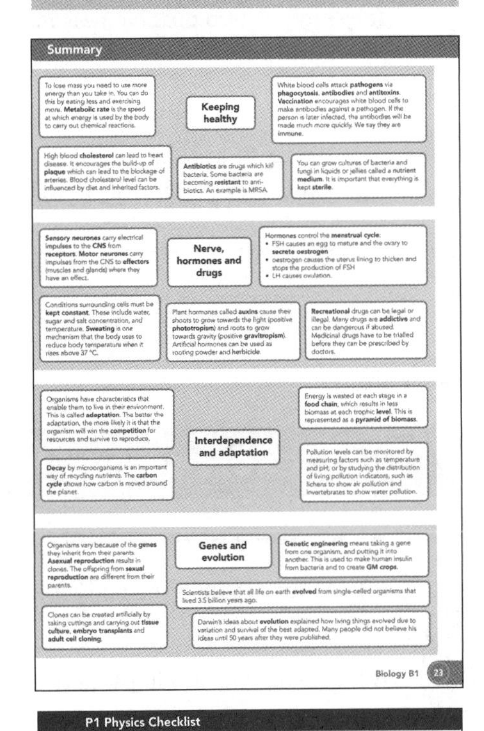

Summary

Keeping healthy

To lose mass you need to use more energy than you take in. You can do this by eating less and exercising more. **Metabolic rate** is the speed at which energy is used by the body to carry out chemical reactions.

White blood cells attack **pathogens** via **phagocytosis**, **antibodies** and **antitoxins**. **Vaccination** encourages white blood cells to make antibodies against a pathogen. If the person is later infected, the antibodies will be made much more quickly. We say they are immune.

High blood **cholesterol** can lead to heart disease. It encourages the build-up of **plaque** which can lead to the blockage of arteries. Blood cholesterol level can be influenced by diet and inherited factors.

Antibiotics are drugs which kill bacteria. Some bacteria are becoming **resistant** to anti-biotics. An example is MRSA.

You can grow cultures of bacteria and fungi in liquids or jellies called a **nutrient medium**. It is important that everything is kept **sterile**.

Nerve, hormones and drugs

Sensory neurones carry electrical impulses to the **CNS** from **receptors**. **Motor neurones** carry impulses from the CNS to **effectors** (muscles and glands) where they have an effect.

Hormones control the **menstrual cycle**:
- FSH causes an egg to mature and the ovary to **secrete oestrogen**
- oestrogen causes the uterus lining to thicken and stops the production of FSH
- LH causes ovulation.

Conditions surrounding cells must be **kept constant**. These include water, sugar and salt concentration, and temperature. **Sweating** is one mechanism that the body uses to reduce body temperature when it rises above 37 °C.

Plant hormones called **auxins** cause their shoots to grow towards the light (positive **phototropism**) and roots to grow towards gravity (positive **gravitropism**). Artificial hormones can be used as rooting powder and **herbicide**.

Recreational drugs can be legal or illegal. Many drugs are **addictive** and can be dangerous if abused. Medicinal drugs have to be trialled before they can be prescribed by doctors.

Interdependence and adaptation

Organisms have characteristics that enable them to live in their environment. This is called **adaptation**. The better the adaptation, the more likely it is that the organism will win the **competition** for resources and survive to reproduce.

Energy is wasted at each stage in a **food chain**, which results in less biomass at each trophic **level**. This is represented as a **pyramid of biomass**.

Decay by microorganisms is an important way of recycling nutrients. The **carbon cycle** shows how carbon is moved around the planet.

Pollution levels can be monitored by measuring factors such as temperature and pH; or by studying the distribution of living pollution indicators, such as lichens to show air pollution and invertebrates to show water pollution.

Genes and evolution

Organisms vary because of the **genes** they inherit from their parents. **Asexual reproduction** results in clones. The offspring from **sexual reproduction** are different from their parents.

Genetic engineering means taking a gene from one organism, and putting it into another. This is used to make human insulin from bacteria and to create **GM crops**.

Scientists believe that all life on earth **evolved** from single-celled organisms that lived 3.5 billion years ago.

Clones can be created artificially by taking cuttings and carrying out **tissue culture**, **embryo transplants** and **adult cell cloning**.

Darwin's ideas about **evolution** explained how living things evolved due to variation and survival of the best adapted. Many people did not believe his ideas until 50 years after they were published.

Biology B1 23

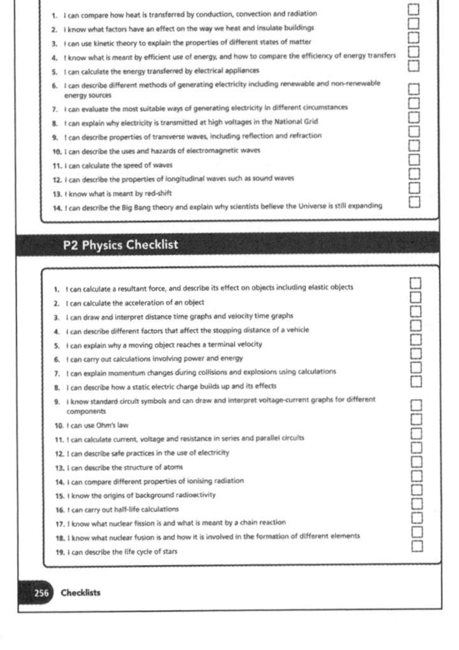

P1 Physics Checklist

1. I can compare how heat is transferred by conduction, convection and radiation
2. I know what factors have an effect on the way we heat and insulate buildings
3. I can use kinetic theory to explain the properties of different states of matter
4. I know what is meant by efficient use of energy, and how to compare the efficiency of energy transfers
5. I can calculate the energy transferred by electrical appliances
6. I can describe different methods of generating electricity including renewable and non-renewable energy sources
7. I can evaluate the most suitable ways of generating electricity in different circumstances
8. I can explain why electricity is transmitted at high voltages in the National Grid
9. I can describe properties of transverse waves, including reflection and refraction
10. I can describe the uses and hazards of electromagnetic waves
11. I can calculate the speed of waves
12. I can describe the properties of longitudinal waves such as sound waves
13. I know what is meant by red-shift
14. I can describe the Big Bang theory and explain why scientists believe the Universe is still expanding

P2 Physics Checklist

1. I can calculate a resultant force, and describe its effect on objects including elastic objects
2. I can calculate the acceleration of an object
3. I can draw and interpret distance time graphs and velocity time graphs
4. I can describe different factors that affect the stopping distance of a vehicle
5. I can explain why a moving object reaches a terminal velocity
6. I can carry out calculations involving power and energy
7. I can explain momentum changes during collisions and explosions using calculations
8. I can describe how a static electric charge builds up and its effects
9. I know standard circuit symbols and can draw and interpret voltage-current graphs for different components
10. I can use Ohm's law
11. I can calculate current, voltage and resistance in series and parallel circuits
12. I can describe safe practices in the use of electricity
13. I can describe the structure of atoms
14. I can compare different properties of ionising radiation
15. I know the origins of background radioactivity
16. I can carry out half-life calculations
17. I know what nuclear fission is and what is meant by a chain reaction
18. I know what nuclear fusion is and how it is involved in the formation of different elements
19. I can describe the life cycle of stars

256 Checklists

Diet and energy

1 (a) Where does the body gets its energy from?

____ [1 mark]

G–E

(b) State one way the body uses up energy.

____ [1 mark]

(c) What happens to the energy the body doesn't use?

____ [1 mark]

2 Susan and Richard are overweight and want to lose mass. They both go on a diet and eat exactly the same food for a week. Richard loses more mass.

D–C

(a) Suggest two possible reasons why.

____

____

____ [2 marks]

(b) Explain why your reasons will lead to Richard losing more mass.

____

____

____ [2 marks]

Diet, exercise and health

1 (a) Which organ in the body makes cholesterol?

____ [1 mark]

G–E

(b) Explain why it is important that your blood cholesterol level is not too high.

____

____ [1 mark]

2 Outline how high cholesterol levels in the blood can lead to a heart attack.

D–C

____

____

____

____

____

____

____ [5 marks]

Pathogens and infections

1 You want to study some bacteria. To do this you need to see them clearly. State what kind of instrument you would need to do this and explain why.

G–E

[2 marks]

2 Explain how viruses and bacteria make us feel ill.

D–C

[3 marks]

Fighting infection

1 (a) What is the function of your immune system?

[1 mark]

G–E

(b) What type of blood cell is part of the immune system?

[1 mark]

2 State two ways that lymphocytes can destroy pathogens.

D–C

[2 marks]

Drugs against disease

G–E

1 (a) State the function of painkillers.

[1 mark]

(b) Kirby has a sore throat. She takes a paracetamol but finds that the pain returns after a while. Explain why.

[2 marks]

D–C

2 Gita has a lung infection.

(a) Her doctor prescribes her some antibiotics. What is an antibiotic?

[1 mark]

(b) After 2 weeks her infection has still not cleared up. The doctor takes a swab of the mucus from her lungs and sends it to a pathology lab, so they can find out which antibiotic to give her. Explain how they would do this.

[4 marks]

Antibiotic resistance

G–E

1 (a) MRSA is an antibiotic-resistant bacteria. What does this mean?

[1 mark]

(b) Why are hospital patients more at risk from MRSA infections?

[1 mark]

D–C

2 Explain why numbers of antibiotic-resistant bacteria have increased.

[4 marks]

Vaccination

1 Explain why you cannot get chicken pox twice.

____

____

____

____________________ [3 marks]

G–E

2 The MMR vaccine is given to children in the UK.

(a) What is the function of this vaccine?

____

____________________ [1 mark]

(b) Explain how it works.

____

____

____________________ [3 marks]

D–C

(c) In the early 21st century, the number of children having the MMR vaccine fell. State the reason for this.

____

____________________ [1 mark]

(d) The number of children having the MMR vaccine has now risen again. Explain why.

____

____________________ [1 mark]

Growing bacteria

1 Paulo used an inoculating loop to spread some *E. coli* bacteria onto a dish containing agar jelly. He sealed the dish and left it in a cool place for the bacteria to multiply.

(a) Name **(i)** the culture **(ii)** the nutrient medium that Paulo used.

____

____________________ [2 marks]

G–E

(b) The inoculating loop he used was sterile. What does this mean?

____

____________________ [1 mark]

2 Why is it important to grow bacteria at a temperature less than 37 °C?

____

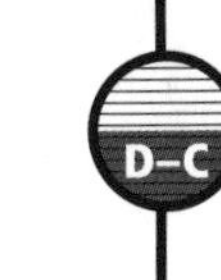

____

____________________ [2 marks]

Co-ordination, nerves and hormones

1 You go to pick up a chocolate bar. How does your brain know where to place your hand?

____

____

____

____ [3 marks]

2 What is the central nervous system (CNS)?

____

____ [1 mark]

Receptors

1 Choose the correct words from the list below to fill in the gaps:

The ______________ in Ben's nose detect ______________ from the lunch cooking in the kitchen.

His stomach is an ______________ and makes digestive juices in response. [3 marks]

taste	effector	light	chemicals	receptors

2 The diagram shows a neurone.

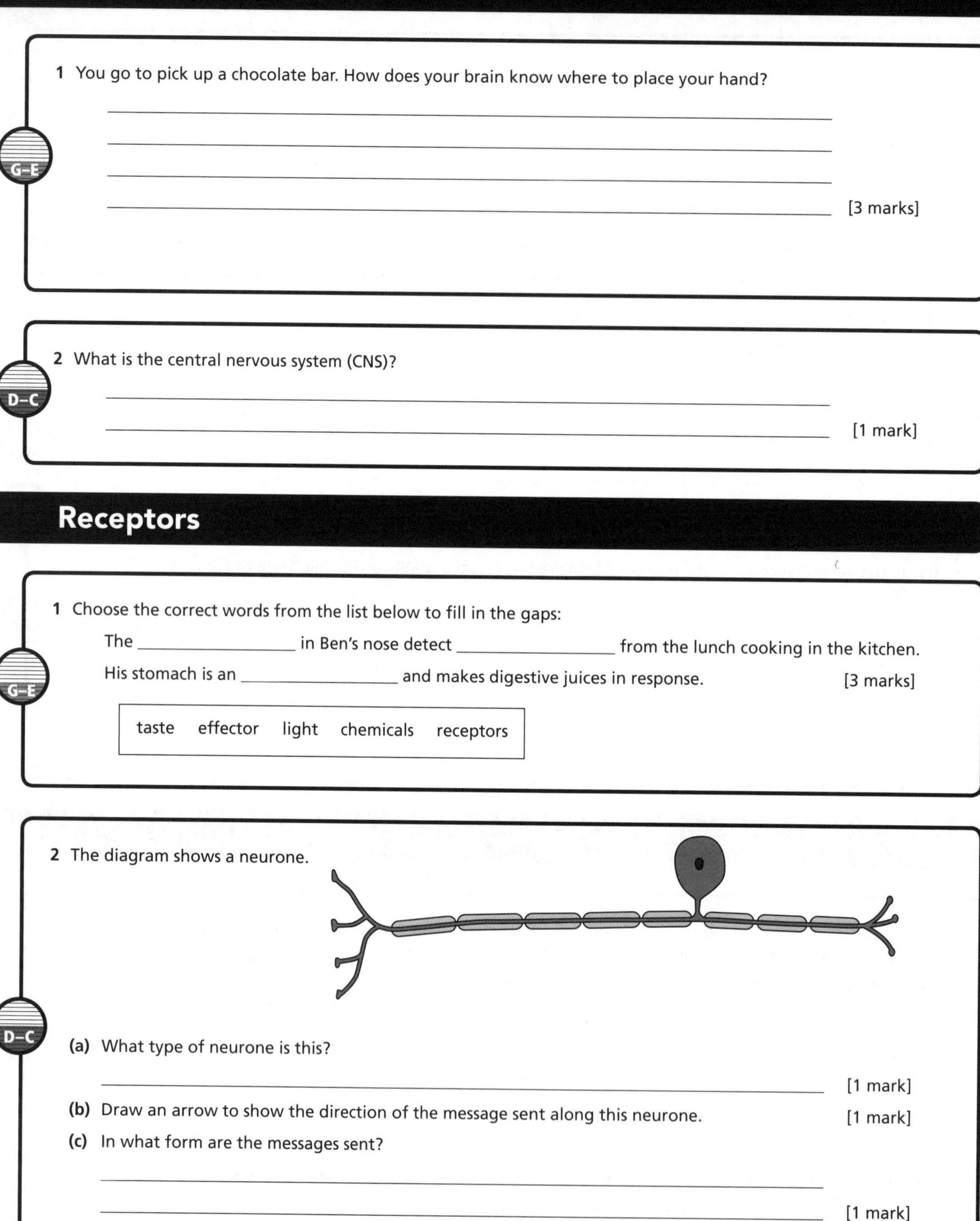

(a) What type of neurone is this?

____ [1 mark]

(b) Draw an arrow to show the direction of the message sent along this neurone. [1 mark]

(c) In what form are the messages sent?

____

____ [1 mark]

Reflex actions

1 (a) Give one example of a reflex action.

[1 mark]

(b) Why do reflex actions have to be fast?

[1 mark]

G–E

2 Your finger touches a sharp pin. Almost immediately, your finger moves away. Describe the sequence of events that have happened to bring this about.

[6 marks]

D–C

Controlling the body

1 (a) Name one condition that must be kept constant in the body.

[1 mark]

(b) Why do conditions need to be kept constant?

[1 mark]

G–E

2 Why is it important that human body temperature is maintained at around 37 °C?

[2 marks]

D–C

Reproductive hormones

G–E

1 (a) Describe the function of the uterus lining.

____

____

____ [2 marks]

(b) What happens to the lining if an egg is not fertilised?

____

____ [1 mark]

D–C

2 Explain how the oestrogen in the contraceptive pill stops a woman from getting pregnant.

____

____

____

____

____

____ [3 marks]

Controlling fertility

G–E

1 (a) Name one hormone found in fertility drugs.

____ [1 mark]

(b) Describe what it does.

____

____

____ [1 mark]

D–C

2 Outline the stages in IVF.

____

____

____

____

____

____

____ [3 marks]

Plant responses and hormones

1 For each of the following examples of tropisms, explain how they help the plant to survive:

(a) shoots growing towards the light.

____

____

______________________________ [1 mark]

(b) roots growing downwards into the soil.

____

____

______________________________ [1 mark]

G–E

2 Tropisms can be positive or negative. Complete each of these statements with the correct word.

(a) A shoot growing towards the light is __________ phototropism. [1 mark]

(b) A shoot growing away from gravity is __________ gravitropism. [1 mark]

D–C

Drugs

1 All drugs have side-effects. What does 'side-effect' mean?

____

____

______________________________ [1 mark]

G–E

2 (a) Name a recreational drug that is:

(i) legal ______________________________

(ii) illegal ______________________________ [2 marks]

(b) Explain why it is difficult to stop smoking.

____

____

______________________________ [2 marks]

D–C

Developing new drugs

1 State one reason why trialling a drug is a very expensive process.

____ [1 mark]

2 The table of data shows the results from a double-blind trial of a flu drug called zanamivir.

	given zanamivir	given a placebo
number of subjects	293	295
mean age in years	19	19
number of days until their temperature went down to normal	2.00	2.33
number of days until they lost all their symptoms and felt better	3.00	2.83
number of days until they felt just as well as before they had flu	4.5	6.3
average score the volunteers gave to their experience of the major symptoms of flu	23.4	25.3

Table 1: Effect of zanamivir and a placebo on soldiers suffering from flu

(a) What is meant by a 'double-blind' trial?

____ [3 marks]

(b) State one control variable in the trial. ______________________________ [1 mark]

(c) Use the data to state one reason why the drug should be distributed.

____ [1 mark]

Legal and illegal drugs

1 **(a)** Class each of these drugs as either legal or illegal: cannabis, alcohol, heroin, nicotine, ecstasy.

____

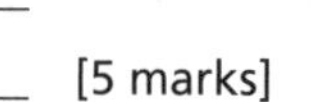

____ [5 marks]

G–E

(b) 'Cannabis is a safe drug.' State two arguments **against** this statement.

____

____ [2 marks]

2 Study the data below and use it to answer the questions that follow.

Year	2000	2001	2002	2003	2004	2005	2006	2007	2008
Men	4483	4938	5069	5443	5431	5566	5768	5732	5999
Women	2401	2561	2632	2721	2790	2820	2990	2992	3032

TABLE 2: The number of deaths from drinking alcohol, in each year from 2000 to 2008, in England and Wales.

(a) Describe two trends as seen in the data.

____

____ [2 marks]

(b) Heroin is considered a much more dangerous drug than alcohol. There were around 900 deaths from heroin in 2008. Describe how this compares with the deaths from alcohol and explain the reason for the difference.

____

____ [2 marks]

Competition

1 Which of the following lists correctly states the resources that plants compete for?

- **A** light, food, space, oxygen
- **B** light, water, space, nutrients
- **C** water, territory, food, nutrients
- **D** space, nutrients, light, oxygen

______________________________ [1 mark]

G–E

2 Suzanne is growing vegetables on her allotment. Explain to her why it is important for her to remove weeds.

____

____

____

______________________________ [2 marks]

D–C

Adaptations for survival

1 Polar bears live in the Arctic where it is very cold. They are adapted to survive there.

(a) State one adaptation a polar bear has and how it helps it to survive in the Arctic.

____

____

______________________________ [2 marks]

(b) Climate change means that the Arctic is getting warmer. Suggest what will happen to the polar bears that live there.

____

____

______________________________ [1 mark]

G–E

2 An image of a wasp and a hoverfly is shown here .

(a) Wasps can sting. How does this help them to survive?

wasp

hoverfly

____

______________________________ [1 mark]

(b) The hoverfly cannot sting. Explain why it looks like the wasp.

____

____

______________________________ [2 marks]

D–C

Environmental change

G–E

1 The Scottish primrose is a plant that only grows in the cold climate in the very north of Scotland. Explain what might happen if temperatures in the UK continue to rise.

___

___

___ [2 marks]

D–C

2 Cockatoos are birds that live in central Australia. They feed on seeds and fruit. Global warming has caused rainfall in their habitat to decrease. Explain how this will affect the cockatoo population.

___

___

___

___

___ [3 marks]

Pollution indicators

G–E

1 **(a)** What is a pollutant?

___

___

___ [1 mark]

(b) State one example of a pollutant.

___ [1 mark]

D–C

2 Cathy works for the Environmental Agency. Her job is to monitor pollution levels. She wants to measure the amount of dissolved oxygen in a stream.

(a) State the instrument she could use to do this.

___ [1 mark]

(b) She finds high levels of oxygen. What does this tell her about the level of pollution in the stream?

___

___ [1 mark]

(c) Give one example of an invertebrate she may find in the stream and explain how this is a further indicator of the pollution levels in the water.

___

___

___ [2 marks]

Food chains and energy flow

1 A food chain that exists in the ocean is: seaweed → limpet → octopus → seal

(a) Choose the organism in the food chain that is an example of a:

(i) producer ______________ (ii) consumer ______________ [2 marks]

G–E

(b) A limpet is a herbivore. What does this mean?

______________________________ [1 mark]

(c) Add correct words in the gaps:

A seal is a ______________, its prey is the ______________ [2 marks]

2 Only a small amount of the energy from light falling on a plant is used for photosynthesis. State one reason for this.

____

______________________________ [1 mark]

D–C

Biomass

1 What is biomass?

____

____

______________________________ [1 mark]

G–E

2 Sketch a pyramid of biomass for the following food chain: Grass → rabbit → fox

D–C

[2 marks]

Decay

1 Name two types of organisms that cause decay.

____

____ [2 marks]

2 June has a compost heap in her garden.

(a) In the summer, she waters the compost. Why?

____

____

____ [3 marks]

(b) What is the function of the air slats at the bottom?

____

____ [2 marks]

(c) She puts the compost onto her plants. Why?

____

____ [2 marks]

Recycling

1 Why is decay an essential process for plants?

____

____

____ [2 marks]

2 To the food chain below, add a label and arrows to show the role of microorganisms. [2 marks]

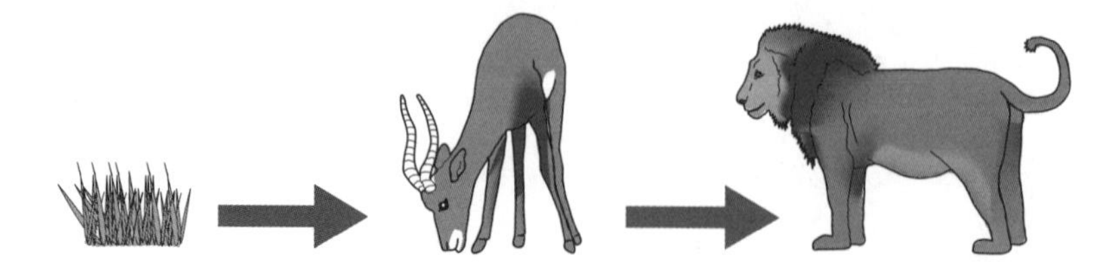

The carbon cycle

1 Name the process that:

(a) releases the carbon locked in the tissues of a dead animal into the air.

____ [1 mark]

(b) passes the carbon in a plant to an animal.

____ [1 mark]

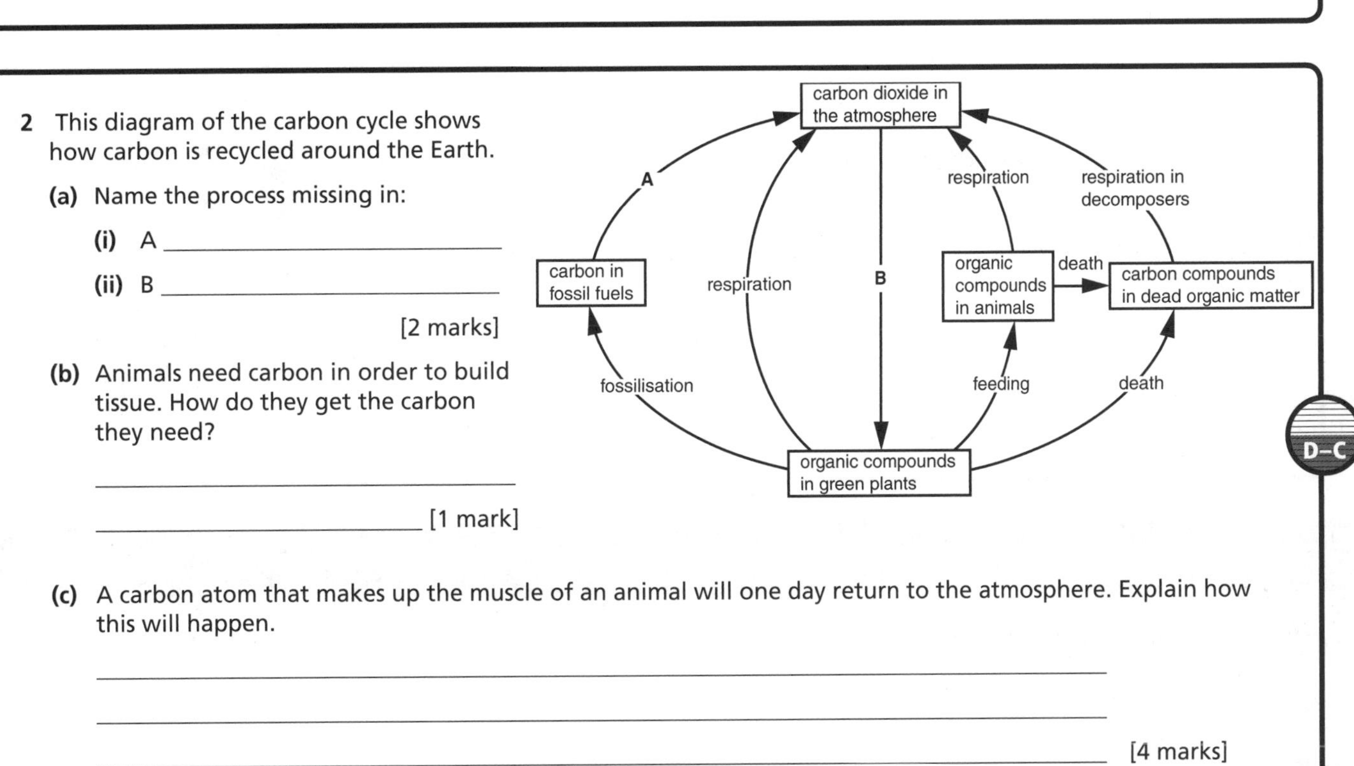

2 This diagram of the carbon cycle shows how carbon is recycled around the Earth.

(a) Name the process missing in:

(i) A ____________________

(ii) B ____________________

[2 marks]

(b) Animals need carbon in order to build tissue. How do they get the carbon they need?

____

____________________ [1 mark]

(c) A carbon atom that makes up the muscle of an animal will one day return to the atmosphere. Explain how this will happen.

____

____

____ [4 marks]

Genes and chromosomes

1 Complete the following sentences:

(a) ____________________ are linked together in chromosomes. [1 mark]

(b) Chromosomes are found in the ____________________ of cells. [1 mark]

(c) During ____________________ chromosomes in the gametes (egg and ____________________) join. [2 marks]

2 Sanjay has genes for blood group A and blood group O. Explain how he got two genes for his blood group.

____ [2 marks]

Reproduction

G–E

1 Choose words to complete the following statements:

(a) The gametes of a male animal are called ______________________ [1 mark]

(b) The gametes of a female animal are called ______________________ [1 mark]

(c) The gametes join together during ______________________ [1 mark]

(d) This is ______________________ reproduction. [1 mark]

D–C

2 This single-celled organism is reproducing by dividing into two.

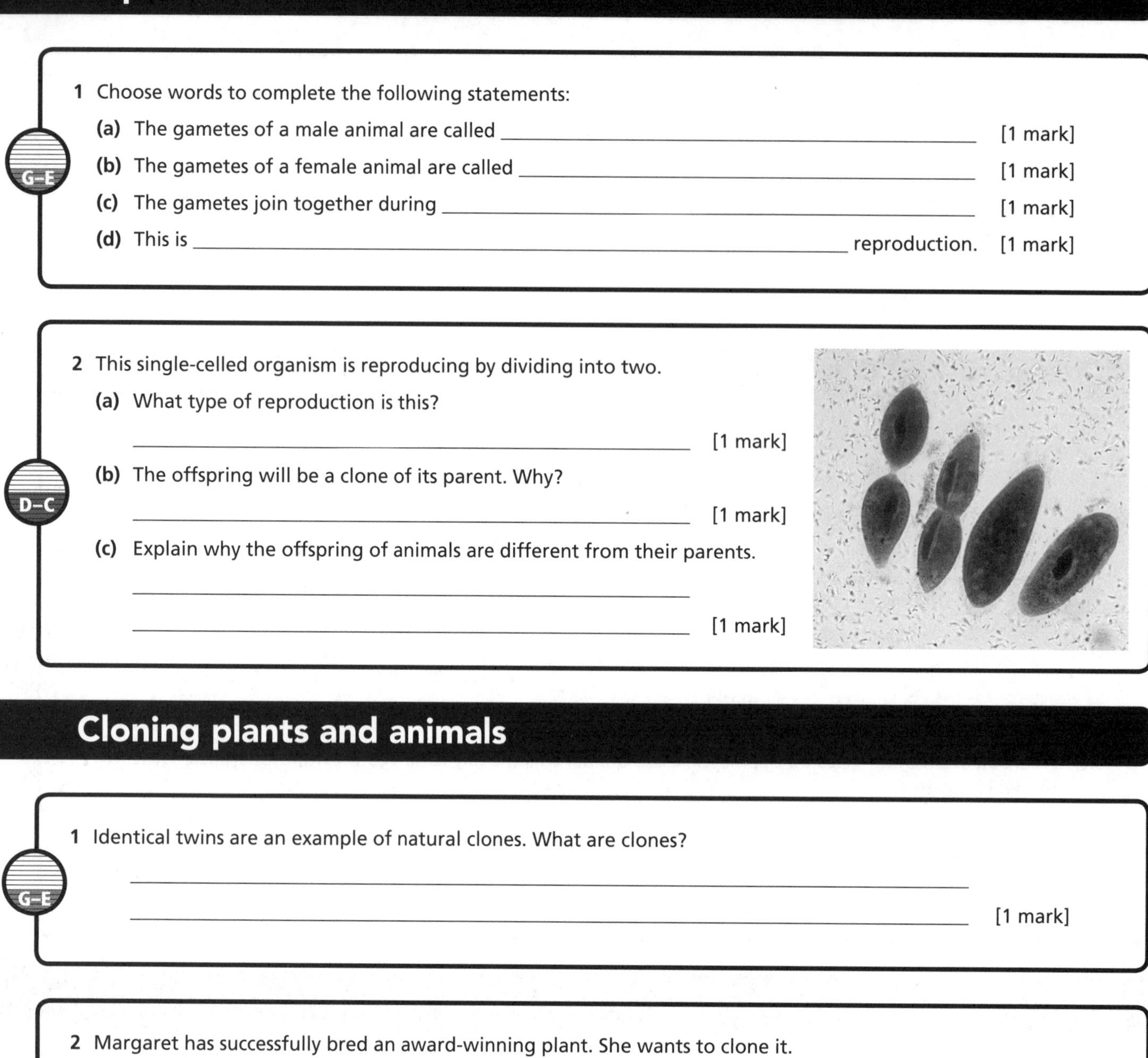

(a) What type of reproduction is this?

______________________ [1 mark]

(b) The offspring will be a clone of its parent. Why?

______________________ [1 mark]

(c) Explain why the offspring of animals are different from their parents.

____

______________________ [1 mark]

Cloning plants and animals

G–E

1 Identical twins are an example of natural clones. What are clones?

____

______________________ [1 mark]

D–C

2 Margaret has successfully bred an award-winning plant. She wants to clone it.

(a) Why does she want to clone it?

____

______________________ [2 marks]

(b) State one method she can use and explain how to carry it out.

____

____

______________________ [3 marks]

Genetic engineering

1 (a) Describe what the term 'genetic engineering' means.

[2 marks]

G–E

(b) Give one example of how genetic engineering can be used.

[1 mark]

2 Human insulin can be produced by bacteria grown in huge vats.

(a) Why are these bacteria known as a GM organisms?

[1 mark]

(b) Outline the procedure used to make GM bacteria that are able to produce human insulin.

D–C

[3 marks]

(c) The culture inside the vat has to be maintained at a temperature of around 37 °C. Why?

[2 marks]

Evolution

1 What does evolution mean?

G–E

[1 mark]

2 Lamarck stated that as a giraffe stretches up to eat leaves from a high tree, its neck gets longer. This characteristic would be passed on to the next generation. Use what you know about how characteristics are inherited to explain why we know this is not true.

D–C

[3 marks]

Natural selection

G–E

1 (a) In a litter of wild rabbits, one of the babies has poor hearing. He will probably not grow up to become an adult. Explain why.

____

______________________________ [2 marks]

(b) How does this reduce the number of rabbits being born with bad hearing?

____

______________________________ [2 marks]

D–C

2 (a) Peppered moths are a pale colour. They like to rest on the trunks of trees. How does the colour of the moth help it to survive?

____

______________________________ [2 marks]

(b) Occasionally there is a mutation and a black moth is born. What is a mutation?

______________________________ [1 mark]

(c) In the past, air pollution in the UK caused the bark of the trees to turn black. Explain how this meant that the numbers of black moths increased.

____

____

______________________________ [3 marks]

Evidence for evolution

G–E

1 Connor found this fossil of a crab at the beach.

(a) What is a fossil?

______________________________ [1 mark]

(b) How are fossils evidence for evolution?

____

____

______________________________ [2 marks]

D–C

2 The bones in the arms of birds, bats and humans are shown here.

(a) Why are they all slightly different?

____

______________________________ [1 mark]

(b) Explain how this is evidence for evolution.

____

____

____

______________________________ [3 marks]

Extended response question

As a seedling grows, it is important that its roots become anchored in the ground. Use what you know about how auxins control cell growth to explain how this happens.

The quality of written communication will be assessed in your answer to this question.

[6 marks]

Atoms, elements and compounds

1 Choose the word from the list that matches the description.

compound **formula** **element** **symbol** **mixture**

(a) It shows the types and numbers of particles in the substance. ____________________ [1 mark]

(b) It is the simplest type of substance. ____________________ [1 mark]

(c) It is made of at least two different atoms. ____________________ [1 mark]

(d) It is used to represent an element. ____________________ [1 mark]

2 Malachite is an ore of copper. It contains both rock, and copper carbonate. When heated with carbon it produces copper, and carbon dioxide.

(a) Name the element obtained from malachite. ____________________ [1 mark]

(b) Name the useful compound in malachite. ____________________ [1 mark]

(c) Explain why you know this substance is a compound.

____________________ [1 mark]

(d) Explain why malachite is a mixture.

____

____________________ [2 marks]

Inside the atom

1 Elements are made up of atoms. Here is a diagram of an atom of beryllium (Be) showing the three subatomic particles.

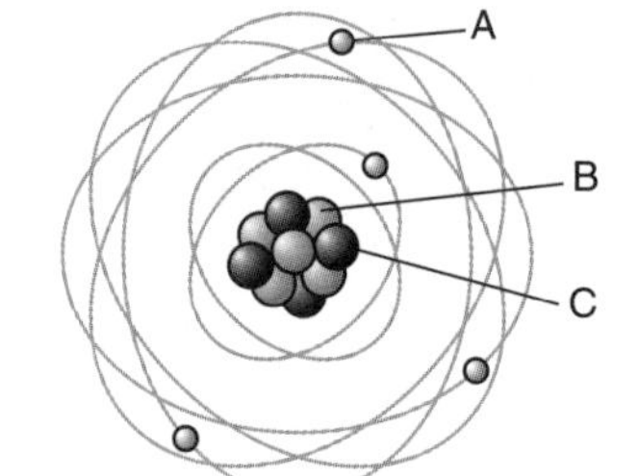

(a) Name particle A. ____________________ [1 mark]

(b) Name particle B. ____________________ [1 mark]

(c) Name particle C. ____________________ [1 mark]

(d) Which particles make up the nucleus?
(Answer using one or more of the letters A, B, C.) ____________ [1 mark]

2 Phosphorus is an element in the periodic table.

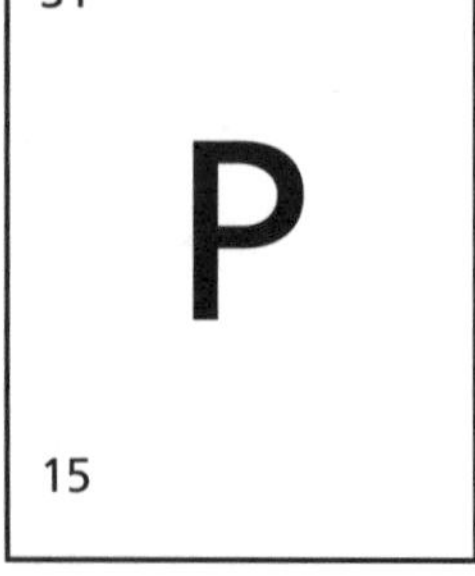

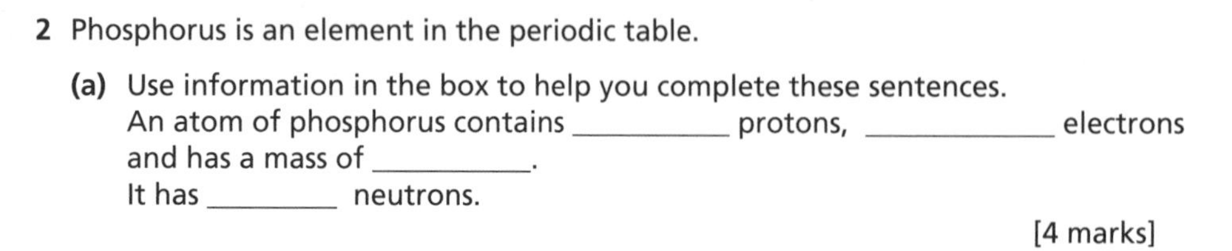

(a) Use information in the box to help you complete these sentences.
An atom of phosphorus contains __________ protons, __________ electrons and has a mass of __________.
It has __________ neutrons.

[4 marks]

(b) Name or give the symbol of an element that:

(i) has the same number of outer electrons as phosphorus. ____________

(ii) has two more outer electrons than phosphorus. ____________

(iii) has one less proton than phosphorus. ____________ [3 marks]

(c) Draw a diagram to show the electronic structure of phosphorus. [2 marks]

Element patterns

1 All known elements are listed on the periodic table. It also gives the atomic number and mass number of each element.

(a) What is the atomic number?

____ [1 mark]

(b) What is the mass number?

____ [1 mark]

(c) Describe how to find the number of electrons in an atom.

____ [1 mark]

G–E

2 (a) How are the elements arranged in the periodic table?

____ [1 mark]

(b) What are horizontal rows of the periodic table called?

____ [1 mark]

(c) What are vertical columns of the periodic table called?

____ [1 mark]

(d) Elements in the same vertical column are sometimes called 'families'. Explain why.

____ [2 marks]

D–C

Combining atoms

1 This is a space filling model of a water molecule (H_2O). The atoms are joined together by covalent bonds.

(a) Describe a covalent bond.

__________________________________ [1 mark]

(b) Draw a dot and cross diagram to show the electron arrangement in the water molecule. [2 marks]

Sodium chloride (NaCl) is a compound where two different elements are present. The atoms transfer electrons to each other.

(c) Give the electronic configuration of a sodium atom.

____ [1 mark]

(d) Describe how the sodium atom becomes a positively charged ion.

____ [2 marks]

(e) Explain how this helps the sodium to form a compound with the chlorine.

____ [3 marks]

Chemical equations

1 Zinc carbonate reacts with hydrochloric acid to make zinc chloride, water and carbon dioxide.

Here is the chemical symbol equation for the reaction.

$ZnCO_3 + 2HCl \rightarrow ZnCl_2 + CO_2 + H_2O$

(a) Write the formula of one of the reactants. ____________________ [1 mark]

(b) How many products are made? ____________________ [1 mark]

(c) How many molecules of hydrochloric acid are needed to make a molecule of carbon dioxide?

____________________ [1 mark]

2 Copper carbonate can be used to make copper. When heated with carbon it produces copper and carbon dioxide.

(a) Complete this word equation

copper carbonate + carbon → copper + __________ [1 mark]

(b) This is the balanced symbol equation for the reaction

$2CuCO_3 + C \rightarrow 2Cu + 3CO_2$

(i) How many molecules of carbon dioxide are produced in the reaction? __________ [1 mark]

(ii) How many atoms of copper can be made for each atom of reacting carbon? __________ [1 mark]

Building with limestone

1 Limestone is a very important raw material for industry. It is used to make many building materials.

(a) What is the chemical name for limestone? ____________________ [1 mark]

(b) Name three building materials that are made from limestone.

____

____

____________________ [3 marks]

(c) Explain why acid rain is a problem for buildings made from limestone.

____

____________________ [2 marks]

2 A company obtains large quantities of limestone by quarrying.
These quarries are often in beautiful parts of the country such as the Peak District.

The company wants to open a new quarry. Describe the advantages and disadvantages, socially and economically for the local population in opening the new quarry.

____

____

____

____________________ [4 marks]

Heating limestone

1 Farmers have heated limestone in limekilns for centuries. They use the calcium oxide produced to help them grow crops.

(a) Which gas is produced when limestone is converted to calcium oxide? ______ [1 mark]

(b) Before the farmers use the calcium oxide, they react it with water. This changes the calcium oxide to another calcium compound sometimes called slaked lime. Name this compound. ______ [1 mark]

(c) Explain why farmers spread slaked lime on their fields.

______

______ [2 marks]

G–E

2 Heating limestone converts it into calcium hydroxide. Calcium hydroxide dissolves in water to make limewater.

(a) Limewater can be used to test for a gas in the air. Which gas is this? ______ [1 mark]

(b) Describe how to use limewater to test for this gas, and what you would see if the gas were present.

______ [2 marks]

(c) Cement is made from limestone.

(i) What substance is limestone heated with to make cement? ______ [1 mark]

(ii) Cement is used to make concrete and mortar. What is the difference between mortar and concrete?

______ [2 marks]

(d) Mortar holds bricks together, concrete can be used to make beams for use in constructing houses. Explain why these two materials made from limestone have these different uses.

______ [2 marks]

D–C

Metals from ores

1 Gold, silver and copper are three metal elements that are found naturally in the ground. The other metals are all found as compounds in rocks. The metal is obtained by either heating the rock with carbon, or using electrolysis.

(a) What do we call rocks that are rich in useful metal compounds? ______ [1 mark]

(b) Name two metals that can be obtained by heating with carbon.

______ [2 marks]

(c) Explain why metals obtained by using electrolysis are more expensive than metals obtained by heating with carbon.

______ [2 marks]

G–E

2 Metals such as iron, copper and zinc are obtained from their ores. Here is a table of some common ores.

name of ore	formula
haematite	Fe_2O_3
bauxite	Al_2O_3
litharge	PbO
zincite	ZnO

(a) Name the ore from the table that can be used to obtain

(i) iron ______ [1 mark]

(ii) zinc ______ [1 mark]

(iii) aluminium ______ [1 mark]

(b) Which ore in the table cannot be obtained by reduction of the ore with carbon. Explain your answer. ______ [2 marks]

(c) What is meant by reduction? ______ [1 mark]

D–C

Extracting Iron

D–C

1 Iron ore (Fe_2O_3) is reduced inside a blast furnace to make iron.

(a) Name the two other substances added to the blast furnace.

(i) ______________________ **(ii)** ______________________ [2 marks]

(b) Hot air is blown through the blast furnace. This produces large volumes of carbon dioxide gas. Explain why.

____

____

____ [3 marks]

(c) Slag is formed in the process. Describe how slag is formed.

____ [1 mark]

Metals are useful

G–E

1 Copper is a very useful metal. Match the property to the use.

unreactive **good conductor of heat** **shiny** **malleable** **good conductor of electricity**

(a) electrical wiring ______________________ [1 mark]

(b) jewellery ______________________ [1 mark]

(c) frying pan ______________________ [1 mark]

(d) coins ______________________ [1 mark]

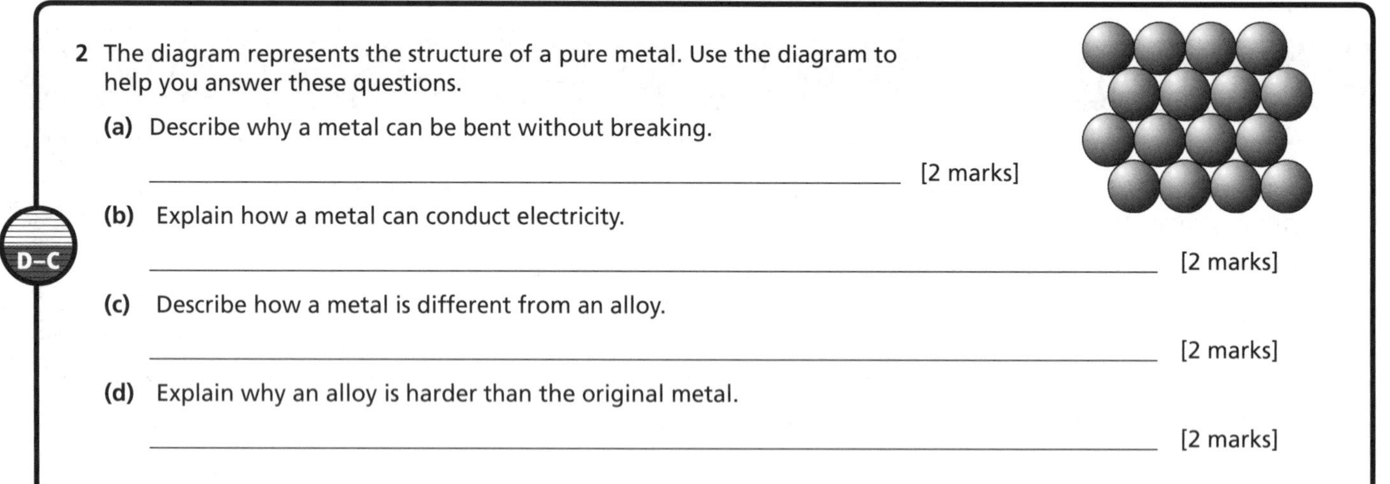

D–C

2 The diagram represents the structure of a pure metal. Use the diagram to help you answer these questions.

(a) Describe why a metal can be bent without breaking.

____ [2 marks]

(b) Explain how a metal can conduct electricity.

____ [2 marks]

(c) Describe how a metal is different from an alloy.

____ [2 marks]

(d) Explain why an alloy is harder than the original metal.

____ [2 marks]

Iron and steel

1 Iron is made by heating iron ore with carbon. The iron made contains about 4% carbon.

(a) What effect does the carbon have on the iron? ______________________ [1 mark]

(b) Iron is converted to steel by reducing the carbon content to 2% or less. The carbon is removed by blowing oxygen gas through the iron. What gas is produced by this process? ____________ [1 mark]

(c) Describe how steel is a more useful material than iron.

____

______________________ [2 marks]

G–E

2 Stainless steel can be made by adding chromium and nickel to iron to make an alloy we call stainless steel.

(a) Give two advantages of using stainless steel instead of iron to make cutlery.

____

______________________ [2 marks]

(b) Vanadium can also be added to iron to make a different stainless steel.
Suggest why having different types of stainless steel is useful.

______________________ [1 mark]

(c) Which part of the periodic table are all these metals found in?

______________________ [1 mark]

D–C

Copper

1 The flow chart shows the stages in the extraction of copper. The copper ore contains sulfur.
Complete the missing information in the flow chart.

removing unwanted rock to ____________ the ore	→	heating the copper ore in air to remove ____________	→	heating in a furnace with ____________ to make ____________ copper.

[4 marks]

G–E

2 Copper is purified using electrolysis of copper sulfate solution.
This diagram shows the process.

(a) What is the anode made from?

______________________ [1 mark]

(b) What is the cathode made from?

______________________ [1 mark]

(c) Where does the pure copper collect? ______________________ [1 mark]

(d) Where do the impurities collect? ______________________ [1 mark]

(e) Describe what happens to copper ions at the cathode.

____

______________________ [2 marks]

D–C

Aluminium and titanium

G–E

1 Look at this table of information about aluminium and titanium.

	aluminium	**titanium**
melting point °C	659	1677
boiling point °C	2447	3277
density g/cm	2.70	4.54
cost £/kg	1.46	4.63
electrical conductivity	good	good
uses	drinks cans jet aircraft bodies and wings overhead power cables	hip replacement joints military jet aircraft and missiles racing bicycle frames

(a) Use information from the table to suggest:

(i) why aluminium is used for overhead power cables.

______________________ [1 mark]

(ii) why titanium is used in military jet aircraft, but not civilian aircraft.

______________________ [1 mark]

(iii) two reasons why aluminium is used for drinks cans.

______________________ [2 marks]

(b) What property of both aluminium and titanium allows them to be used as drinks cans and as hip replacement joints? ______________________ [1 mark]

D–C

2 Aluminium is extracted from its ore by electrolysis.

(a) What is meant by electrolysis? ______________________ [1 mark]

(b) Name the main aluminium ore. ______________________ [1 mark]

(c) (i) Why is aluminium an expensive metal?

______________________ [2 marks]

(ii) Explain how dissolving the aluminium ore in cryolite reduces the cost of aluminium manufacture.

______________________ [3 marks]

Metals and the environment

G–E

1 Aluminium is made from the ore bauxite, which is found in the Amazon Rainforests in Brazil.

Aluminium melts at 659 °C. Aluminium is extracted from bauxite by electrolysis at a temperature of 900 °C.

(a) Explain why it is cheaper to recycle an aluminium can than to make a new one.

______________________ [1 mark]

(b) Suggest two reasons why recycling an aluminium can is better for the environment than making a new one from bauxite.

______________________ [2 marks]

D–C

2 Many metals are recycled.

(a) Evaluate the economic benefits of recycling metals against using new metal. Your answer should suggest two advantages and two disadvantages of recycling metals.

______________________ [6 marks]

(b) Describe the environmental disadvantages of mining metal ores.

______________________ [3 marks]

A burning problem

1 (a) Name three fossil fuels. ______ [1 mark]

(b) Biofuels are renewable energy sources. What is meant by a renewable energy source?

______ [1 mark]

(c) Fossil fuels are made up of hydrocarbons. Which two elements are present in a hydrocarbon?

______ [1 mark]

(d) When a hydrocarbon burns it produces two compounds. Name the two compounds.

______ [2 marks]

G–E

2 In 2010, an Icelandic volcano erupted sending large quantities of ash into the atmosphere, and gases such as sulfur dioxide and carbon dioxide. Jet aircraft were prevented from flying over a large part of Europe.

(a) Explain how the ash may cause global dimming.

______ [2 marks]

(b) Describe how the volcanic eruption may lead to more acid rainfall.

______ [2 marks]

(c) Suggest how the eruption could lead to global warming.

______ [1 mark]

D–C

Reducing air pollution

1 Some fossil fuels contain sulfur.

At oil refineries, sulfur is removed from petrol and diesel before they are sent to filling stations.

(a) Explain why sulfur is removed from petrol and diesel before being burnt.

______ [2 marks]

(b) Sulfur compounds are not removed from coal. Instead coal fired power stations remove the sulfur after combustion.

(i) Name the sulfur compound that has to be removed from the waste gases in a power station.

______ [1 mark]

(ii) Name the substance that removes this sulfur compound.

______ [1 mark]

G–E

2 Car exhaust systems are fitted with catalytic converters.

(a) Why are cars fitted with a catalytic converter?

______ [1 mark]

(b) What environmental benefit is gained by using catalytic converters.

______ [1 mark]

(c) To reduce vehicle emissions, some cars are designed to use ethanol instead of petrol as a fuel.

Explain why ethanol is a green fuel. ______ [3 marks]

D–C

Crude oil

1 Crude oil is separated into useful substances such as petrol and diesel.

The table shows the boiling ranges of several of these substances.

substance	mean number of carbon atoms in molecules	boiling point range (°C)
liquid petroleum gas	3	-42 to -0.5
petrol	8	20 to 100
paraffin	13	170 to 240
diesel	17	240 to 330

G–E

(a) Why is a boiling range used, rather than a single boiling point? ____________ [1 mark]

(b) Which of the substances will be hardest to ignite?

____________ [1 mark]

(c) Which of the substances will have the least mean number of atoms in its molecules? ____________ [1 mark]

(d) How does the boiling range vary with mean number of carbon atoms?

____________ [1 mark]

2 Crude oil is a mixture of many different hydrocarbons.

D–C

(a) What is meant by *hydrocarbon*?

____________ [1 mark]

(b) Here is a diagram of how the crude oil is separated into more useful substances.

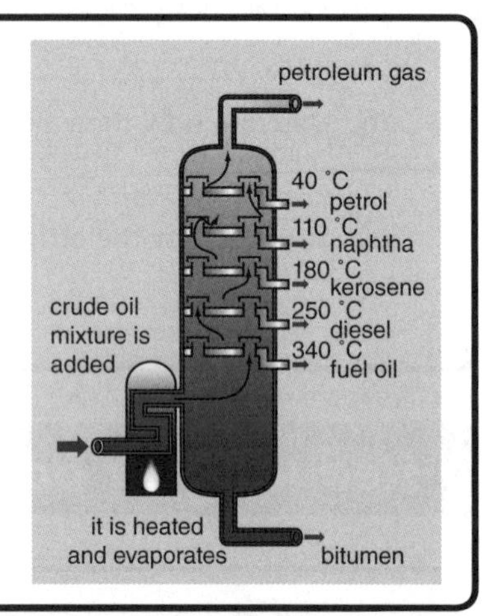

(i) Name the separation process. ____________ [1 mark]

(ii) Name the substance with the highest boiling point. ____________ [1 mark]

(iii) Name the substance with the lowest boiling point. ____________ [1 mark]

(c) Explain how the process separates crude oil into different useful substances.

____________ [4 marks]

Alkanes

1 Alkanes are hydrocarbon molecules where each atom is joined to other atoms by covalent bonds. Here is a diagram of a methane molecule.

```
   H
   |
H—C—H
   |
   H
```

G–E

(a) What is the formula of the methane molecule? ____________ [1 mark]

(b) How many bonds does a carbon atom make in methane? ____________ [1 mark]

(c) What is a covalent bond? ____________ [1 mark]

(d) Sketch a diagram of ethane (C_2H_6)

[1 mark]

2 Here is the general formula for an alkane. $C_n H_{2n+2}$

(a) Explain how you could use this to work out the formula of octane which has 8 carbon atoms.

____________ [2 marks]

D–C

(b) As alkanes have the same basic structure, they have similar properties that change as the molecule gets bigger. Explain how each of these properties changes as the alkane molecule gets bigger:

(i) flammability? ____________

(ii) boiling point? ____________

(iii) viscosity? ____________ [3 marks]

Cracking

1 Cracking converts long hydrocarbon chains into useful smaller chains such as petrol and diesel.

(a) Why is it important to crack large hydrocarbons to make more petrol and diesel?

____ [1 mark]

(b) As well as diesel and petrol, the cracking makes this molecule.

ethene

(i) What is the formula of ethene? _______________ [1 mark]

(ii) How is ethene different to ethane? _______________ [1 mark]

(iii) What type of useful substances are made from ethene? _______________ [1 mark]

G–E

2 Cracking hydrocarbon molecules is a very important part of an oil refinery's work. The longer alkane molecules are broken down into shorter molecules.

decane → octane + molecule X

(a) Write the formula of decane.

_______________ [1 mark]

(b) Name molecule X.

_______________ [1 mark]

(c) What does the = symbol mean in the structural diagram of molecule X? _______________ [1 mark]

(d) Why is molecule X a very useful chemical?

____ [2 marks]

Alkenes

1 The general formula for alkenes is C_nH_{2n}. Propene is an alkene, it has the formula C_3H_6.

(a) What is meant by a general formula? _______________ [1 mark]

(b) Butene has four carbon atoms. What formula would butane have? Explain your answer.

____ [2 marks]

(c) Alkenes have a double bond. What is a double bond?

____ [1 mark]

G–E

2 Ethene is the simplest alkene

(a) How are alkenes different from alkanes?

____ [1 mark]

(b) Use the diagram to explain why alkenes are more reactive than alkanes.

____ [2 marks]

(c) Alkenes are said to be unsaturated hydrocarbons. What do chemists mean by unsaturated?

____ [1 mark]

D–C

Making ethanol

G–E

1 Ethanol is a member of the alcohol family. It has the formula of C_2H_5OH.

(a) Draw the structural formula of ethanol. [1 mark]

(b) Ethanol can be produced by adding yeast to sugar dissolved in water.

(i) What do we call the process where yeast converts sugar to ethanol? ________________ [1 mark]

(ii) Why does the yeast and sugar solution have to be kept warm?

____ [1 mark]

D–C

2 Ethanol (C_2H_5OH) can be made by two methods: fermentation from sugars, or from ethene (C_2H_4) obtained from crude oil.

(a) What is added to ethene to produce ethanol? ________________ [1 mark]

(b) Ethanol made from sugars is a renewable biofuel. What is a biofuel?

____ [1 mark]

(c) Suggest two advantages and two disadvantages of using sugar to produce ethanol for use as a car fuel.

____ [4 marks]

Polymers from alkenes

G–E

1 The diagram shows how poly(propene) is made. The process uses a catalyst.

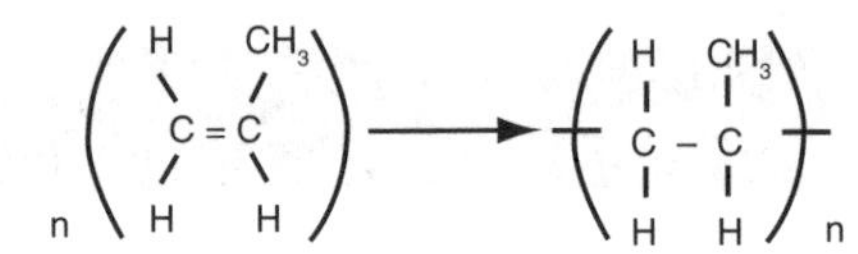

(a) What is the name given to large molecules like poly(propene)? ________________ [1 mark]

(b) In the diagram what does 'n' mean? ________________ [1 mark]

(c) Why is a catalyst used? ________________ [1 mark]

(d) Molecules like styrene can make polymers. Name the polymer made by styrene.

____ [1 mark]

2 Alkenes such as ethene can be used to make polymers such as poly(ethene).

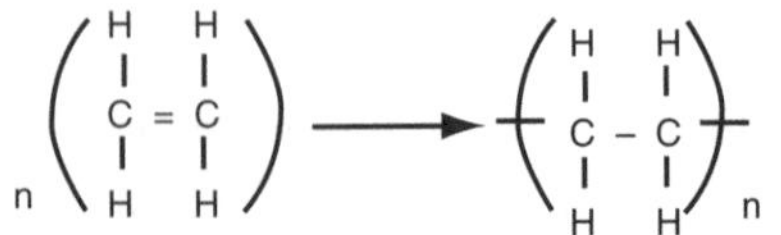

(a) What is meant by a polymer?

____ [1 mark]

(b) What word do we use to describe the small molecule used to make the polymer?

____ [1 mark]

(c) Explain how the ethene molecules join together to make poly(ethene).

____ [2 marks]

Designer polymers

1 Chemists have made a large number of different polymers.

(a) Suggest why we need a large number of different polymers. ______ [1 mark]

(b) Teflon is a very slippery polymer. It has many uses. One is to coat the inside of artificial hip joints.

(i) Explain why Teflon is used in a hip joint. ______ [1 mark]

(ii) What other property allows Teflon to be used safely inside your body? ______ [1 mark]

(iii) Artificial hip joints need replacing every 10 years. Suggest why the joint might need replacing.

______ [1 mark]

G–E

2 Plastic polymers are easily moulded into shape, and are low density (lightweight), waterproof and resistant to acids and alkalis. Increasingly, car and aircraft bodies are being made from designer plastic polymers such as carbon fibre-reinforced plastic.

(a) Give three reasons why these expensive plastics are being used in cars and aircraft.

______ [3 marks]

(b) Gore-tex® is a designer fabric, it lets water vapour through but not liquid water.
Explain why it is a good choice from which to make an outdoor coat.

______ [2 marks]

(c) Suggest why the use of designer polymers is likely to increase in the future.

______ [1 marks]

D–C

Polymers and waste

1 Most plastics and polymers are not biodegradable. When they are no longer needed and thrown away, this causes problems. Many are sent to landfill sites.

(a) What is meant by biodegradable? ______ [1 mark]

(b) Give two reasons why plastics and polymers should not be sent to landfill sites.

______ [2 marks]

(c) Plastics often have this type of symbol printed into them.

1
PET

(i) What does the symbol mean? ______ [1 mark]

(ii) The number in the middle of the symbol can be different. What does the number tell you?

______ [1 mark]

G–E

2 There are three ways to dispose of polymer waste. It can be sent to landfill sites, it can be burnt in incinerators or it can be recycled.

(a) There are problems with all three methods of disposal. Suggest a disadvantage of:

(i) putting the polymer waste in a landfill site. ______

(ii) incinerating the polymer waste. ______

(iii) recycling the polymer waste. ______ [3 marks]

(b) Evaluate which of the three methods of disposal provides the most effective use of the limited resources available to make polymers. Give reasons for your answer.

______ [3 marks]

D–C

Oils from plants

G–E

1 Vegetable oils are useful products from plants. They are used in foodstuffs, cosmetics and fuels.

Complete the flow chart by filling in the missing spaces to show how olive oil is produced, using these words:

dissolved **crushed** **liquid** **distillation** **filtered** [5 marks]

Olives are ________ → the ________ obtained is then ________ → the remaining pulp is then ________ in a solvent → the solvent is then removed from the olive oil by ________

D–C

2 Vegetable oils are oils that are made from plants such as sunflower, maize, olives, almonds and walnuts. They have many uses, such as in foodstuffs, cosmetics, and as fuels such as biodiesel.

(a) Why are plant oils suitable for use as fuels? ________ [1 mark]

(b) From which parts of the plant is the plant oil obtained? ________ [1 mark]

(c) Describe how the plant oil is obtained from the plant.

________ [2 marks]

(d) Cooking food in plant oils raises the energy content of the food, and changes the flavour of the food. Explain why cooking a potato in plant oil makes the potato taste different from when it is cooked in water.

________ [3 marks]

Biofuels

G–E

1 Sugar cane is a really useful biofuel plant. The sugar can be converted to ethanol (C_2H_5OH) which can be used instead of petrol as a green fuel. The leftover canes can be turned into a solid biofuel to burn to produce electricity.

(a) Biofuels are renewable. What does renewable mean? ________ [1 mark]

(b) Where does the energy stored in sugar cane come from? ________ [1 mark]

(c) Sugar cane is termed to be carbon neutral. What does carbon neutral mean?

________ [1 mark]

(d) Sugar cane contains no sulfur. Explain why power stations burning biofuels made from sugar cane produce no acid rain.

________ [1 mark]

D–C

2 Biofuels are fuels made from animal or plant materials. They are considered to be 'greener' than fossil fuels.

(a) Evaluate the economic and environmental benefits of using biofuels. Your answer should suggest two advantages and two disadvantages.

____

____

________ [6 marks]

(b) Plant oils are far more viscous than diesel. Describe how they can be modified to be used in diesel engines.

____

________ [2 marks]

Oils and fats

1 Animal fats are saturated fats. Plant oils are unsaturated fats. Animal fats are solid at room temperature. Plant oils are usually liquid at room temperature.

(a) What is meant by the term 'saturated fat'? ______________________ [1 mark]

(b) Explain why animal fats are solid at room temperature. ______________________ [1 mark]

(c) Spreadable butter is made by mixing saturated butter with an unsaturated vegetable oil.

(i) Why is it easier to spread? ______________________ [1 mark]

(ii) It is claimed to be healthier than pure butter. Suggest why.

______________________ [2 marks]

G–E

2 Low-fat spreads and margarines are made from plant oils. Butter is made from milk produced by cows. Some students tested butter, hard margarine, soft margarine and low-fat spread. They tested each spread for saturated fat by dissolving a sample of each in a little ethanol, adding 2 cm^3 of orange bromine water and timing how long the reaction took.

(a) What would you expect to see happen? ______________________ [1 mark]

(b) Explain why the test is not a fair one. ______________________ [2 marks]

Here are their results.

Spread	Mean time for the reaction to complete in seconds
butter	2
hard margarine	4
soft margarine	7
low-fat spread	8

(c) Which spread has the most saturated molecules present? Explain your answer.

______________________ [2 marks]

(d) Explain the difference between a 'saturated' fat and an 'unsaturated' fat.

______________________ [2 marks]

D–C

Emulsions

1 Salad cream is made from a mixture of egg yolk, vegetable oil, mustard and vinegar. It is an emulsion.

(a) What is meant by an emulsion? ______________________ [1 mark]

(b) Mustard is an emulsifier. Explain why it is important to add an emulsifier to salad cream.

______________________ [1 mark]

(c) Suggest why adding insufficient emulsifier could make it hard to sell the salad cream.

______________________ [2 marks]

G–E

2 In 2010, there was a large oil spillage off the coast of the USA between New Orleans and Florida. The crude oil did not mix with the seawater but floated on top of it. The crude oil and seawater were mixed together using a dispersant or detergent, to make an emulsion.

(a) What word do we use to describe two liquids that don't mix together? ______________________ [1 mark]

(b) What type of substance is the dispersant or detergent? ______________________ [1 mark]

(c) When making salad cream, egg yolk is added to a mixture of oil, water, vinegar and powdered mustard.

(i) Name the aqueous substances in the mixture. ______________________ [1 mark]

(ii) What is the purpose of the egg yolk? ______________________ [1 mark]

D–C

Earth

1 This is a diagram of the Earth's structure.

Complete the table below.

Part	Name	Its structure
A		solid rock
B		
C		

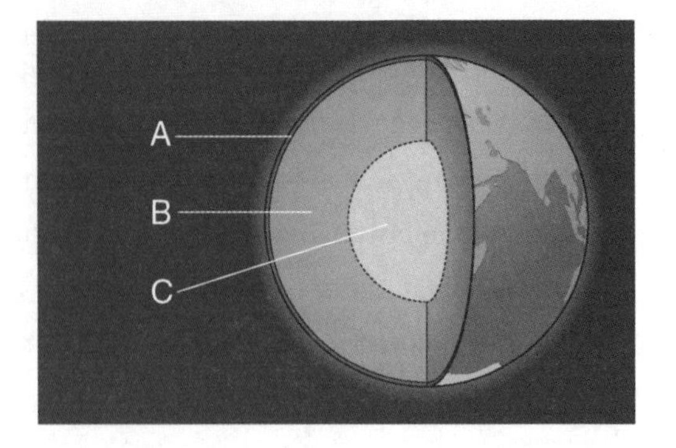

[5 marks]

Continents on the move

1 The diagram shows the continent of Pangea 200 million years ago. What happened to Pangea has been explained using the theory of continental drift.

(a) What is the theory of continental drift?

____

______________________________ [2 marks]

(b) Explain what has happened to Pangea over the last two hundred million years.

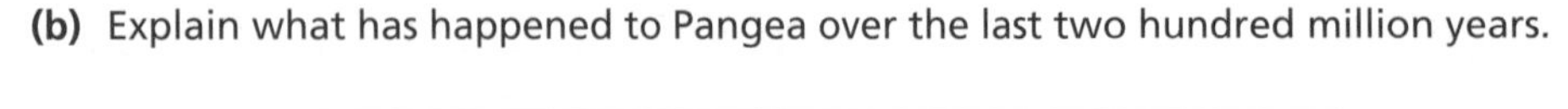

____

______________________________ [2 marks]

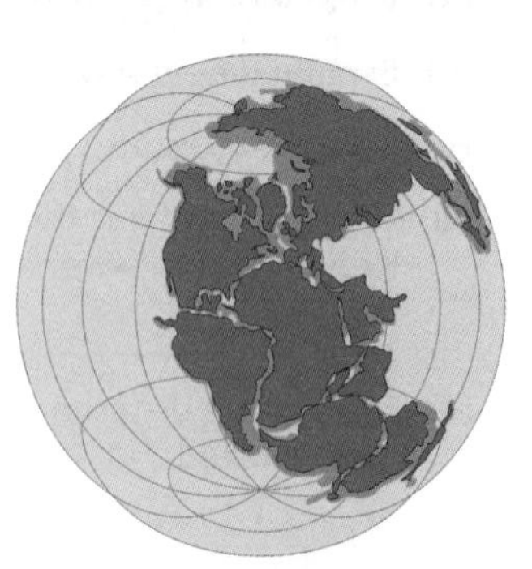

200 million years ago

2 Alfred Wegener suggested in 1915 that originally all the continents were joined together in one continent called Pangea. Most scientists thought he was wrong.

(a) Explain why most scientists disagreed with Wegener's theory.

____

______________________________ [2 marks]

(b) Wegener showed that South America could fit together with Africa like jigsaw pieces. Give two other pieces of evidence that were found which persuaded scientists to support his theory.

____

______________________________ [2 marks]

(c) New York is slowly moving away from London by a few centimetres each year. Describe how this is happening, using Wegener's theory.

____

______________________________ [2 marks]

Earthquakes and volcanoes

1 Every year Britain experiences over 200 earthquakes, but we don't usually feel them.

(a) Suggest why we don't notice Britain's earthquakes. ______________________ [1 mark]

(b) What causes an earthquake? ______________________ [2 marks]

(c) Explain why scientists find them very hard to accurately predict.

______________________ [2 marks]

G–E

2 This diagram shows the main tectonic plates of the Earth's crust.

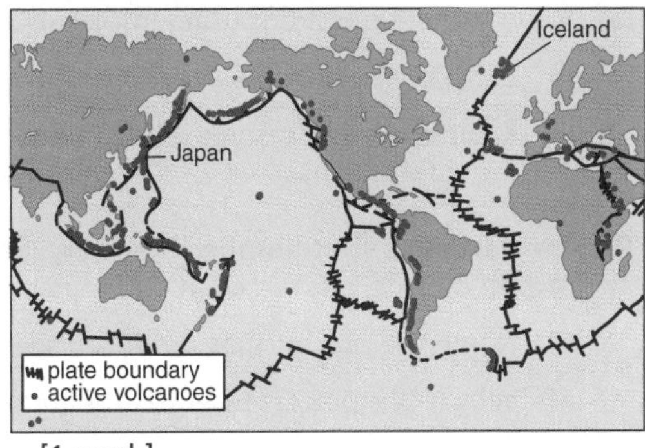

(a) What is a tectonic plate? ______________________ [1 mark]

(b) Where would you expect to find volcanoes?

______________________ [1 mark]

(c) Explain why volcanoes are found in these areas.

______________________ [2 marks]

(d) In Spring 2011, a tsunami devastated the east coast of Japan.

(i) What do we mean by a tsunami? ______________________ [1 mark]

(ii) How was the tsunami caused? ______________________ [2 marks]

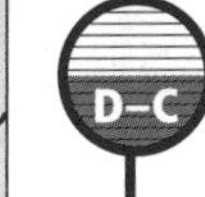

The air we breathe

1 This pie chart shows the composition of the air.

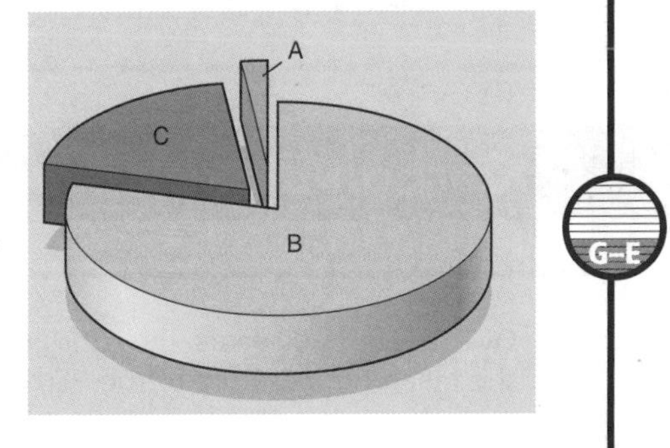

(a) Which letter represents nitrogen? ______________________ [1 mark]

(b) Which letter represents oxygen. ______________________ [1 mark]

(c) The third section is a mixture of mainly argon, carbon dioxide, and water vapour. The amount of carbon dioxide and water vapour varies.

(i) Suggest why the amount of carbon dioxide changes.

______________________ [1 mark]

(ii) Suggest why the amount of water vapour changes.

______________________ [1 mark]

G–E

2 The composition of the air has not always been the same since the Earth formed. It has changed.

(a) Where do we think the early atmosphere came from?

______________________ [1 mark]

(b) This graph shows how three gases in the air have changed since the Earth formed. Use information from the graph to help you answer these questions.

% of atmosphere by mass

CO_2 N_2 O_2

A B C D E

(i) How old is the Earth? ______________________ [1 mark]

(ii) Which letter, A to E, shows when plants were first able to photosynthesise? ______________________ [1 mark]

(iii) When did life start to live on the land? ______________________ [1 mark]

(c) Explain why life could exist only in the sea until ozone appeared in the atmosphere.

______________________ [2 marks]

D–C

The atmosphere and life

1 The table below has some information about three planets in the solar system.

	Venus	Earth
mean daytime temperature (°C)	470	18
main gases in the atmosphere	carbon dioxide sulfuric acid clouds	nitrogen oxygen
state of water on the planet	gas	liquid
living organisms	no	yes

(a) Answer these questions using information from the table.

(i) Explain why water is a liquid on Earth but not Venus. ____________________ [1 mark]

(ii) Explain why spaceships sent to Venus must be made of very unreactive metal.

____________________ [1 mark]

(b) The Earth's early atmosphere had very large amounts of carbon dioxide. Over millions of years the carbon dioxide was removed and replaced with oxygen.

(i) Name the process that produced the oxygen. ____________________ [1 mark]

(ii) Why is this process essential for animals? ____________________ [1 mark]

2 Primitive life first evolved about 3.4 billion years ago. No one knows how it evolved, although there are many different theories. Most of the theories suggest that life formed in the seas.

(a) Explain why scientists do not know how life began. ____________________ [1 mark]

(b) Suggest why most theories believe that life began in the seas. ____________________ [1 mark]

(c) Experiments to prove how life began usually concentrate on devising a method to make amino acids. Explain why. ____________________ [2 marks]

Carbon dioxide levels

1 Over the last 200 years, the concentration of carbon dioxide in the air has increased slightly. At the same time, the mean temperature of the Earth has risen by 1 °C. Most scientists think there is a link between these two facts.

G–E

(a) Why has the concentration of carbon dioxide in the air increased over the last two hundred years?

____________________ [1 mark]

(b) Give two possible effects of this global warming.

____________________ [2 marks]

2 The diagram below shows the carbon cycle.
Some of the processes have been replaced by letters.

(a) (i) Name process A. ____________________

(ii) Name process B. ____________________

(iii) Name process C. ____________________

(iv) Name process D. ____________________ [4 marks]

carbon dioxide in air
D
D
A
B
carbon in animals
carbon in plants
bio-fuels
fossil fuels
feeding
C
waste and death
carbon in decomposers
with air
without air

(b) Explain why process D appears twice. ____________________ [1 mark]

(c) Explain why burning biofuels is said to be 'carbon neutral'. ____________________ [2 marks]

Extended response question

Biofuels are fuels made from plant materials. In the UK in 2010, petrol and diesel had to contain at least 5% biofuel.

- Petrol and diesel made from crude oil are non-renewable energy sources.
- Biofuels are renewable sources of energy.
- Burning petrol, diesel and biofuels releases carbon dioxide into the atmosphere.
- Carbon dioxide is thought to cause global warming.
- Much of the carbon dioxide released into the atmosphere dissolves in the sea.

Use the information above, and your knowledge and understanding, to give the positive and negative environmental impacts of increasing the percentage of biofuels in petrol and diesel.

The quality of written communication will be assessed in your answer to this question.

[6 marks]

Energy

G–E

1 Explain where the energy is transferred to from:

(a) a moving car ______ [1 mark]

(b) a hot bath ______ [1 mark]

(c) a stretched bow just as an arrow is released.

______ [1 mark]

D–C

2 (a) Describe the energy transfers taking place when a hairdryer is turned on.

______

______

______ [4 marks]

(b) Where is the energy finally transferred to? ______ [1 mark]

(c) A student investigates how quickly hairdryers transfer energy by timing how long it took for each hairdryer to dry a piece of damp cloth. Here are their results:

Model of hairdryer	Time needed to dry damp cloth
Traveller	2 minutes 30 seconds
Olympus1000	1 minute 20 seconds
Hairdry	2 minutes 10 seconds

Explain which hairdryer transferred energy quickest.

______ [2 marks]

(d) State two variables that the student should control during this investigation.

______ [2 marks]

Infrared radiation

G–E

1 Two cars are parked on a hot day in a car park.

(a) How does energy from the Sun reach the cars? ______ [1 mark]

(b) Explain why the black car heats up faster than the white car.

______ [2 marks]

D–C

2 Solar panels are fitted on house roofs and are used to heat water.

(a) Explain which is the best colour for the manufacturer to choose for the solar panels.

______

______ [3 marks]

(b) Why are solar panels normally fitted to south-facing roofs in the UK?

______ [2 marks]

(c) Solar panels should be insulated to avoid heat losses at night or during winter. Explain why.

______

______ [3 marks]

Kinetic theory

1 A bowl of ice cream is left at room temperature and melts.

(a) What change of state occurs? ______________________________ [1 mark]

(b) How you can tell that the state of matter has changed?

______________________________ [1 mark]

G–E

2 (a) Explain what melting means.

______________________________ [1 mark]

(b) A piece of ice left on a plate starts to melt. Explain the process of melting by writing about particles.

____

______________________________ [4 marks]

(c) Why do the plate and its surroundings cool down when ice melts?

______________________________ [2 marks]

(d) Explain why the ice melts more quickly in a warmer room.

______________________________ [1 mark]

D–C

Conduction and convection

1 A plastic spoon and a metal spoon are left in a hot drink.

(a) How is heat transferred through the spoons?

______________________________ [1 mark]

(b) Why does the metal spoon get hot quicker than the plastic spoon?

______________________________ [2 marks]

(c) The air above the drink warmed more quickly than the air around the sides of the drink. Explain why.

______________________________ [2 marks]

G–E

2 (a) Explain why convection cannot take place in solids.

______________________________ [2 marks]

(b) Use your ideas about heat transfers to explain how the thermos flask keeps a drink hot.

____

____

____

____

____

______________________________ [6 marks]

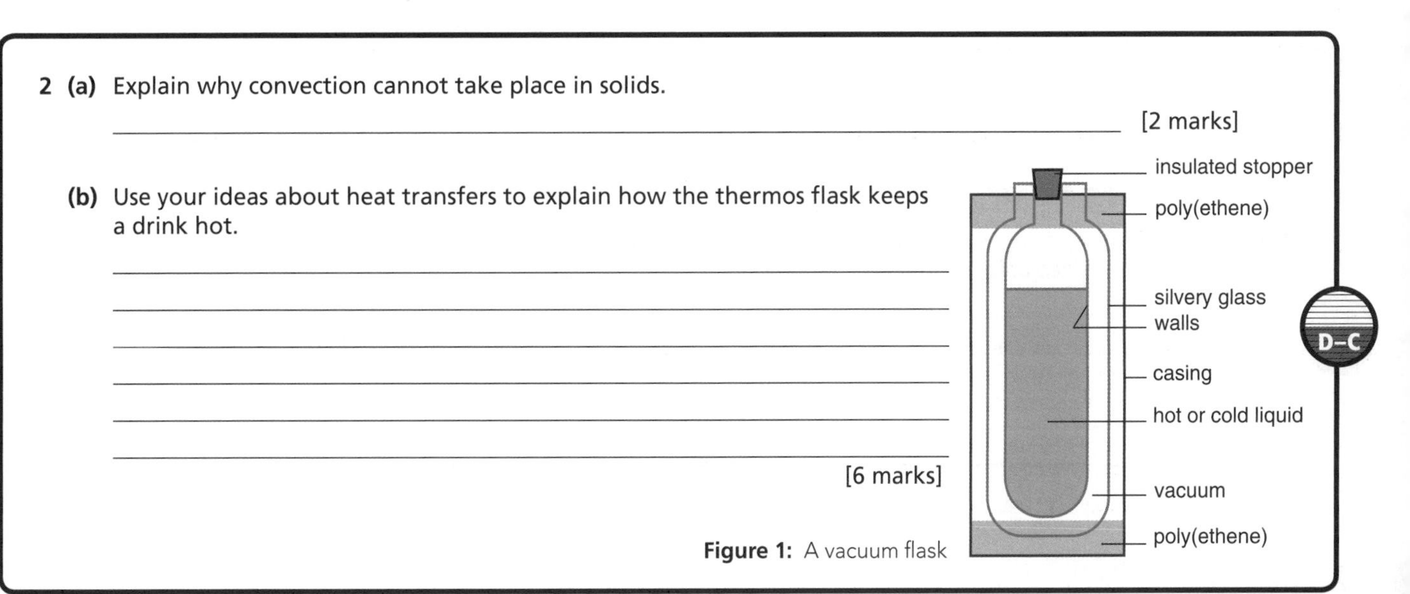

Figure 1: A vacuum flask

D–C

Evaporation and condensation

G–E

1 (a) Write down the differences between evaporation and condensation.

____

______________________________ [2 marks]

(b) In the morning, dew can be found on surfaces outside. The moisture originally came from the surrounding air. What process causes dew to form?

______________________________ [1 mark]

D–C

2 Suna hangs her washing on a washing line outside.

(a) What is meant by evaporation?

______________________________ [1 mark]

(b) Write down two differences between evaporation and boiling.

____

______________________________ [2 marks]

(c) Write down two weather conditions that make evaporation take place more quickly. Give a reason why each weather condition affects the rate of evaporation.

(i) Condition 1______________

Why it increases the rate of evaporation:

____

______________________________ [2 marks]

(ii) Condition 2______________

Why it increases the rate of evaporation:

____

______________________________ [2 marks]

Rate of energy transfer

G–E

1 Hot soup is served at lunchtime. It is important that the soup cools down slowly. Write down three features of a container that would keep the soup warm for as long as possible.

____

____

______________________________ [3 marks]

D–C

2 When a potato is cut into smaller pieces, it cooks more quickly. Explain why this happens using these terms in your answer: surface area; conduction; radiation.

____

____

____

______________________________ [4 marks]

Insulating buildings

1 An architect is choosing materials for a house extension.

(a) Why should he choose materials with a low U-value?

______________________________ [2 marks]

G–E

(b) Explain why double-glazing has a lower U-value than a single, thick pane of glass.

______________________________ [2 marks]

2 A homeowner paid for an energy survey and report for their home. The report suggested several ways to reduce energy losses in the home.

(a) Suggest one reason why the homeowner was prepared to pay for a report.

______________________________ [1 mark]

Here are four suggestions made by the consultant.

Change	Estimated cost	Estimated annual saving
Change all light bulbs to energy efficient bulbs	£80	£95
Increase loft insulation to a depth of 15 cm	£200	£120
Set central heating thermostat on to a lower temperature setting	£0	£40
Put the central heating onto an automatic timer switch	£15	£20

D–C

All of these measures are put in place in the first year.

(b) Write down the overall savings over five years. ______________________________ [3 marks]

(c) Loft insulation reduces heat losses because it slows down convection and conduction. Explain whether it has a high or low U-value.

______________________________ [2 marks]

(d) State two other methods to reduce heat losses in the home. Explain how each method reduces heat losses.

____

______________________________ [4 marks]

Specific heat capacity

1 Water has a large specific heat capacity. Oil has a low specific heat capacity.

(a) Will water or oil reach a higher temperature if they absorb the same amount of heat? ______ [1 mark]

G–E

A coolant is used to absorb heat from engines and keep them cool.

(b) Explain whether water or oil is the best choice for a coolant.

______________________________ [2 marks]

2 **(a)** Explain what is meant by specific heat capacity. ______________________________ [1 mark]

(b) The specific heat capacity of aluminium is 900 J/kg °C. Calculate the temperature rise when a 250 g block of aluminium is supplied with 1800 J of energy. Write down the equation you use, then clearly show how you calculate your answer.

D–C

____

______________________________ [3 marks]

Energy transfer and waste

G–E

1 (a) Whenever anything happens, energy is conserved. What does this mean about the energy before and afterwards?

____

______________________________ [2 marks]

(b) How does a Sankey diagram show conservation of energy?

____

______________________________ [2 marks]

D–C

2 The picture shows a Sankey diagram for a petrol engine.

(a) Use the diagram to write down the energy equation for the petrol engine.

______________________________ [2 marks]

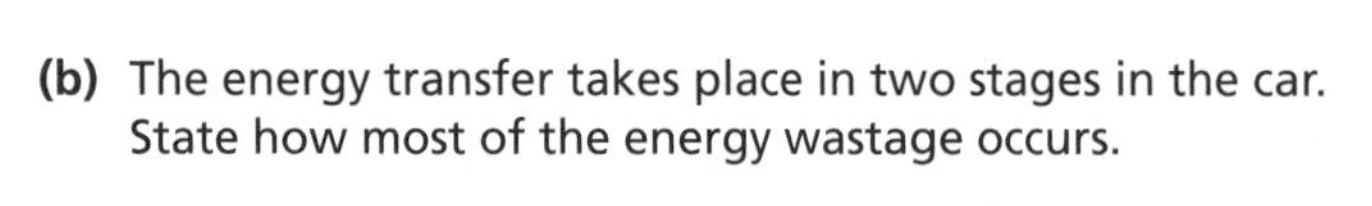

(b) The energy transfer takes place in two stages in the car. State how most of the energy wastage occurs.

______________________________ [1 mark]

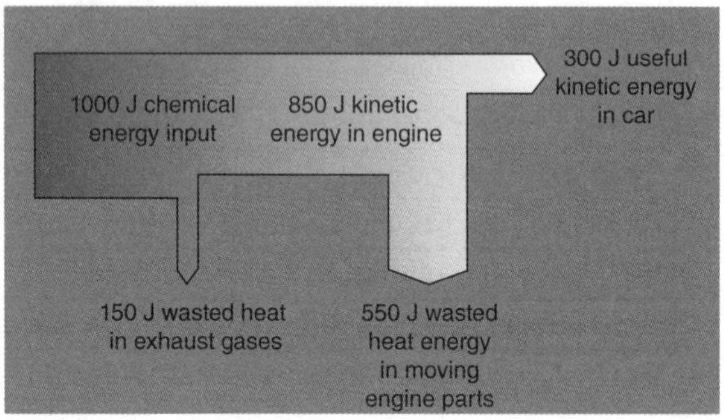

1000 J chemical energy input
850 J kinetic energy in engine
300 J useful kinetic energy in car
150 J wasted heat in exhaust gases
550 J wasted heat energy in moving engine parts

Figure 1: Sankey diagram for a petrol engine

(c) Describe how conservation of energy is shown on a Sankey diagram.

____

____

______________________________ [3 marks]

Efficiency

G–E

1 A shopkeeper tells a customer that one electric drill is more efficient than a different electric drill. They both cost the same.
Why should the customer buy the more efficient drill?

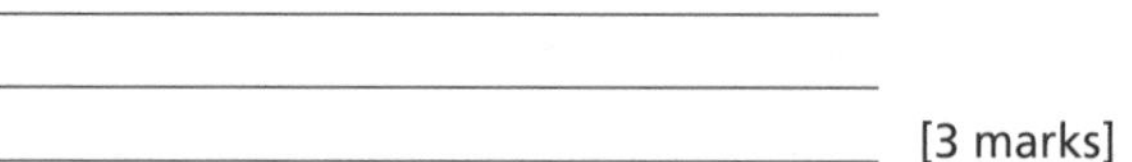

____

____

______________________________ [3 marks]

D–C

2 The diagram shows the energy transfer for an electric fan.

(a) Use the Sankey diagram to calculate the efficiency of the electric fan.

____

______________________________ [2 marks]

(b) Explain whether all new electrical appliances should display Sankey diagrams for people to compare their energy efficiency.

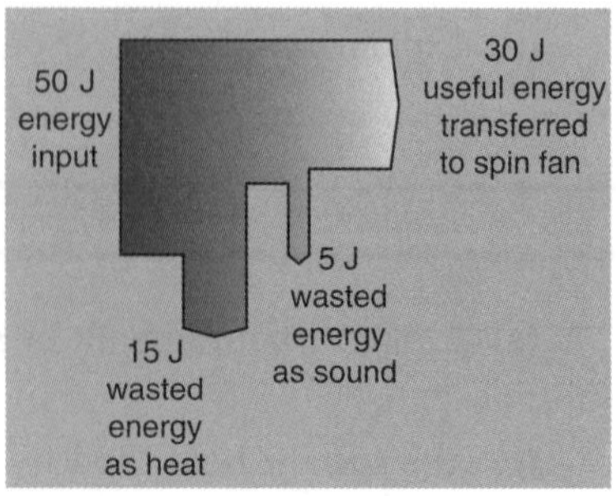

____

____

______________________________ [3 marks]

50 J energy input
30 J useful energy transferred to spin fan
5 J wasted energy as sound
15 J wasted energy as heat

Figure 2: Sankey diagram for an electric fan

Electrical appliances

1 Match the phrases A, B, C and D with the numbers 1–4 in the sentences.

A – store energy B – chemical C – different energy forms D – do not store lots of energy

Electrical energy has many uses because it can be changed into ___1___
Batteries are used where there is no mains electricity because they ___2___
Batteries are not used in electric heaters because they ____3____
Batteries store ____4___ energy [4 marks]

G–E

2 A company has designed a wind-up mobile phone battery recharger.

(a) Describe the useful energy changes that take place when a person winds up the recharger and charges up the battery.

________________ -> ____________________ -> ____________________ [3 marks]

(b) Explain one advantage of;

(i) using a wind-up battery recharger

____

____ [2 marks]

(ii) using a battery in the mobile phone.

____

____ [2 marks]

D–C

Energy and appliances

1 An electric kettle has a power of 2.2 kW.

(a) How much energy does it transfer each second?

____ [1 mark]

(b) Why does a more powerful kettle heat 500 ml of water quicker than a less powerful kettle?

____

____ [2 marks]

G–E

2 Calculate the energy transferred when an 850 W microwave oven is switched on for 15 minutes.

____

____

____ [3 marks]

D–C

The cost of electricity

1 An oven is switched on for 2 hours. The power of the oven is 3 kW. How much energy does the oven transfer?

____

____ [2 marks]

2

	Filament bulb	Energy efficient bulb
Power	100 W	16 W
Cost when new	30p	£5.00
Lifetime	1 year	5 years

(a) Each bulb is used for 1500 hours per year. Use this information to calculate the energy used by the filament bulb in a year in kilowatt-hours.

____ kWh [2 marks]

(b) If each kilowatt-hour costs 15p, calculate the total cost of using the filament bulb over five years.

____

____ [3 marks]

(c) The equivalent cost if energy efficient bulbs are used is less than £19. Explain why the use of energy efficient bulbs is being encouraged for environmental reasons.

____

____ [2 marks]

Power stations

1 Match the parts of a power station, A, B, C and D, with the job that it does (1–4).

Part of power station: A – boiler, B – turbine, C – generator, D – step-up transformer

Job:

1 – make the generator spin ______________________

2 – increase the voltage ______________________

3 – change kinetic energy into electric energy ______________________

4 – heat water to create steam ______________________ [4 marks]

2 **(a)** Describe the useful energy transfers taking place in a gas-fired power station.

____

____

____ [4 marks]

(b) Explain why a combined heat and power station is more efficient than a power station that just generated electricity.

____

____

____ [4 marks]

Renewable energy

1 (a) Explain why wind turbines do not always generate electricity.

______________________________ [1 mark]

(b) Explain why some hydroelectric power stations cannot be used all the time.

______________________________ [1 mark]

G–E

(c) Write down one disadvantage of using biomass power stations.

______________________________ [1 mark]

2 Both Iceland and Norway use renewable sources of energy to generate electricity. Iceland is a volcanic island and Norway is a mountainous country with many valleys and fast flowing rivers. Suggest which is the most suitable renewable energy resource for each country to use to generate electricity, explaining your answers.

____

____

____

______________________________ [4 marks]

Electricity and the environment

1 People living on a small island use renewable energy sources only.

(a) Write down three renewable energy sources.

____

____

______________________________ [3 marks]

G–E

(b) Write down one advantage of using renewable energy sources on the island.

______________________________ [1 mark]

(c) Write down one disadvantage of using renewable energy sources on the island.

______________________________ [1 mark]

2 There are plans to build a new power station near a large city. The choice is between a coal fired power station or wind turbines set on nearby hills in a local beauty spot.

Give one environmental advantage and one environmental disadvantage for each of these schemes.

Type of station	Advantage of the scheme	Disadvantage of the scheme
Coal fired		
Wind turbines		

[4 marks]

Making comparisons

1 Our power stations are getting older.

(a) Write down one advantage and one disadvantage of replacing the older power stations with newer, more efficient power stations.

Advantage: ______________________________

Disadvantage: ______________________________ [2 marks]

G–E

(b) Write down three things to consider before choosing which new power station to build.

____

____

______________________________ [3 marks]

2 (a) Explain two factors that affect the demand for electricity during the day.

____

____

____

______________________________ [3 marks]

D–C

(b) A reliable energy source generates electricity whenever it is needed.
Put these energy sources in order of reliability, starting with the least reliable first.
gas, wind, waves, tidal

______________________________ [3 marks]

The National Grid

1 Step-up transformers are used in the National Grid.

(a) What does a step-up transformer do?

______________________________ [1 mark]

G–E

(b) Why are step-up transformers used in the National Grid?

____

______________________________ [2 marks]

2 The National Grid distributes electricity throughout the UK. Overhead high voltage transmission lines transmit electricity across the country at voltages up to 400 000 V. Distribution lines transmit electricity at voltages up to 33 000 V throughout an area.

(a) Why is less energy wasted when electricity is transmitted at higher voltages?

____

______________________________ [2 marks]

D–C

(b) Calculate the current in a high voltage transmission line. The power of the line is 200 000 000 W.

____

____

______________________________ [3 marks]

What are waves?

1 Waves can be electromagnetic or mechanical

(a) What do both types of wave transfer?

______________________________ [1 mark]

(b) Write down one difference between these types of wave.

______________________________ [1 mark]

G–E

(c) Write down whether each type of wave is mechanical or electromagnetic.

(i) sound ______________

(ii) light ______________

(iii) gamma ______________

(iv) water ______________ [4 marks]

2 Describe two differences and two similarities between transverse waves and longitudinal waves.

Difference 1 ______________________________

Difference 2 ______________________________

Similarity 1 ______________________________

Similarity 2 ______________________________ [4 marks]

D–C

Changing direction

1 (a) What effect is the diagram showing ?

______________________________ [1 mark]

(b) Why is this effect useful when broadcasting radio programmes?

____

______________________________ [2 marks]

G–E

2 Two people had a conversation in a room. The door was open and the conversation could be clearly heard on the other side of the wall. Use your ideas of diffraction to explain why.

____

____

______________________________ [3 marks]

D–C

Sound

1 Complete the sentences using the words:
amplitude, frequency

(a) A loud, low-pitched rumble has a large ________________ and a small ________________

(b) The quiet high-pitched whistle has a small ______________ and a large ________________ [4 marks]

2 (a) A musician plays a note with a frequency of 440 Hz. Another musician plays a slightly higher pitched note. Write down a possible frequency of this note.

____ [1 mark]

(b) How does the wavelength of a sound change when it becomes higher pitched?

____ [1 mark]

(c) What changes when a sound becomes louder?

____

____ [2 marks]

Light and mirrors

1 Alisha shines a torch at a mirror.

(a) What happens to the light when it reaches the mirror?

____ [1 mark]

(b) How can you tell what direction the light will travel after it reaches the mirror?

____

____ [2 marks]

2 Describe how the image of a child in a plane mirror compares with the real child by choosing the correct words.

The image is *upright/inverted.*
It is *real/virtual*.
It is the *same distance behind/further behind* the mirror than the object is in front. [3 marks]

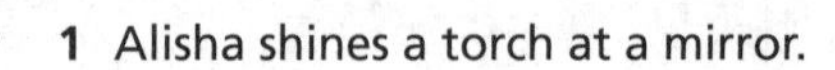

Using waves

1 Write down two things that are the same for all electromagnetic waves.

____

____ [2 marks]

2 Match the type of electromagnetic wave with these uses.

Type of electromagnetic radiation	Use
Radio wave	Photography
Microwave	Remote control
Infrared radiation	BBC TV broadcasts
Visible light	Satellite TV

[4 marks]

The electromagnetic spectrum

1 The table shows the members of the electromagnetic spectrum. Fill in the names missing on the diagram.

Radio waves	Microwaves	A	B	Ultraviolet	C	Gamma rays

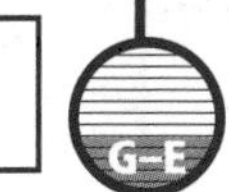

[3 marks]

2 Radio waves can be detected from galaxies, such as the Milky Way. If the frequency of radio waves is measured as 2 700 000 kHz, calculate the wavelength of the radiation. Electromagnetic waves travel at 300 000 km/s

____

____

____

____ [4 marks]

Dangers of electromagnetic radiation

G–E

1 Match the words in A, B, C and D with the numbers 1–4 in the sentences.

A absorb, B burn, C ionise, D kill

Cells __1__ many types of radiation that can damage them.
High doses of gamma rays __2__ cancer cells.
Low doses of X-rays and gamma rays __3__ cells, damaging them.
Infrared radiation and microwaves __4__ cells. [4 marks]

D–C

2 Rachel put sun block on her skin before she went out in the Sun.

(a) What type of electromagnetic radiation does sun block protect her from?

______________________________ [1 mark]

(b) Write down three forms of harm caused by too much exposure to this type of radiation.

____
____
______________________________ [3 marks]

(c) Write down two other ways that Rachel could reduce her exposure to radiation from the Sun.

____
____
____
______________________________ [4 marks]

Telecommunications

G–E

1 Which type of electromagnetic waves are used for:

(a) mobile phone calls ______________________________

(b) satellite TV broadcasts ______________________________

(c) signals sent through fibre optic cables. ______________________________ [3 marks]

D–C

2 The Olympic Games brings together sportsmen and women from all over the world. Events from the Olympics are transmitted to audiences worldwide. Explain how electromagnetic waves can be used to allow viewers to see live transmissions of the London Olympics in Australia.

____
____
______________________________ [3 marks]

Cable and digital

1 **(a)** Write down one difference between digital and analogue signals.

____ [1 mark]

(b) Describe how digital signals travel through optical fibres.

____

____ [2 marks]

G–E

2 **(a)** What is the name given to the type of signal shown in figure A?

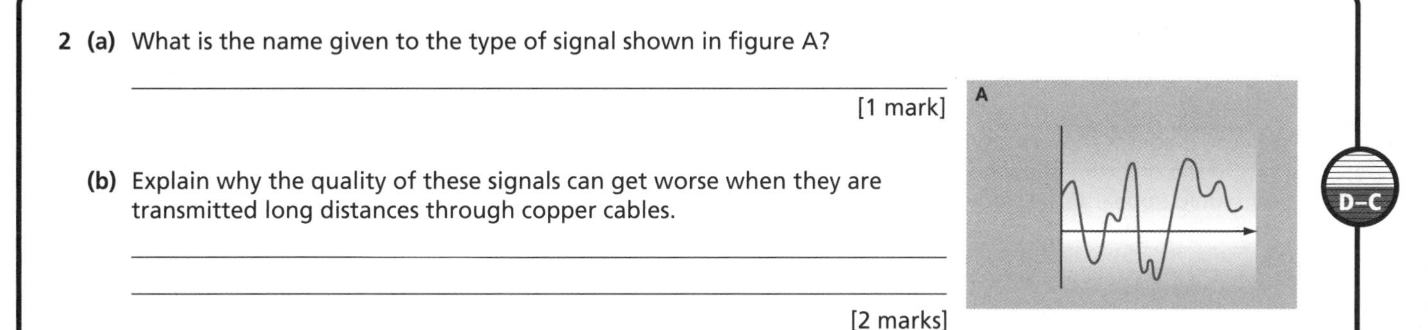

____ [1 mark]

(b) Explain why the quality of these signals can get worse when they are transmitted long distances through copper cables.

____

____ [2 marks]

D–C

Searching space

1 Describe two reasons why astronomers need telescopes to find out more about the Solar System.

____

____ [2 marks]

G–E

2 **(a)** Use the table to explain why we need different telescopes to detect different objects in space.

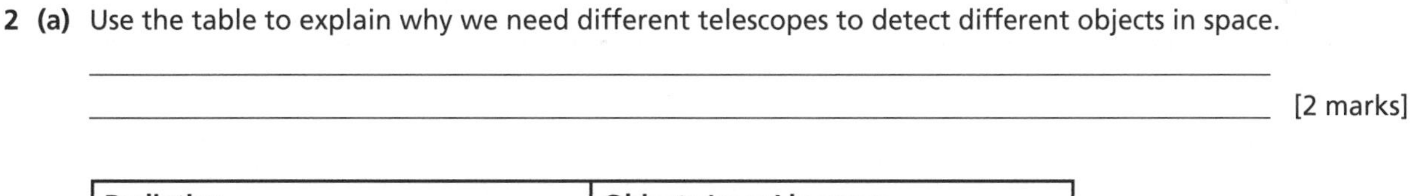

____

____ [2 marks]

Radiation	Objects 'seen' in space
gamma ray	neutron stars
X-ray	neutron stars
ultraviolet	hot stars, quasars
visible	stars
infrared	red giants
far infrared	protostars, planets
radio	pulsars

D–C

(b) Telescopes that detect radiation with long wavelengths need large receiving dishes because the radio signals from objects in space are very weak. Explain whether a radio telescope is larger or smaller than an optical telescope that detects visible light.

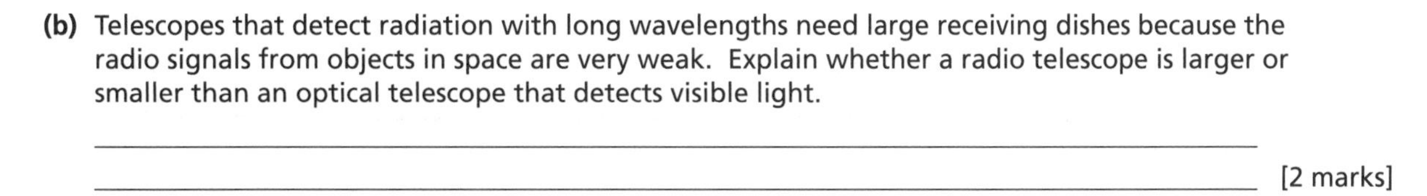

____

____ [2 marks]

Waves and movement

G–E

1 (a) Choose the correct words to complete the sentences:
When a police car drives towards you, the pitch of its siren is higher/lower because the wavelength of the signal is longer/shorter.

[2 marks]

(b) What happens to the pitch of the note if the car slows down?

______________________________ [1 mark]

D–C

2 (a) When a car passes a person, its sound appears to change pitch. What is the name of this effect?

______________________________ [1 mark]

(b) The diagram shows sound waves coming from a car. On the diagram, show how the sound wave would appear if the car was travelling away from the person.

[2 marks]

Origins of the Universe

G–E

1 The pattern of light from distance stars has changed, and is redder than before.

(a) What is this effect called?

______________________________ [1 mark]

(b) The effect is greater when the stars are further away. What does this tell us about more distant stars?

____

______________________________ [2 marks]

D–C

2 Scientists have developed several models of the start of the Universe. One model is the Steady State theory. This theory describes a universe that is constantly expanding. Matter is being constantly created. The Big Bang theory is another theory that is more widely accepted.

(a) Describe the Big Bang theory of the creation of the Universe.

____

____

______________________________ [3 marks]

(b) Describe the evidence that supports the Big Bang theory.

____

____

______________________________ [3 marks]

Extended response question

Electromagnetic waves are used for communication. Explain why different electromagnetic waves are used to communicate in different situations.

The quality of written communication will be assessed in your answer to this question.

[6 marks]

Animal and plant cells

G–E

1 You were formed when the nucleus from a sperm and egg fused together to produce a zygote.

(a) Name a cell mentioned in the sentence above.

______________________________ [1 mark]

(b) Name a cell part (organelle) mentioned.

______________________________ [1 mark]

D–C

2 **(a)** What type of cell is shown in the diagram?

______________________________ [1 mark]

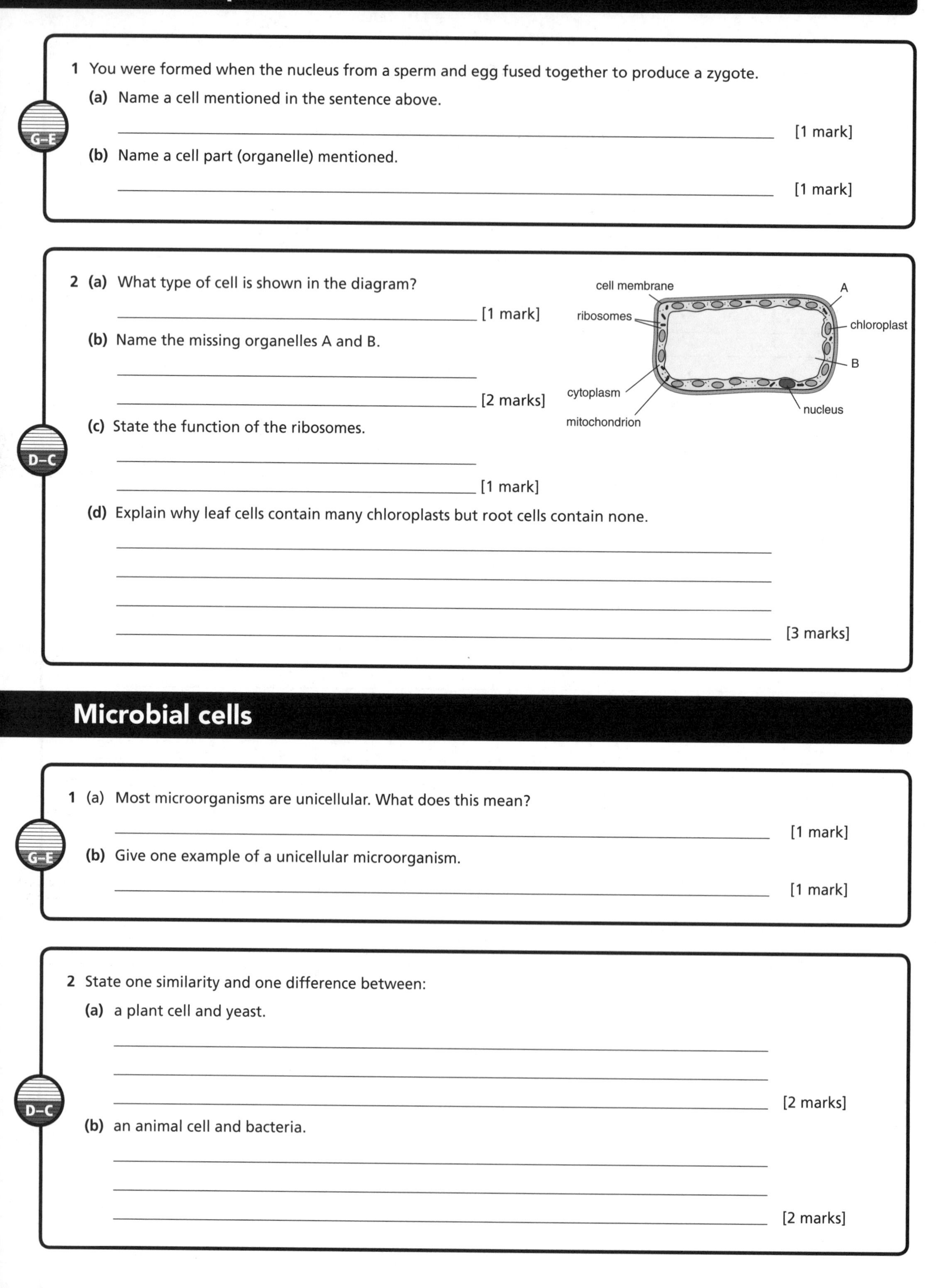

(b) Name the missing organelles A and B.

____

______________________________ [2 marks]

(c) State the function of the ribosomes.

____

______________________________ [1 mark]

(d) Explain why leaf cells contain many chloroplasts but root cells contain none.

____

____

____

______________________________ [3 marks]

Microbial cells

G–E

1 (a) Most microorganisms are unicellular. What does this mean?

______________________________ [1 mark]

(b) Give one example of a unicellular microorganism.

______________________________ [1 mark]

D–C

2 State one similarity and one difference between:

(a) a plant cell and yeast.

____

____

______________________________ [2 marks]

(b) an animal cell and bacteria.

____

____

______________________________ [2 marks]

Diffusion

1 Complete the sentence by filling in the gaps with the correct words.

Diffusion is the spreading out of ______________________ from a ______________________ concentration to a ______________________ concentration. [3 marks]

G–E

2 A cake is cooking in the kitchen. Use ideas about diffusion to explain how the smell travels around the house.

____

____

____

____

____ [4 marks]

D–C

Specialised cells

1 (a) Sperm cells are specialised. What does this mean?

____

____

____ [1 mark]

(b) Bacteria do not have specialised cells. Why not?

____

____

____ [2 marks]

G–E

2 State one adaptation of a red blood cell and explain why it has this adaptation.

____

____

____ [2 marks]

D–C

Tissues

G–E

1 Classify each of these as a cell, tissue or organ.

(a) muscle

______________________________ [1 mark]

(b) neurone

______________________________ [1 mark]

(c) sperm

______________________________ [1 mark]

(d) stomach

______________________________ [1 mark]

D–C

2 (a) What type of tissue is shown in the diagram?

______________________________ [1 mark]

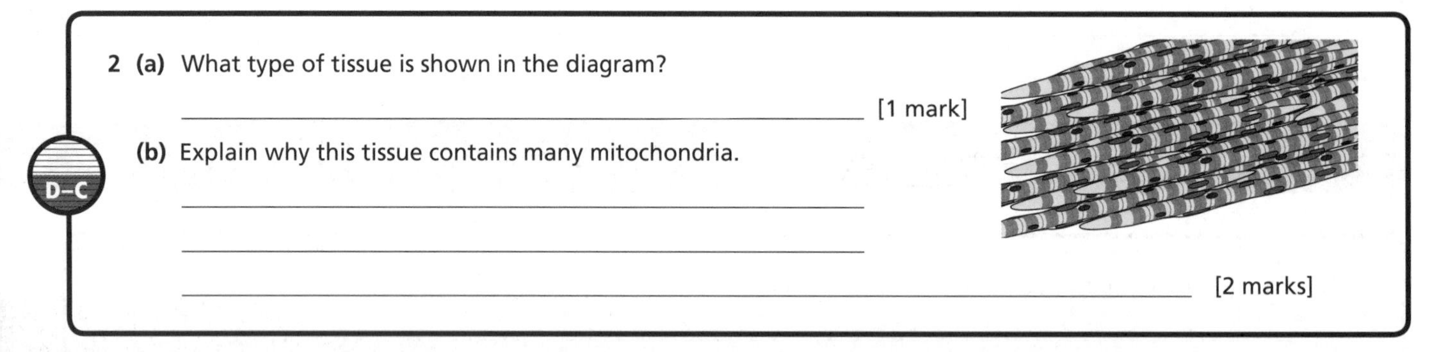

(b) Explain why this tissue contains many mitochondria.

____

____

______________________________ [2 marks]

Animal tissues and organs

G–E

1 Explain why the skin is classed as an organ.

____

____

______________________________ [1 mark]

D–C

2 Match the digestive system organ to its function.

	Organ		Function
1	stomach	A	contractions of muscle in walls mixes food with digestive juices
2	liver	B	produces digestive juices
3	small intestine	C	is where water is absorbed from the undigested food, producing faeces
4	pancreas	D	produces bile
5	large intestine	E	is where the absorption of soluble food occurs

______________________________ [5 marks]

Plant tissues and organs

1 State one function of each of the following plant organs.

(a) leaf ______________________________ [1 mark]

(b) root ______________________________ [1 mark]

(c) stem ______________________________ [1 mark]

G–E

2 The diagram shows a cross-section of a leaf.

(a) State the function of each of these tissues:

(i) upper epidermis

____

______________________________ [2 marks]

(ii) mesophyll.

______________________________ [1 mark]

(b) Name the tissue that carries water to the leaves.

______________________________ [1 mark]

D–C

Photosynthesis

1 A rose is a plant which carries out photosynthesis.

(a) Why is photosynthesis an essential process for the rose?

______________________________ [1 mark]

(b) The flowers of the rose are white. Can the petal cells carry out photosynthesis? Give a reason for your answer.

____

______________________________ [2 marks]

G–E

2 Tyrone kept some geranium plants in a dark cupboard for two days. He then added iodine to their leaves. Iodine turns black if it comes into contact with starch.

(a) Predict what he would see.

______________________________ [1 mark]

(b) Explain the reasons behind your prediction.

____

____

______________________________ [3 marks]

D–C

(c) He also tested the leaves of some variegated (green and white) geranium leaves that had been kept in a sunny place. This is what he saw.
Explain why he got these results.

went black

stayed orange

____

____

______________________________ [4 marks]

Limiting factors

G–E

1 Plants grow faster in the summer than in the winter. Explain one reason why this is so.

___

___

___ [2 marks]

D–C

2 An experiment was set up as shown in the diagram. The lamp was moved towards the plant and the average number of bubbles released by the plant per minute were counted, averaged and recorded at each distance. The results are shown in the table below.

collected gas
inverted test tube
bubbles of gas
beaker
water
inverted funnel
water–weed

(a) Name the gas in the bubbles.

___ [1 mark]

(b) Use the results to describe how light intensity affects the rate of photosynthesis.

___ [1 mark]

(c) Describe what happens to the rate of photosynthesis once the lamp gets closer than 15 cm away.

___ [1 mark]

(d) Explain why this has happened.

___

___

___ [2 marks]

Distance between the lamp and plant (cm)	50	45	40	35	30	25	20	15	10	5
Number of bubbles of oxygen released per minute	1	2	5	10	16	32	54	56	56	56

The products of photosynthesis

G–E

1 (a) Choose the correct word from the list below to fill in the gaps.

Glucose is a store of ___ Plants carry out a reaction called ___ which releases it. [2 marks]

energy photosynthesis respiration sugar

(b) Plants can convert glucose into different substances. Name a substance which:

(i) makes cell walls ___

(ii) is a store of energy ___

(iii) is needed to make new cytoplasm ___ [3 marks]

D–C

2 Peas are high in protein. Explain how the pea plant makes this protein.

___

___

___ [3 marks]

Distribution of organisms

1 State two reasons why many animals cannot live at the top of tall mountains.

[2 marks]

G–E

2 Explain why the distribution of:

(a) plants in a cave is low

[2 marks]

D–C

(b) egg wrack seaweed at the top of a beach, far from the sea, is low.

[2 marks]

Using quadrats to sample organisms

1 (a) Why do scientists use samples to study the distribution of organisms living in a habitat?

[1 mark]

G–E

(b) State one sampling technique.

[1 mark]

2 Rory wanted to estimate the number of snails on his lawn. He threw a quadrat onto it. You can see the result on the right.

(a) Rory's lawn has an area of 2 m^2. Estimate how many snails there are on the entire lawn.

[2 marks]

(b) Explain how he could increase the validity of his results.

[2 marks]

0.5m

0.5m

D–C

Proteins

1 State **two** functions of proteins in the body.

____

____

____ [2 marks]

G–E

2 The image below represents part of a protein molecule.

D–C

(a) What are the individual units called?

____ [1 mark]

(b) Why is it important that these units are arranged in the correct order?

____

____ [2 marks]

Enzymes

1 Choose the correct words below to fill in the gaps. [4 marks]

Enzymes are ______________ molecules which ______________

chemical reactions inside living organisms. They are biological ______________ .

They are ______________ . This means they each only control one type of reaction.

carbohydrate	catalysts	fussy	protein	slow down	specific	speed up

G–E

2 Emma carried out an investigation where she added amylase to starch and measured how long it took before all the starch disappeared. She repeated the experiment at different temperatures.

(a) In this reaction what is the:

(i) substrate?

____ [1 mark]

(ii) product?

____ [1 mark]

D–C

(b) Name her independent variable.

____ [1 mark]

(c) State one control variable she should use.

____ [1 mark]

Enzymes and digestion

1 Describe the role of enzymes in digestion.

____

____

____ [2 marks]

2 Fill in the gaps in the table. [6 marks]

Enzyme group	Substrate	Product	Where enzyme is produced
amylase	(a)__________	sugars	(b)__________ glands in the mouth and pancreas
(c)__________	protein	amino acids	pancreas, (d)__________ and small intestine
(e)__________	lipids (fats and oils)	fatty acids and (f)__________	pancreas and small intestine

Enzymes at home

1 Enzymes can be used in biological washing powders. How are these enzymes manufactured?

____

____

____ [1 mark]

2 **(a)** Harry was eating a greasy burger and spilt fat down his T-shirt. Explain why washing it in biological washing powder will help remove the fat stain.

____

____ [3 marks]

(b) He decided to wash the T-shirt at 60 °C to make sure the stain came out. Explain why this is not a good idea.

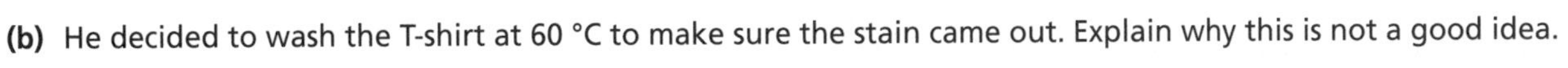

[3 marks]

Enzymes in industry

1 Meat tenderiser contains enzymes which break down the fibrous proteins in the meat. What type of enzyme is found in meat tenderiser?

____ [1 mark]

2 Manufacturers of sports drinks often make sugar syrup for their drinks from starch solution.

(a) State how they turn the starch solution into sugar syrup.

____ [1 mark]

(b) Why don't they just use sugar syrup?

____ [1 mark]

D–C

(c) Explain why slimming foods may contain fructose instead of glucose.

____

____ [3 marks]

(d) Name the enzyme used to turn glucose into fructose.

____ [1 mark]

Aerobic respiration

1 For each of these statements state whether they are true or false.

(a) Starchy foods like pasta are high in energy.

____ [1 mark]

G–E

(b) The energy stored in food is released in a reaction called respiration.

____ [1 mark]

(c) Only animals carry out respiration.

____ [1 mark]

2 **(a)** Why do all your cells need a supply of oxygen?

____

____ [2 marks]

(b) Describe how oxygen from the air reaches all your cells.

____

____ [2 marks]

Using energy

1 State **two** ways that humans use energy.

____

____

____

____ [2 marks]

G–E

2 If you use more energy than you take in you will lose body mass.

(a) Andrew goes running every day. How does this help him to lose body mass?

____

____ [2 marks]

D–C

(b) Explain how running on a cold day will help him to lose more mass.

____

____ [2 marks]

Anaerobic respiration

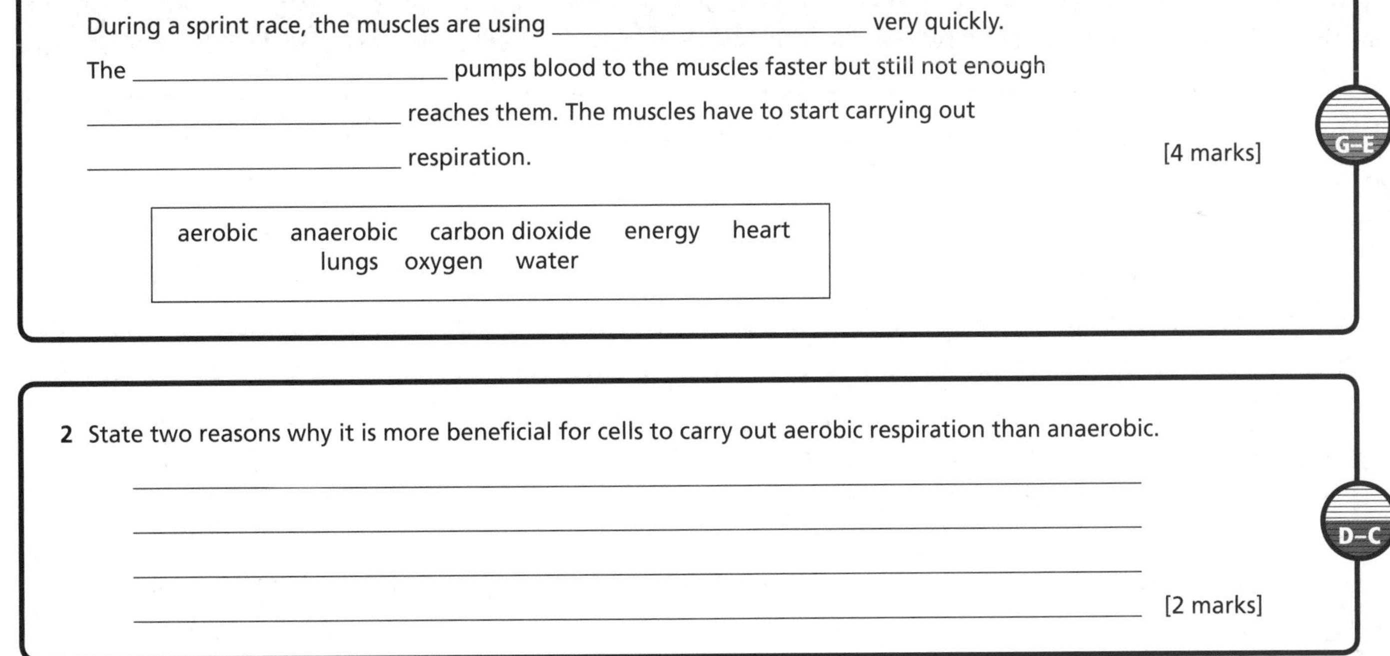

1 Choose the correct words from below to complete this paragraph.

During a sprint race, the muscles are using ____________________ very quickly.

The ____________________ pumps blood to the muscles faster but still not enough

____________________ reaches them. The muscles have to start carrying out

____________________ respiration. [4 marks]

G–E

aerobic	anaerobic	carbon dioxide	energy	heart
lungs	oxygen	water		

2 State two reasons why it is more beneficial for cells to carry out aerobic respiration than anaerobic.

____

____

____

____ [2 marks]

D–C

Cell division – mitosis

G–E

1 You were once a single cell called a zygote. Describe what happened in order to create the millions of cells that make you up today.

____

____

____

____ [3 marks]

D–C

2 Describe why:

(a) mitosis results in two daughter cells.

____

____ [1 mark]

(b) the daughter cells are genetically identical to the parent cell.

____

____ [1 mark]

(c) skin cells frequently undergo mitosis.

____

____ [1 mark]

Cell division – meiosis

G–E

1 For each of these human cells state how many chromosomes it contains.

(a) sperm ____________________

(b) egg ____________________

(c) zygote ____________________ [3 marks]

D–C

2 A horse has 64 chromosomes in its body cells. This diagram shows the stages that happen during meiosis in a female horse.

parent cell
1
2 3
4 5 6 7
daughter cells

(a) Where in the female horse's body does meiosis take place?

____ [1 mark]

(b) State the number of chromosomes in each of the numbered cells.

____ [3 marks]

(c) Explain why meiosis is essential for successful sexual reproduction.

____

____

____ [2 marks]

Stem cells

1 Choose correct words to complete the following statements about stem cells.

(a) Stem cells are cells that have the potential to do any job. They are ______________________ [1 mark]

(b) Stem cells then change into the different types of cell that have a specific job. This is called

______________________ [1 mark]

2 Parkinson's disease is caused by the death of cells in the brain. Explain how stem cells could be used in the future to treat it.

____

____

____

______________________ [3 marks]

Genes, alleles and DNA

1 The answer to each of the following questions is either DNA, genes or chromosomes. Choose the correct one.

(a) Many of these are found on one chromosome.

______________________ [1 mark]

(b) These are arranged in pairs inside the nucleus of most cells.

______________________ [1 mark]

(c) The shortened name of deoxyribonucleic acid.

______________________ [1 mark]

2 DNA fingerprinting can be used to find out who a child's father is. Use the DNA fingerprint below to:

(a) state who the father is.

______________________ [1 mark]

(b) explain how the DNA fingerprint is evidence that he is the father.

____

______________________ [2 marks]

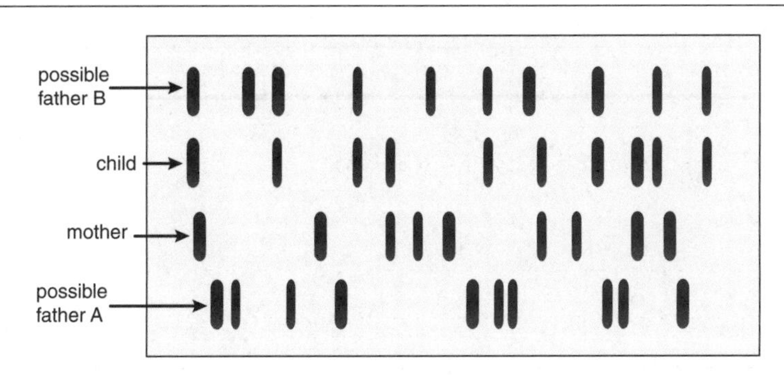

Mendel

G–E

1 What did Mendel's experiments on inheritance involve?

[1 mark]

D–C

2 Explain why scientists did not accept Mendel's ideas until after his death.

[1 mark]

How genes affect characteristics

G–E

1 State the sex chromosomes in:

(a) an egg cell

[1 mark]

(b) a male body cell

[1 mark]

(c) a female body cell.

[1 mark]

D–C

2 **(a)** Natasha has two alleles for hair colour. Why has she got two?

[1 mark]

(b) She has blonde hair. The blonde hair allele is recessive (b). What pair of alleles has she got?

[1 mark]

(c) Her friend Nisha has brown hair. The brown hair allele is dominant over the blonde hair allele. What possible alleles has Nisha got?

[2 marks]

Inheriting chromosomes and genes

1 Explain why it is the father that determines the sex of the child.

G–E

[3 marks]

2 A long-haired guinea pig was bred with a short-haired one and produced 12 short-haired offspring.

(a) Which allele is most likely to be dominant?

[1 mark]

(b) What alleles will the offspring have?

[1 mark]

(c) Draw a genetic diagram to show the cross between one of the offspring and a long-haired guinea pig.

D–C

[3 marks]

(d) What is the ratio of homozygous to heterozygous genotypes?

[1 mark]

How genes work

1 **(a)** What is the shape of a DNA molecule?

[1 mark]

(b) Use the words below to fill in the gaps.

[3 marks]

G–E

______ are lengths of DNA that code for characteristics. DNA carries a code called the ______ code. It provides instructions to the cell about which type of ______ it should make.

carbohydrates chromosomes DNA genes genetic nucleus proteins

2 Erin has brown eyes. She inherited the brown eye allele from her mother. Describe how this allele produces brown eyes.

D–C

[4 marks]

Genetic disorders

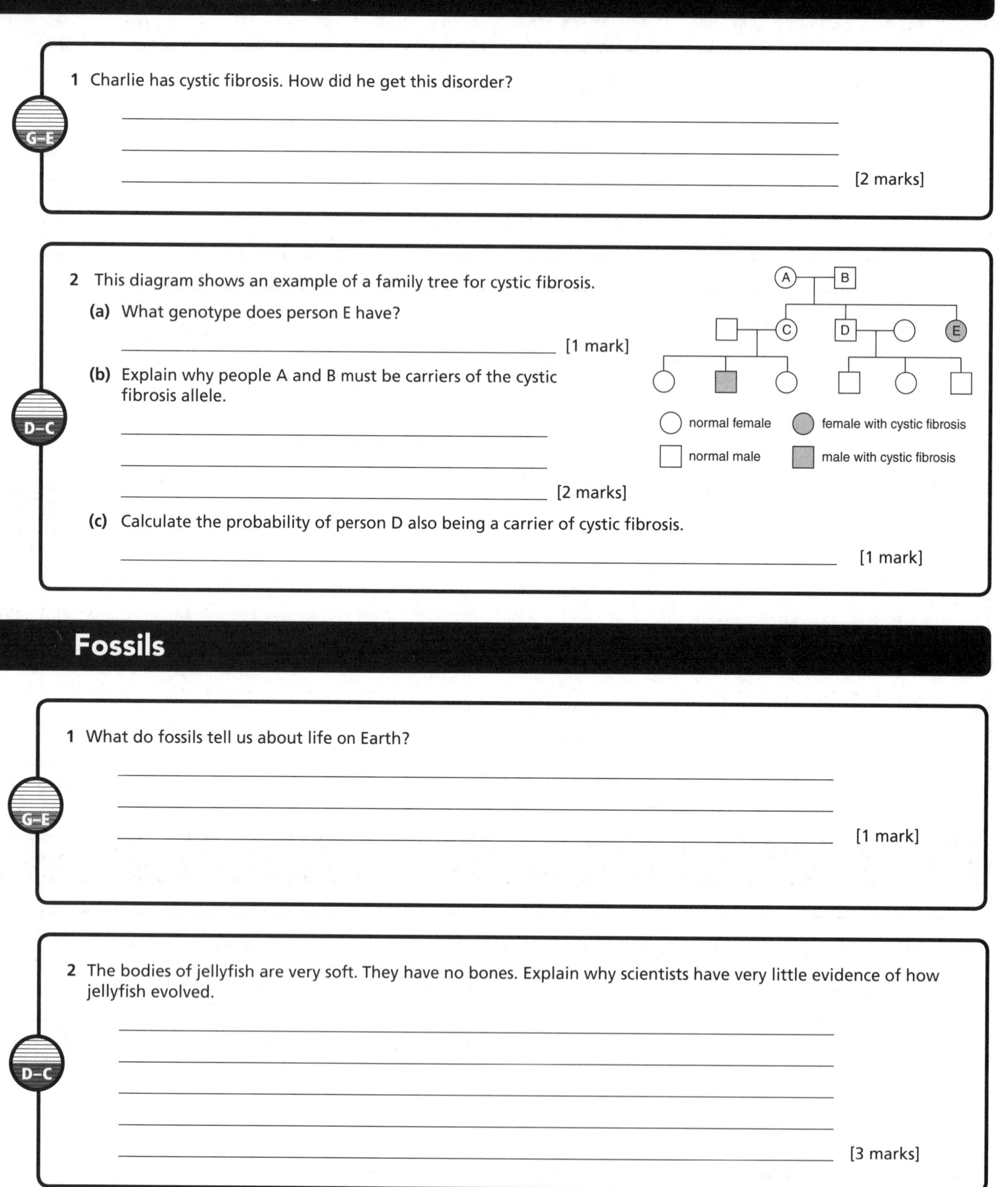

1 Charlie has cystic fibrosis. How did he get this disorder?

____

____

______________________________ [2 marks]

2 This diagram shows an example of a family tree for cystic fibrosis.

(a) What genotype does person E have?

______________________________ [1 mark]

(b) Explain why people A and B must be carriers of the cystic fibrosis allele.

____

____

______________________________ [2 marks]

(c) Calculate the probability of person D also being a carrier of cystic fibrosis.

______________________________ [1 mark]

Fossils

1 What do fossils tell us about life on Earth?

____

____

______________________________ [1 mark]

2 The bodies of jellyfish are very soft. They have no bones. Explain why scientists have very little evidence of how jellyfish evolved.

____

____

____

____

______________________________ [3 marks]

Extinction

1 Many species of plants and animals have become extinct in recent times. Many of these extinctions can be blamed on people.

(a) What does it mean when a species is extinct?

[1 mark]

(b) State two ways that people can cause the extinction of species.

[2 marks]

G–E

2 The Croatian dace is a small fish found only in one small stream in Croatia. People introduced trout into the stream. State two possible reasons why Croatian dace are in danger of becoming extinct.

[2 marks]

D–C

New species

1 A male donkey and a female horse can breed together to form a mule. Mules are sterile (they cannot reproduce). How is this evidence that horses and donkeys are different species?

[1 mark]

G–E

2 Charles Darwin visited the Galapagos Islands off the coast of South America. He found that on each island there was a different species of finch. We now know that they all descended from one species found on the mainland. Explain the process that led to the evolution of these many different species of finch.

[5 marks]

D–C

Extended response question

Control systems are used to keep human temperature at around 37 °C. Explain why it is dangerous for our body temperature to go much higher. Use what you know about enzymes in your answer.

The quality of written communication will be assessed in your answer to this question.

[6 marks]

Investigating atoms

1 (a) Complete this table (**i** to **iv**) about the mass and charge on subatomic particles.

Particle	Relative mass	Relative charge
proton	**(i)**	+1
(iii)	1	**(ii)**
electron	**(iv)**	−1

[4 marks]

28
Si
14

(b) Use the data in the box to help you answer these questions.

(i) How many protons does this element have? ______________________

(ii) How many neutrons does this element have? ______________________

(iii) Write the electronic structure of this element. ______________________ [3 marks]

Mass number and isotopes

1 Atoms are made up of three subatomic particles, protons, neutrons and electrons. They also have a relative atomic mass.

(a) What is the atomic number of an element? [1 mark]

(b) What is the mass number of an element? [1 mark]

(c) Explain why an atom of an element always has the same number of protons and electrons. [2 marks]

2 Many elements have different isotopes. Potassium has two isotopes: potassium-39 and potassium-40.

(a) What is an isotope?

______________________ [1 mark]

(b) Describe how potassium-39 is different from potassium-40.

______________________ [2 marks]

(c) In a sample of potassium atoms, 90 were found to be potassium-39, and 10 were potassium-40. Calculate the relative atomic mass of potassium.

____

______________________ [2 marks]

Compounds and mixtures

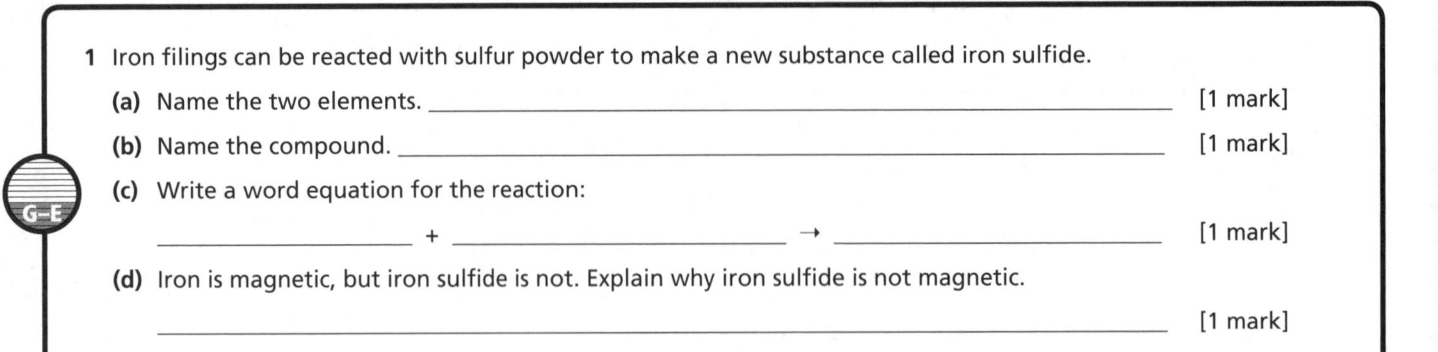

1 Iron filings can be reacted with sulfur powder to make a new substance called iron sulfide.

(a) Name the two elements. ______________________________ [1 mark]

(b) Name the compound. ______________________________ [1 mark]

(c) Write a word equation for the reaction:

______________ + ______________ → ______________ [1 mark]

(d) Iron is magnetic, but iron sulfide is not. Explain why iron sulfide is not magnetic.

______________________________ [1 mark]

2 Barium chloride solution is used to test for the presence of compounds containing sulfate ions. Some barium chloride was reacted with aluminium sulfate solution.

$$3BaCl_2(aq) + Al_2(SO_4)_3(aq) \rightarrow 3BaSO_4(s) + 2AlCl_3(aq)$$

(a) What does (aq) mean? ______________________________ [1 mark]

(b) Calculate the relative formula mass (M_r) of barium chloride.

(Relative atomic masses: Ba = 137, Cl = 35.5) ______________________________ [2 marks]

(c) What is the mass of one mole of barium chloride? ______________________________ [1 mark]

Electronic structure

1 Magnesium and calcium are in Group 2 of the periodic table.

The diagrams show the electronic arrangement of magnesium and calcium.

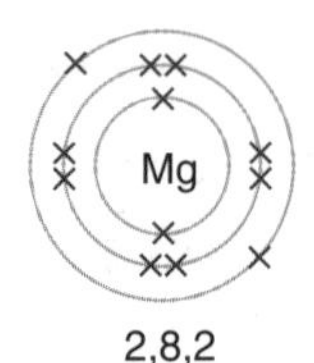

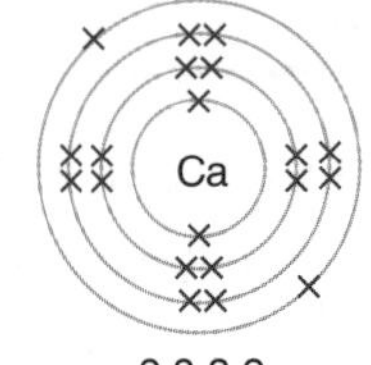

(a) Name another element in Group 2. ______________________________ [1 mark]

(b) How many outer electrons would you expect the element you have named to have? ______________________________ [1 mark]

(c) Explain why all the elements have similar chemical reactions.

______________________________ [2 marks]

2 The diagram shows the electronic structure of a sodium atom.
Draw similar diagrams for the following atoms, writing the electronic structure under each diagram:

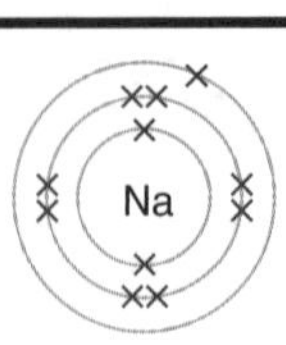

(a) fluorine [2 marks]

(b) helium [2 marks]

(c) aluminium [2 marks]

(d) sulfur [2 marks]

(Atomic numbers: Al = 13, S = 16, F = 9, He = 2)

Ionic bonding

1 An ion forms when an atom gains or loses electrons and becomes charged. Complete the sentences below choosing one word from each box.

Metal atoms [lose / gain] electrons to become [positively / negatively] charged ions. Gaining an electron turns the atom into a [positive ion / negative ion], losing an electron turns the atom into a [positive ion / negative ion]. They gain or lose electrons so that the ion has [a full / an empty / a noble gas] electronic structure. [5 marks]

2 The compound potassium fluoride (KF) is being investigated by some scientists. They have discovered the following facts about potassium fluoride.

property	
melting point	857 °C
boiling point	1502 °C
conduction as a solid	no
conduction as a solution	yes

(a) The scientists think potassium fluoride has ionic bonds. Give two reasons from the table for their conclusion.

____

_______________________ [2 marks]

(b) A potassium atom has the electronic structure of 2,8,8,1. A fluorine atom has the electronic structure 2,7.

(i) Describe the electronic structure of a potassium ion.

_______________________ [1 mark]

(ii) Describe the electronic structure of a fluoride ion.

_______________________ [1 mark]

(c) Explain how the two ions form the compound potassium fluoride.

_______________________ [2 marks]

Alkali metals

1 Sodium is an element. It conducts electricity, is stored under oil, and reacts violently with water.

(a) Which piece of information in the question suggests sodium is a metal.

_______________________ [1 mark]

(b) Why is it stored under oil?

_______________________ [1 mark]

(c) Sodium floats on water when it reacts. What does this tell you about the density of sodium.

_______________________ [1 mark]

2 Lithium, sodium and potassium are the first three elements of Group 1 or the alkali metals. They all have a single outer electron, and all react easily with water to form hydrogen gas and the metal hydroxide. Potassium has a violent reaction with water, and lithium just floats and fizzes in water.

(a) Write a word equation for the reaction of sodium with water. _______________________ [1 mark]

(b) Explain why all the Group 1 metals form positive ions with a 1+ charge.

_______________________ [2 marks]

(c) Caesium is another Group 1 metal. It is lower in the group than potassium. Suggest how it would react with water.

Explain your answer. _______________________ [2 marks]

Halogens

1 Complete the table about the properties of the halogens.

Name	Formula	Boiling point °C	state at room temperature	colour
fluorine	F_2		gas	pale yellow
chlorine		–34	gas	
	Br_2	58		dark red
iodine	I_2	183	solid	dark grey

[5 marks]

2 The diagram shows the electronic arrangement of a chlorine atom.

(a) Complete the second diagram to show the electron arrangement of a chloride ion. [1 mark]

(b) What is the charge on a chloride ion? ________________ [1 mark]

(c) Name the element with the same electronic structure as a chloride ion. ________________ [1 mark]

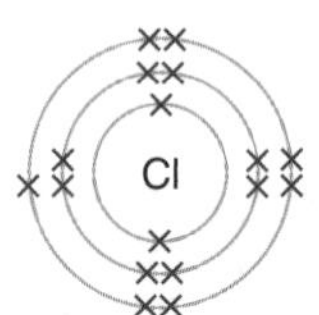

chlorine atom   chloride ion

Ionic lattices

1 The table gives some information about four substances.

compound	melting point °C	electrical conductivity as a solid	electrical conductivity in water
P	1418	no	yes
Q	1704	no	yes
R	2447	yes	does not dissolve
S	79	no	no

Use information from the table and your knowledge to answer the questions.

(a) Which substance(s) might be metal(s)? Explain your answer.

________________ [2 marks]

(b) Which substance(s) are probably covalent compound(s)? Explain your answer.

________________ [2 marks]

(c) Which substance(s) are probably ionic compound(s)? Explain your answer.

________________ [2 marks]

2 The diagram shows part of a sodium chloride crystal lattice.

(a) How many chloride ions are attracted to each sodium ion? [1 mark]

(b) Sodium chloride has a melting point of approximately 800 °C, but the molecule hydrogen chloride is a gas at room temperature. Explain why.

________________ [2 marks]

(c) Two electrodes, one positive and one negative, are placed in some molten sodium chloride, and a current passed through. Describe how the sodium and chloride ions react.

________________ [3 marks]

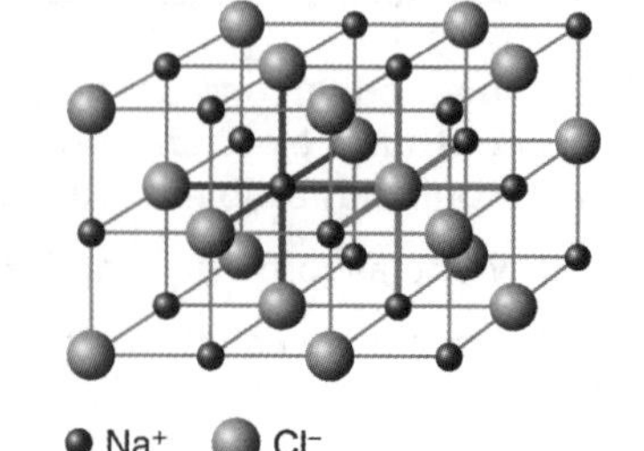

Covalent bonding

1 Complete the following sentences about bonding, using the words given below.

electrons **non-metal** **metal** **pair** **protons** **ion**

When two ______________ atoms join together they form covalent bonds. The outer shells of __________ overlap and form covalent bonds. A covalent bond is a shared ______________ of ______________. [4 marks]

2 Look at this diagram of two chlorine atoms.

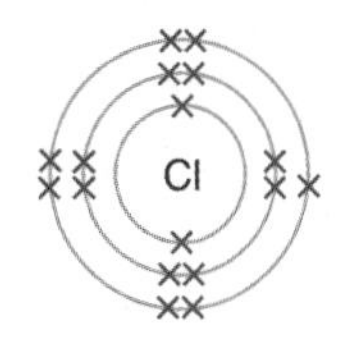

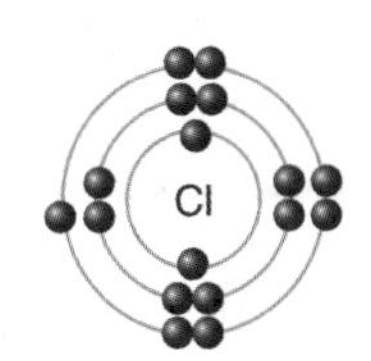

(a) Draw a diagram to show how the two atoms will make a molecule of chlorine gas (Cl_2).

[2 marks]

(b) The atoms are joined together by a covalent bond. What is a covalent bond?

____ [1 mark]

(c) Explain why non-metal elements form molecules with covalent bonds.

____ [2 marks]

Covalent molecules

1 Look at the diagram of the covalent molecule of ammonia.

H N H H

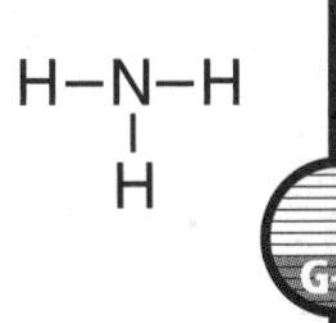

(a) What is the formula of ammonia?

______________________________ [1 mark]

(b) Explain why ammonia is a gas at room temperature.

______________________________ [1 mark]

(c) Explain why ammonia cannot conduct electricity.

____ [1 mark]

2 The table shows some data about the halogens, Group 7.

halogen	boiling point in °C	diameter of molecule in nm
fluorine	–220	0.28
chlorine	–34	0.39
bromine	58	0.46
iodine	114	0.53

(a) Room temperature is 20 °C. Which of the halogens are gases at room temperature? ______________________ [1 mark]

(b) How many outer electrons has a bromine atom? ___ [1 mark]

(c) Describe the trend in boiling points as the halogen molecule gets larger.

____ [1 mark]

(d) The boiling point of each halogen is affected by the intermolecular forces between the molecules. Describe, with reasons, how the intermolecular forces between halogen molecules change in the group.

____

____ [3 marks]

Covalent lattices

G–E

1 The diagram shows the structures of diamond and graphite, two covalent lattices of carbon.

graphite diamond

(a) How many covalent bonds does each carbon atom have in:

(i) diamond? ________________ [1 mark]

(ii) graphite? ________________ [1 mark]

(b) Both diamond and graphite melt at over 4000 °C. Explain why they both have very high melting points. [2 marks]

D–C

2 (a) Diamond is one of the hardest known substances. Describe how its structure enables it to be very hard.

____

____ [2 marks]

(c) Graphite is often used as a lubricant. Describe how its structure enables it to be a good lubricant.

____

____ [2 marks]

Polymer chains

G–E

1 Polymers are very large molecules made from monomers. Here are two representations of polymer chains.

-A-A-A-A-A-A-A-A-A-A-A-A-A-A-A-A-A-A-A-A-A-A-A-A-

polymer M

-A-B-A-B-A-B-A-B-A-B-A-B-A-B-A-B-A-B-A-B-A-B-A-B-

polymer N

(a) What is a monomer?

____ [1 mark]

(b) How is polymer A different from polymer B?

____

____ [2 marks]

(c) Both diagrams only show part of each polymer chain. Why?

____ [1 mark]

D–C

2 A bottle manufacturer has two different types of plastic bottle. One bottle is made from a thermosetting polymer, the other from a thermosoftening polymer. The thermosetting polymer is cheaper to produce.

(a) Describe the difference between thermosetting and thermosoftening polymers.

____

____ [2 marks]

(b) Suggest an environmental benefit from using the thermosoftening polymer bottle.

____ [1 mark]

(c) Describe how the structure of the thermosetting polymer differs from the thermosoftening polymer.

____

____ [2 marks]

Metallic properties

1 The table shows some of the properties of metals and non-metals. Fill in the missing spaces.

Metals	Non-metals
shiny	
good electrical conductor	
	brittle
	poor conductor of heat

[4 marks]

2 The diagram on the right shows the metal lattice of a pure metal such as titanium.

When making a shape-memory alloy, titanium metal is mixed with nickel.

(a) Draw a diagram in the box on the far right to show the effect of adding nickel atoms to the titanium metal lattice. [2 marks]

(b) What is a shape-memory alloy?

_______________________________________ [1 mark]

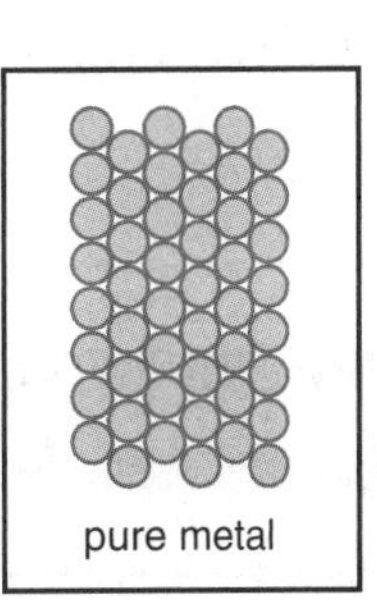

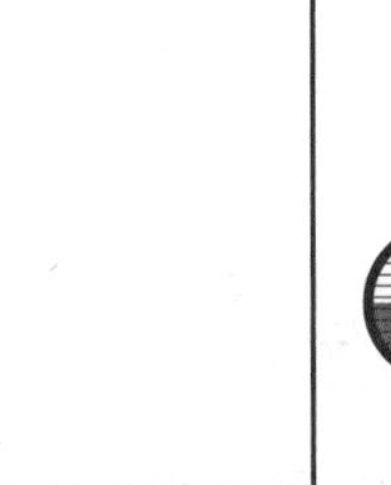

(c) Describe how a shape-memory alloy is useful in making a dental brace.

_______________________________________ [2 marks]

Modern materials

1 Many people now buy spectacles that have photochromic lenses. Photochromic lenses are an example of a 'smart material'.

(a) What do we mean by a 'smart material'?

_______________________________________ [1 mark]

(b) The frames of the spectacles can also be made from 'smart' alloy. Suggest why.

_______________________________________ [1 mark]

(c) What 'smart' property does a thermochromic material have?

_______________________________________ [1 mark]

G–E

2 Titanium(IV) oxide is used in sunscreen. The large solid particles used are white in colour and, if applied thickly, coat the skin in a white paste. Chemists are developing nano-sized particles of titanium(IV) oxide to overcome this problem.

(a) How many atoms are there in a nanoparticle?

_______________________________________ [1 mark]

(b) How large are nanoparticles?

_______________________________________ [1 mark]

(c) Explain why sunscreens' nanoparticles of titanium(IV) oxide will be an improvement on current sunscreens.

____

_______________________________________ [2 marks]

D–C

Identifying food additives

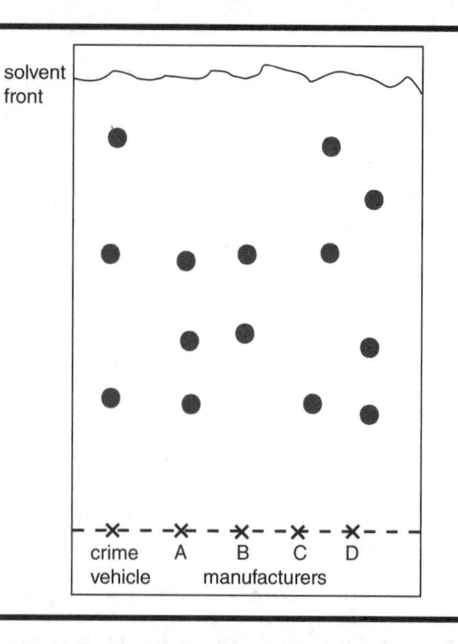

1 Some crime scene investigators used chromatography to process some paint found at the scene of a serious hit-and-run accident. They wanted to know the manufacturer of the vehicle. Here is the chromatogram they produced.

D–C

(a) How many different chemicals does the sample from the hit-and-run car contain?

______________________________ [1 mark]

(b) All the spots are red. How many different red pigments are used by all the different manufacturers?

______________________________ [1 mark]

(c) Which manufacturer made the car?

______________________________ [1 mark]

Instrumental methods

1 This diagram shows a gas chromatography system.

sample
long coiled chromatogram tube
fused silica
solvent gas
detector
GC column
Gas chromatograph oven

(a) Why is the instrument in an oven?

______________________________ [1 mark]

(b) Why is the solvent a gas?

______________________________ [1 mark]

(c) What is the function of the detector?

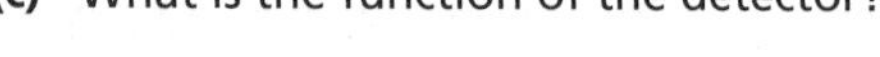

______________________________ [1 mark]

2 Gas Chromatography–Mass Spectrometry (GC-MS) is an instrumental method, used to determine the identity of different chemicals.

(a) What is an instrumental method?

______________________________ [1 mark]

(b) Why are instrumental methods used?

______________________________ [2 marks]

(c) How does a mass spectrometer help analyse the results of the gas chromatogram?

______________________________ [2 marks]

Making chemicals

1 Complete these sentences using the words provided.

products environmental calculate economic hazards reactants

To make a chemical you need to know what ______________ are needed. You need to ______________ the masses needed of each one so the process is ______________. You should consider the ______________ of the reactants and ______________, and their ______________ problems. [6 marks]

2 Chemicals can be made by a variety of different reactions such as neutralisation, oxidation, reduction, and precipitation.

(a) What is a neutralisation reaction? ______________ [1 mark]

(b) What is a precipitation reaction? ______________ [1 mark]

(c) What is an oxidation reaction? ______________ [1 mark]

(d) Why are oxidation and reduction reactions sometimes described as redox reactions?

______________ [2 marks]

D–C

Chemical composition

1 A chemical company sells a NPK fertiliser.

Different fertilisers have different NPK values.

(a) What does NPK mean? ______________ [1 mark]

(b) What does the 20–15–12 mean on the sack?

____

______________ [2 marks]

(c) Explain how this information could be useful to a farmer.

____

______________ [2 marks]

G–E

2 A student heated some copper oxide using this apparatus to reduce some copper oxide to copper metal.
The student weighed the 'boat', then the copper oxide and 'boat' before heating the copper oxide strongly. When cooled, the student then weighed the copper produced in the 'boat'.

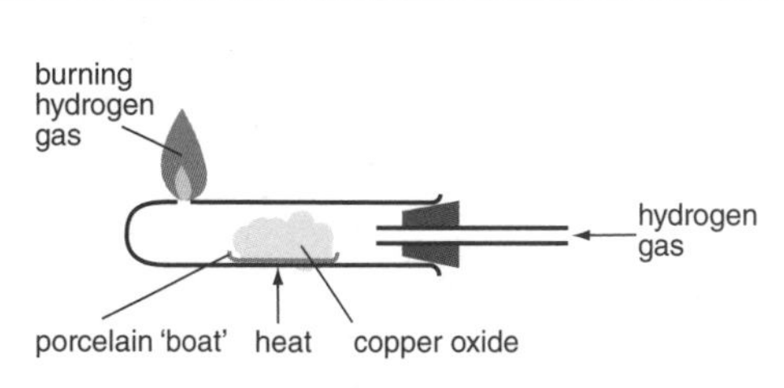

Mass of boat = 15.44 g,
boat and copper oxide = 19.42 g,
boat and copper produced = 18.62 g

(a) What was the mass of copper oxide used?

______________ [1 mark]

(b) What was the mass of copper produced?

______________ [1 mark]

(c) What was the mass of oxygen removed from the copper oxide?

______________ [1 mark]

(d) Use the masses to calculate the formula of copper oxide. Show your working.
(Relative atomic masses, (A_r): Cu = 63.5, O = 16)

______________ [3 marks]

Quantities

1 The reaction for the thermal decomposition of calcium carbonate to calcium oxide is shown by this equation.

$$CaCO_3(s) \rightarrow CaO(s) + CO_2(g)$$

(Relative atomic masses, (A_r): Ca = 40, C = 12, O = 16)

(a) Calculate the relative formula mass of calcium carbonate.

______________________________ [1 mark]

(b) Calculate the relative formula mass of calcium oxide.

______________________________ [1 mark]

How much product?

1 A student decided to make some copper sulfate crystals. The student calculated that the theoretical yield of the reaction would be 2.6 g. When the copper sulfate made was weighed, the actual yield was 1.9 g.

(a) What is meant by 'theoretical yield'?

______________________________ [1 mark]

(b) What is meant by 'actual yield'?

______________________________ [1 mark]

(c) Give one reason why the theoretical yield and actual yield were different.

______________________________ [1 mark]

2 A student planned to make some copper sulfate by reacting copper oxide with sulfuric acid.

$$CuO(s) + H_2SO_4(aq) \rightarrow CuSO_4(aq) + H_2O(\ell)$$

(Relative atomic masses, (A_r): Cu = 63.5, H = 1, S = 32, O = 16)

The student used 1.59 g of copper oxide with an excess of sulfuric acid. The student weighed the copper sulfate made, and found he had made 2.2 g of copper sulfate.

(a) What is meant by an excess?

______________________________ [1 mark]

(b) Calculate the theoretical yield of copper sulfate.

____

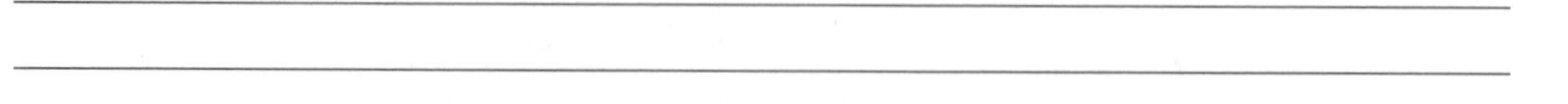

______________________________ [3 marks]

(c) Calculate the percentage yield obtained.

______________________________ [1 mark]

Reactions that go both ways

1 Litmus solution is red in acid, and purple in alkaline solutions. Adding alkali to red litmus makes it turn purple, and adding acid to purple litmus makes it turn red. The reaction of litmus solution with an acid and alkali is reversible.

(a) What do we mean by reversible reaction? ______________________ [1 mark]

(b) Why is it useful for the colour change of an indicator to be reversible?

______________________ [1 mark]

2 Cobalt chloride paper can be used to test for the presence of water. It uses this reversible reaction:

$$\underset{\text{pink}}{CoCl_2{\cdot}6H_2O(s)} \rightleftharpoons \underset{\text{blue}}{CoCl_2(s) + 6H_2O(\ell)}$$

(a) What does this symbol $\rightleftharpoons$ mean? ______________________ [1 mark]

(b) Describe the colour change when water is added to cobalt chloride paper.

______________________ [1 mark]

(c) Explain why cobalt chloride paper may be dried and used again.

______________________ [2 marks]

Rates of reaction

1 The rate of a chemical reaction can be altered by changing one of the conditions of the reaction. State whether each of these changes will increase, decrease or have no effect on the rate of a chemical reaction.

(a) decreasing the temperature ______________________ [1 mark]

(b) changing the beaker the reaction is carried out in ______________________ [1 mark]

(c) adding a catalyst ______________________ [1 mark]

(d) using larger pieces of solid ______________________ [1 mark]

(e) increasing the concentration of one reactant ______________________ [1 mark]

2 A group of students wanted to measure the rate of reaction when calcium carbonate dissolves in hydrochloric acid.

$$CaCO_3(s) + HCl(aq) \rightarrow CaCl_2(aq) + H_2O(\ell) + CO_2(g)$$

They used the apparatus shown in the diagram. They placed 20 g of calcium carbonate in a conical flask then added 25 cm^3 of dilute hydrochloric acid.

(a) Describe the measurements they need to take to follow the rate of the reaction.

______________________ [2 marks]

Another group of students measured the volume of gas produced, and plotted a graph of their results.

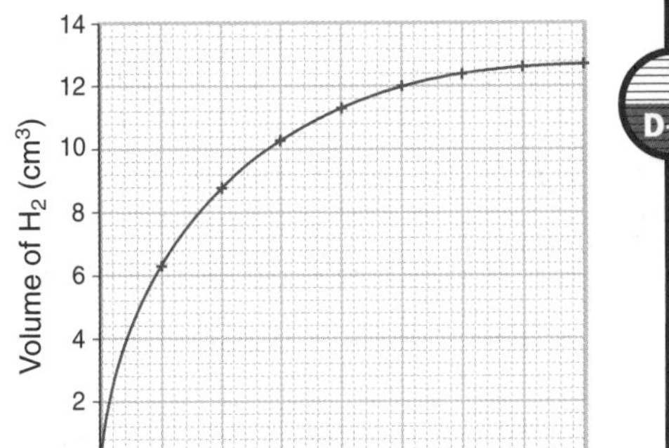

(b) Explain why the graph is a curve.

____

______________________ [2 marks]

(c) What is the maximum volume of carbon dioxide obtained from the reaction?

______________________ [1 mark]

Collision theory

1 Reactants must collide to react. Use collision theory to explain these observations.

(a) Powdered calcium carbonate reacts faster with hydrochloric acid than the same mass in lumps.

____ [1 mark]

(b) Diluted acids always react slower than more concentrated acids.

____ [1 mark]

(c) Increasing the temperature that a reaction is carried out at always increases the rate of reaction.

____ [1 mark]

2 Magnesium ribbon reacts with dilute hydrochloric acid like this:

$$Mg(s) + 2HCl(aq) \rightarrow MgCl_2\,(aq) + H_2(g)$$

(a) If a 2 g piece of magnesium ribbon was reacted with an excess of hydrochloric acid, what would happen to the rate of reaction if the temperature was increased? Explain your answer.

____ [2 marks]

(b) If a 2 g piece of magnesium ribbon was cut into six pieces then reacted with an excess of hydrochloric acid, what would happen to the rate of reaction? Explain your answer.

____ [2 marks]

(c) If a 2 g piece of magnesium ribbon was reacted with a more dilute solution of hydrochloric acid, what would happen to the rate of reaction? Explain your answer.

____ [2 marks]

Adding energy

1 Copper oxide reacts very slowly with hydrochloric acid to make copper chloride. If the mixture is gently heated the reaction rate increases.

(a) Explain why warming the mixture increases the rate of the reaction.

____ [2 marks]

(b) Suggest why the mixture is only warmed, and not heated strongly.

____ [1 mark]

2 A student planned to make some iron sulfate by reacting 2 g of iron with an excess of sulfuric acid .

$$Fe(s) + H_2SO_4(aq) \rightarrow FeSO_4(aq) + H_2(g)$$

The student carried out the reaction at two different temperatures. The results are shown on the graph.

Volume of H_2 (cm^3) / Time (min) — 30°C, 20°C

(a) What is the maximum volume of hydrogen gas the student can obtain? ____________________ [1 mark]

(b) The student decided to repeat the experiment at 40 °C.

(i) Sketch on the graph what the results should look like. [2 marks]

(ii) Explain your curve, using collision theory.

____ [2 marks]

Concentration

1 Two identical pieces of magnesium ribbon are dissolved in dilute nitric acid and concentrated nitric acid.

(a) Which piece of magnesium ribbon will dissolve first?

____ [1 mark]

(b) Explain why this piece dissolved first.

____ [1 mark]

2 The graph shows the volume of gas produced when 3 g of aluminium reacted with hydrochloric acid at three different concentrations.

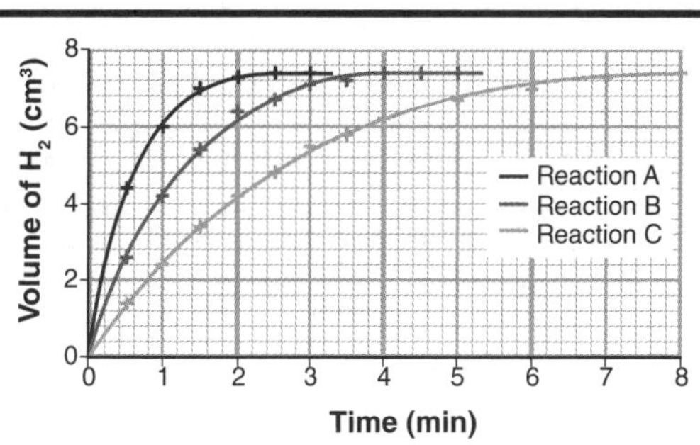

(a) Which graph, A, B or C, shows the reaction with the highest concentration of hydrochloric acid? _______ [1 mark]

(b) Which graph, A, B or C, shows the reaction with the slowest rate of reaction? Explain your answer. ___________ [2 marks]

(c) Explain why all three curves reach the same maximum height.

____

____ [2 marks]

(d) How long did it take reaction B to finish? _______________________ [1 mark]

Size matters

1 Both coal and coal dust are highly flammable. When mining coal underground it is important to keep the quantity of coal dust in the air to the minimum.

Suggest why 2 kg of coal dust in the air is far more dangerous in a mine than a 2 kg lump of coal.

____

____ [2 marks]

1 A student looked at burning iron in a Bunsen flame. The student tried burning a small nail, some iron wool, and some iron filings in a Bunsen flame. Here are the observations of the experiment.

type of iron	observations
nail	glowed red, did not burn, did not change size
iron wool held by tongs	bright red glow that spread through the wool, as it spread the wool disappeared, brown bits appeared on the heat mat
iron filings sprinkled into flame	bright sparks as filings sprinkled onto flame, brown bits on heat mat

(a) Which of the three types of iron reacted fastest? _______________________ [1 mark]

(b) Explain your answer using collision theory.

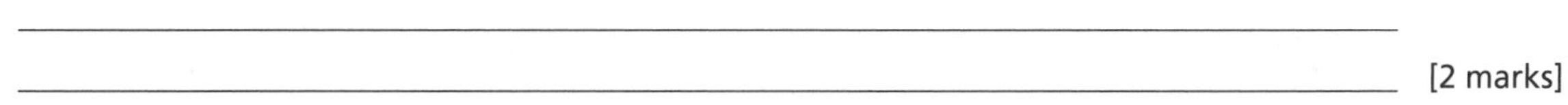

____ [2 marks]

(c) Which of the types of iron did not react? What evidence is in the table for this?

____

____ [2 marks]

Clever catalysis

G–E

1 Ammonia is made from hydrogen and nitrogen with the help of an iron catalyst. It originally used a ruthenium catalyst.

(a) Where in the periodic table are ruthenium and iron to be found? ______________________ [1 mark]

(b) What is a catalyst? ______________________ [1 mark]

(c) Suggest why iron is used rather than ruthenium in the making of ammonia today.

______________________ [1 mark]

D–C

2 Hydrogen peroxide decomposes slowly to form water and oxygen. It can be speeded up by using manganese(IV) oxide as a catalyst.
A student wanted to prove that manganese(IV) oxide was a catalyst, so did an experiment. Here are the results.

volume of hydrogen peroxide in cm^3	mass of manganese(IV) oxide at the start in grams	mass of manganese(IV) oxide at the end in grams	time for 50 cm^3 of oxygen gas to be produced in seconds.
100	0.0	0.0	250
100	2.3	2.3	50

(a) By how much did the manganese(IV) oxide affect the rate of reaction? Explain your answer.

____

______________________ [2 marks]

(b) What evidence is there that the manganese(IV) oxide is a catalyst? Explain your answer.

____

______________________ [2 marks]

(c) Suggest how the manganese(IV) oxide may have affected the reaction rate.

____

______________________ [2 marks]

Controlling important reactions

G–E

1 Ammonia is a very useful gas. It is used to make fertilisers. Controlling the rate of the reaction is very important to maximise the ammonia made, whilst keeping the costs economic. Ammonia is made by a reversible reaction between hydrogen and nitrogen.

(a) Write a word equation for the manufacture of ammonia.

______________________ [1 mark]

(b) The reaction uses an iron catalyst. What does this suggest about the rate of the reaction between nitrogen and hydrogen? ______________________ [1 mark]

D–C

2 Ammonia is made by a reversible reaction between hydrogen and nitrogen, using an iron catalyst.

(a) At 450 °C only 20% of the nitrogen and hydrogen become ammonia. Explain why a temperature of 450 °C is used for the reaction.

____

______________________ [2 marks]

(b) Explain why a pressure of 200 atmospheres is used for the reaction?

____

______________________ [2 marks]

The ins and outs of energy

1 A student was asked to dissolve 2 g of calcium oxide in water, and record the temperature change. The student then repeated the experiment using 2 g of ammonium nitrate and the same volume of water. Here are the results.

compound dissolved	temperature of water in °C	temperature of solution in °C	change in temperature in °C
calcium oxide	19	25	
ammonium nitrate	18	11	

(a) Complete the table to show the temperature changes. [2 marks]

(b) Which reaction was an exothermic reaction? Explain why.

____ [2 marks]

2 A student dissolved 5 g of ammonium nitrate in 50 cm³ of water. The temperature of the solution fell by 6 °C.

(a) What type of reaction is this? ____________________ [1 mark]

(b) Explain why the temperature dropped.

____ [2 marks]

(c) The diagram shows a self-heating can of coffee. Which two chemicals must react for the coffee to start to heat up?

____ [1 mark]

(d) Where does the heat energy come from to heat the coffee?

____ [2 marks]

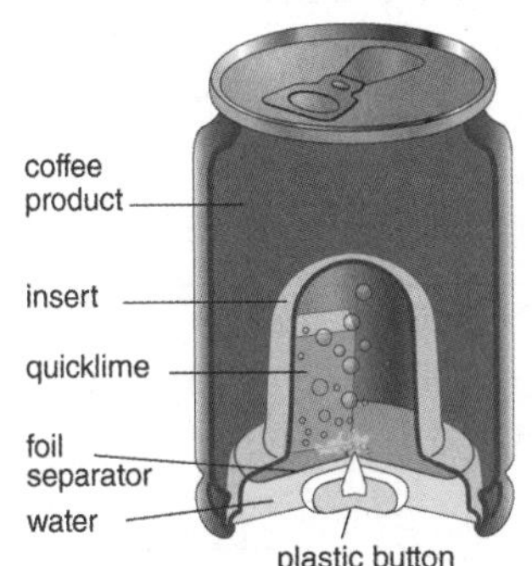

Acid–base chemistry

1 **(a)** Complete these two word equations to name the ions formed by the acid and alkali.

hydrochloric acid + water → chloride ion + ____________ ion

sodium hydroxide + water → sodium ion + ____________ ion [2 marks]

(b) Explain why water is formed during a neutralisation reaction.

____ [1 mark]

2 Alkalis and bases can both neutralise an acid.

(a) Describe the difference between an alkali and a base.

____ [1 mark]

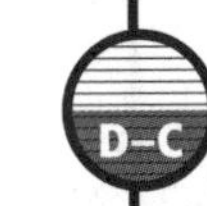

(b) When sodium hydroxide solution reacts with nitric acid it forms two new compounds. Complete the equation to show the two products.

sodium hydroxide + nitric acid → ____________ +  [1 mark]

(c) Copper carbonate can act as a base and react with hydrochloric acid. Complete the equation to show the products formed.

copper carbonate + hydrochloric acid → ____________ + ____________ + ____________ [2 marks]

Making soluble salts

1 Salts can be made by reacting different substances with different acids.
Complete these word equations, filling in the missing words.

G–E

(a) nitric acid + ______________________ → magnesium nitrate + ______________________

(b) hydrochloric acid + ______________________ → potassium chloride + water

(c) lithium oxide + sulfuric acid → ______________________ + water [4 marks]

2 A student wanted to make some cobalt sulfate. The student planned to use cobalt metal, but a friend said this would be too hazardous and suggested using cobalt oxide instead.

(a) Name the acid needed to make cobalt sulfate. ______________________ [1 mark]

(b) The student added some cobalt oxide to the acid.

(i) What should the student do to help the cobalt oxide and acid to react?

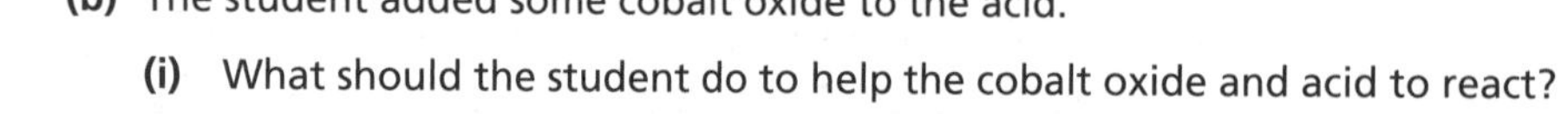

______________________ [1 mark]

D–C

(ii) Suggest two ways that the student could use to show all the acid had been used up.

____

______________________ [2 marks]

(c) How could the student make sure that the cobalt sulfate made was pure, and had no cobalt oxide left in it?

______________________ [1 mark]

(d) How could the student obtain a solid sample of cobalt sulfate crystals?

____

______________________ [2 marks]

Insoluble salts

1 Insoluble salts are made by mixing together two soluble salts which produce a salt that is insoluble.

G–E

(a) What does insoluble mean?

______________________ [1 mark]

(b) This chemical equation describes the making of lead iodide, an insoluble salt.

$2KI(aq) + Pb(NO_3)_2(aq) \rightarrow 2KNO_3(aq) + PbI_2(s)$

Write the formula of the compound that is the precipitate.

______________________ [1 mark]

2 Calcium carbonate obtained from limestone or chalk is too impure to be used in kitchen cream cleaning liquids. Instead it is obtained by using this reaction.

$$CaCl_2(aq) + Na_2CO_3(aq) \rightarrow CaCO_3(s) + 2NaCl(aq)$$

(a) Name the products in the reaction. ______________________ [1 mark]

(b) After the reaction is complete, how could you separate the calcium carbonate from the solution?

______________________ [1 mark]

(c) Explain how you could ensure that the calcium carbonate would be pure when dried.

______________________ [1 mark]

Ionic liquids

1 Ionic compounds cannot conduct electricity when they are solid. If they are dissolved in water then they can conduct electricity.

(a) Why can ionic compounds conduct electricity when they are dissolved in water?

______________________________ [2 marks]

G–E

(b) As well as when dissolved in water, when else can ionic compounds conduct electricity?

______________________________ [1 mark]

2 Potassium dichromate ($K_2Cr_2O_7$) is a yellow ionic solid that dissolves in water to form potassium ions (K^+) and dichromate ions ($Cr_2O_7^{2-}$). The yellow colour is caused by the dichromate ions. If an electric current is passed through a filter paper soaked in water with a pipette of potassium dichromate added as shown on the right, this is what happens.

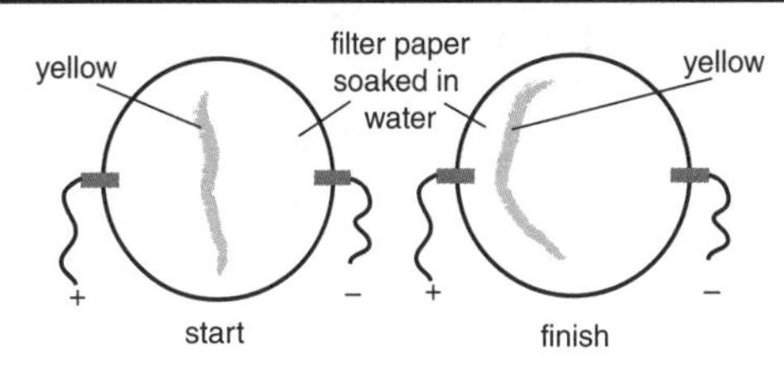

(a) What is the name given to the process of splitting a compound using electricity?

______________________________ [1 mark]

D–C

(b) Suggest why the yellow colour travels towards the positive electrode or anode.

______________________________ [2 marks]

(c) Explain what will happen to a potassium ion when it reaches the negative electrode.

______________________________ [2 marks]

Electrolysis

1 A teacher demonstrated the electrolysis of sodium chloride solution (brine) to the class, using this apparatus.

At the positive electrode, a gas was given off that smelled like the swimming pool. At the negative electrode, a gas was also produced that the teacher thought was hydrogen, and the solution turned blue.

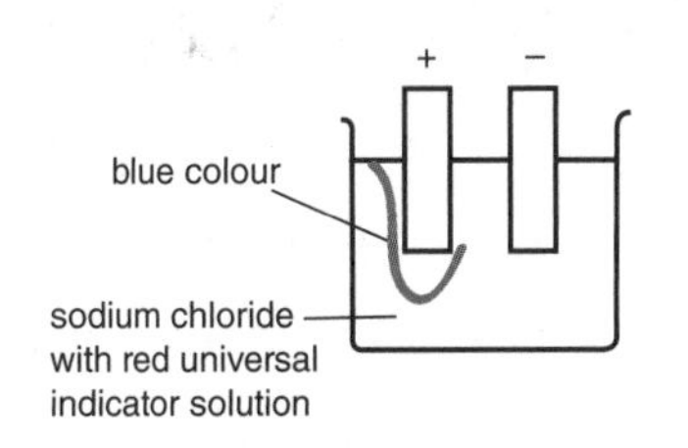

(a) Name the gas produced at the positive electrode.

______________________________ [1 mark]

G–E

(b) How could the teacher show the gas at the negative electrode was hydrogen?

______________________________ [1 mark]

(c) Name the substance that turned the sodium chloride solution blue at the negative electrode.

______________________________ [1 mark]

2 The diagram shows how copper is purified using electrolysis.

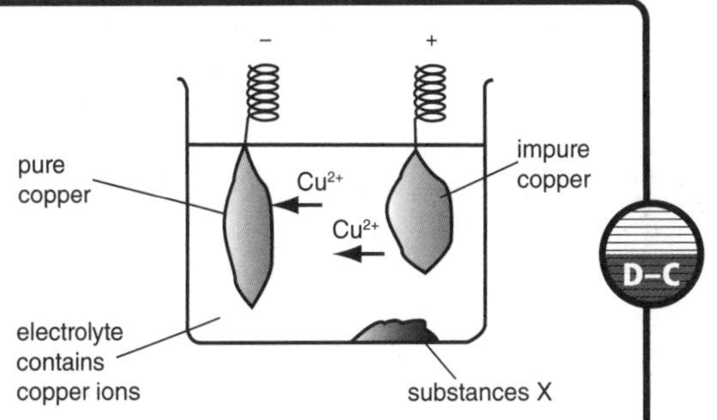

(a) Which electrode has the impure copper on? ______________ [1 mark]

(b) What happens to the impure copper at this electrode?

______________________________ [2 marks]

D–C

(c) Explain why the electrolyte solution has to contain copper ions.

______________________________ [1 mark]

(d) What are substances X? ______________ [1 mark]

Sodium hydroxide is made by the electrolysis of brine (sodium chloride) at room temperature. Sodium chloride is an ionic compound containing sodium ions (Na^+) and chloride ions (Cl^-).

Sodium hydroxide can also be made by the electrolysis of molten sodium chloride and calcium chloride at 600 °C, and then reacting the sodium produced with water.

The figure shows the apparatus used to electrolyse the brine.

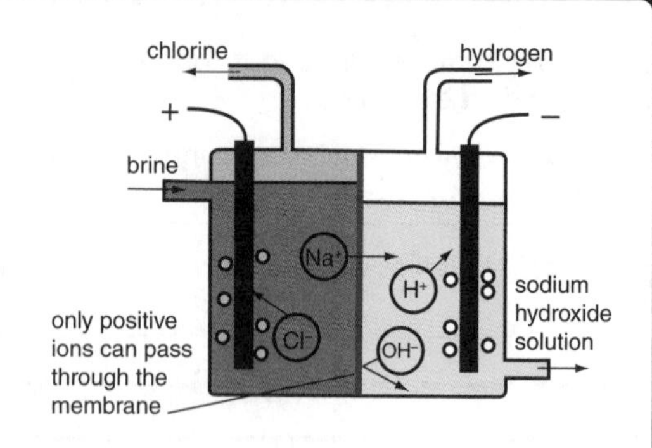

Use the information given, and your knowledge and understanding of this process, to answer this question.

Explain, as fully as you can, why sodium hydroxide is made by the electrolysis of brine rather than from molten sodium chloride.

The quality of written communication will be assessed in your answer to this question.

[6 marks]

See how it moves

1 The distance–time graph shows the motion of a bus.

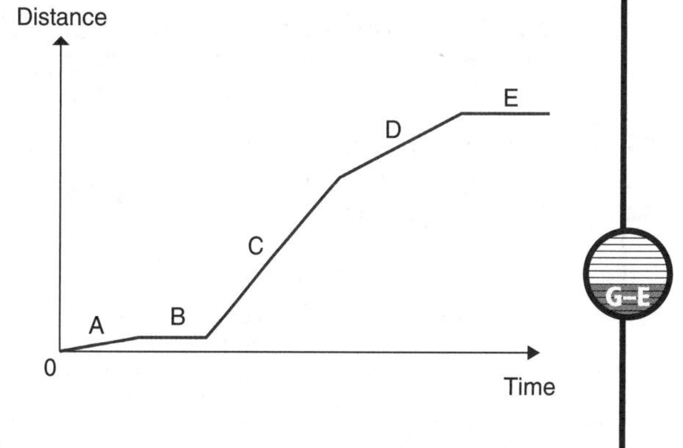

(a) Describe the motion of the bus at:

(i) section A ______________________

(ii) section B ______________________

[2 marks]

(b) In which section did the bus travel fastest?

____

[1 mark]

(c) Describe the appearance of the graph if the bus is accelerating.

____

______________________ [2 marks]

2 (a) Calculate the speed of the car in the first hour using the distance–time graph.

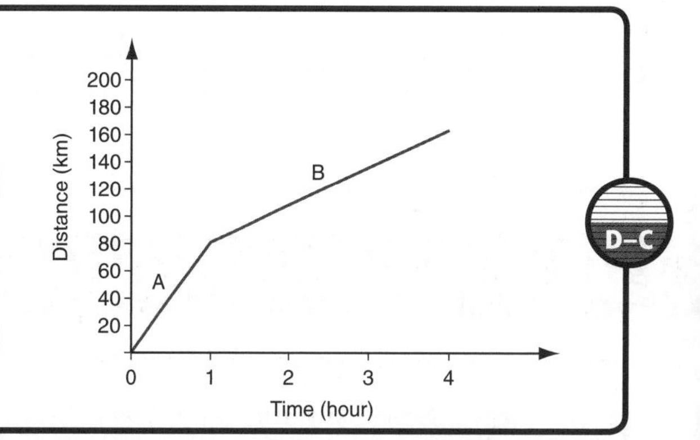

____

[2 marks]

(b) What is the average speed of the car over the whole journey?

____

[2 marks]

Speed is not everything

1 The velocity–time graph shows the motion of a falling object. Use the graph to describe how the object's speed changes:

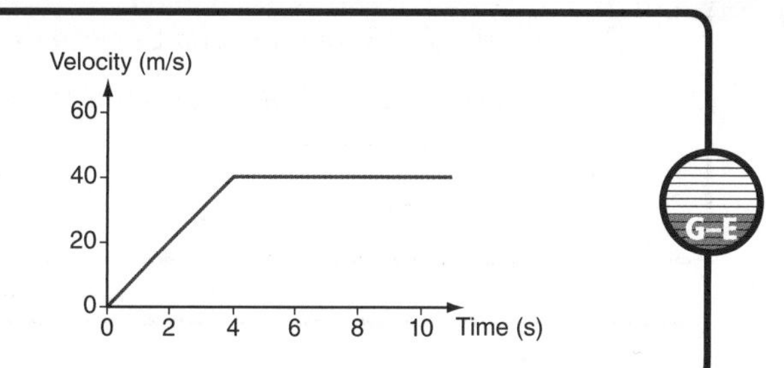

(a) During 0–4 seconds.

(b) During 4–10 seconds. [2 marks]

2 (a) Explain what is meant by velocity.

____

______________________ [2 marks)

(b) Explain why the velocity of a car changes as it drives round a corner at a constant speed.

____

______________________ [2 marks]

(c) The car initially travels at 40 m/s. After braking, the car slows down and stops 5 seconds later. What is its acceleration?

____

______________________ [3 marks]

Forcing it

1 The diagram shows the forces acting on a car.

(a) Calculate the resultant force acting on the car.

[2 marks]

G–E

(b) Is the car accelerating, decelerating or travelling at a steady speed?

[1 mark]

air resistance 400 N

forward force from engine 1000 N

400 N friction

2 A van is trying to pull another car out of a ditch. Both vehicles cannot move.

(a) Describe how the force from the van compares with the force from the car.

[2 marks]

D–C

(b) A tractor is used to pull the car out of the ditch.
How does the force from the tractor compare with the force from the van?

[1 mark]

Forces and acceleration

1 A car is stopped at traffic lights. The lights change and the car starts to move forwards.

(a) How did the forces acting on the car change?

[2 marks]

G–E

The car drives at a steady speed then slows down at a junction.

(b) How did the forces acting on the car change?

[2 marks]

2 The table shows results from an experiment to investigate the link between force and acceleration.

Force	Acceleration
1 N	2 m/s^2
2 N	4 m/s^2
3 N	6 m/s^2
4 N	7 m/s^2
5 N	10 m/s^2

(a) Describe the pattern shown in the results.

[2 marks]

D–C

(b) Explain whether the results have been taken over a large enough range.

[2 marks]

(c) Why should results from an experiment be repeated?

[2 marks]

Balanced forces

1 The diagram shows a tug of war between two teams. Neither team is winning so nobody is moving.

(a) Write down two pairs of balanced forces that act on the person labelled X

(i) Pair 1 ____________________ and ____________________

(ii) Pair 2 ____________________ and ____________________ [4 marks]

(b) Describe how forces change when the team on the right hand side starts to win by pulling the other team towards them.

____________________ [2 marks]

X

G–E

2 A person blows up a balloon. When the balloon is released, it moves away, before falling to the floor.

(a) Explain what provides the force on the balloon when it moves.

____________________ [1 mark]

(b) Why does the balloon fall to the floor?

____________________ [1 mark]

(c) Why doesn't the balloon fall through the floor?

____________________ [2 marks]

D–C

Stop!

1 **(a)** Stopping distance can be split into two parts. What are these called?

____________________ [2 marks]

(b) Nicole is taking her first driving lesson. Why will her thinking distance be longer than her driving instructor's?

____________________ [1 mark]

(c) A car has more energy if it travels faster. How does this affect its stopping distance?

____________________ [2 marks]

G–E

2 The table shows how thinking distance changes with speed.

Thinking distance (m)	0	6	9	12
Speed (km/h)	0	32	48	64

(a) Describe how thinking distance changes with speed.

____________________ [2 marks]

(b) A driver drinks a glass of wine and sometime later their thinking time is calculated as 1.0 seconds. Calculate their thinking distance when travelling at 9 m/s.

____________________ [1 mark]

(c) A speed of 9 m/s is the same as 32 km/h. Use information from the table to describe the effect of the drink on the thinking distance of a driver.

____________________ [1 mark]

D–C

Terminal velocity

1 A lorry travels along the motorway at 80 km/h. A car travels in town at 40 km/h. Describe two reasons why the lorry has more air resistance than the car.

____

____ [2 marks]

2 (a) What is meant by terminal velocity?

____ [1 mark]

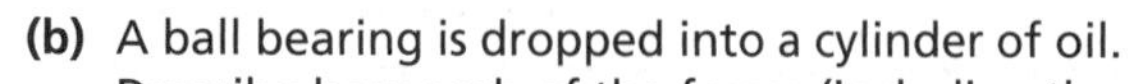

(b) A ball bearing is dropped into a cylinder of oil.
Describe how each of the forces (including the resultant force) changes:

(i) when the ball bearing is first released in the oil.

____

____ [2 marks]

(ii) when the ball bearing reaches terminal velocity.

____

____ [2 marks]

Forces and elasticity

1 Which objects are storing elastic potential energy?

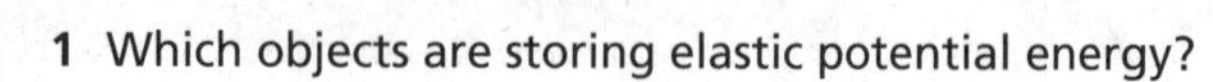

(a) a stretched rubber band ____

(b) a piece of string pulled tight ____

(c) a piece of plasticine ____

(d) a squashed spring in a chair seat ____

2 Alex set up the experiment shown in the diagram to test the hypothesis that the extension of a spring depends on the force applied to it. The extension of the spring is its change in length.

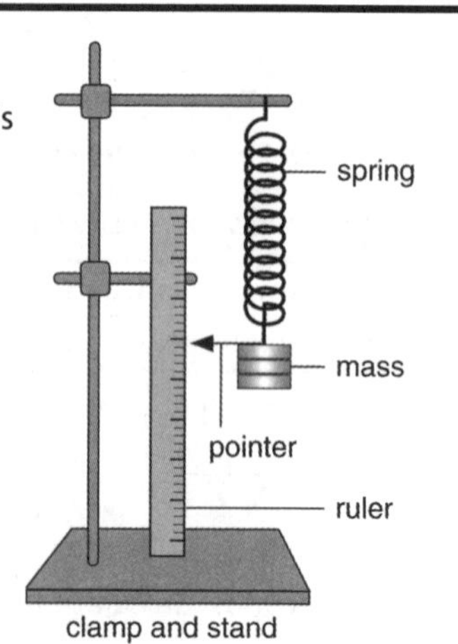

(a) What is the independent variable? ______________________ [1 mark]

(b) What is the dependent variable? ______________________ [1 mark]

(c) Explain why it is important that Alex writes down all the data he collects, and not just the result of calculations.

____ [1 mark]

(d) Explain two steps Alex should take to make sure his results are valid.

____

____ [2 marks]

Energy to move

1 An aeroplane has kinetic energy.

(a) What energy store was its kinetic energy transferred from?

______________________________ [1 mark]

(b) Where is its kinetic energy transferred to when it is flying?

______________________________ [1 mark]

(c) Where is its kinetic energy transferred to when it lands on the runway and stops?

______________________________ [1 mark]

G–E

2 Complete these sentences:

(a) A moving car has kinetic energy transferred from ____________ energy in its fuel. This energy is transferred to the surroundings in the form of ____________ energy and ____________ energy.

[3 marks]

(b) Explain why the energy transferred to the surroundings every second is less for a vehicle travelling on a smooth road than if it is travelling on a rough road at the same speed.

____

______________________________ [2 marks]

D–C

Working hard

1 Energy is transferred when a force makes something move. Is energy transferred in these examples?

(a) a person stretches a rubber band and releases it ____________ [1 mark]

(b) a person holds a picture against a wall ____________ [1 mark]

(c) a candle burns ____________ [1 mark]

(d) a person uses a stairlift to travel up a floor ____________ [1 mark]

G–E

2 Muhammad did 12 J of work when he lifted an apple.

(a) How much energy was transferred to the apple?

______________________________ [1 mark]

(b) What is the name of the force he was working against when he lifted the apple up?

______________________________ [1 mark]

(c) Muhammad dragged a box of apples 2 m along the ground. He measured the force needed as 25 N. How much work did he do?

____

____

______________________________ [3 marks]

D–C

Energy in quantity

G–E

1 A person climbs onto a diving board. They dive into the water.

(a) Where has the diver got most gravitational potential energy? Explain your answer

____

____ [2 marks]

(b) Where has the diver got most kinetic energy? Explain your answer.

____

____ [2 marks]

D–C

2 A box is stored on a shelf in a warehouse.

(a) Calculate the gravitational potential energy stored by a 6 kg box placed on a shelf 0.8 m above the ground.

____

____

____ [3 marks]

(b) How much energy does the box have when it is lifted to a shelf that is 0.6 m higher?

____

____

____ [3 marks]

Energy, work and power

G–E

1 An electric motor can do 6000 J of work in 10 seconds when it is lifting some sacks.

(a) What is its power?

____

____ [2 marks]

(b) If a more powerful setting is used, how will the time taken to lift the sacks change?

____ [1 mark]

D–C

2 A man uses a motor to lift some machinery. The motor lifts 400 kg of machinery by 3 m in 60 seconds.

(a) What is the power of the motor?

____

____

____

____ [4 marks]

(b) Explain what could change if the man uses a more powerful motor.

____

____

____

____ [4 marks]

Momentum

1 A ball has a mass of 0.5 kg.

(a) What is its momentum when it is not moving?

____ [1 mark]

G–E

(b) What is its momentum when a kick makes it move at 3 m/s?

____ [2 marks]

2 The diagram shows two trolleys travelling towards each other and colliding. The two trolleys stick together after the collision.

Inelastic collision

2 kg → 3 m/s 2 kg → O

before

2 kg 2 kg → V_1

after

D–C

(a) Calculate the total momentum before the collision.

____ [3 marks]

(b) Write down the momentum after the collision.

____ [1 mark]

(c) Calculate the speed that the joined up trolleys moves off with.

____ [3 marks]

Static electricity

1 A piece of plastic is rubbed on a person's jumper. The plastic becomes negatively charged.

(a) What charge is left on the jumper? ________________________ [1 mark]

G–E

(b) Describe one way to show someone that the plastic is charged.

____ [2 marks]

2 Owen rubs a balloon on some cloth and the balloon becomes charged.

(a) Explain how rubbing the balloon makes it negatively charged.

____

____ [3 marks]

(b) What charge does the cloth have after it is rubbed on the balloon?

____ [1 mark]

D–C

(c) How can the balloon be used to find the charge of another object?

____

____ [3 marks]

Moving charges

G–E

1 Two identical balloons are rubbed on the same piece of cloth. They both become charged.

(a) Will they have the same charge or a different charge?

____ [1 mark]

(b) The balloons are each suspended from a piece of thread. Explain if they will repel or attract each other when they are brought close together.

____

____ [2 marks]

D–C

2 A negatively charged balloon can stick on an uncharged painted wall.

(a) What is the name of this effect? _______________________________ [1 mark]

(b) Explain why the balloon can stick on an uncharged painted wall.

____

____ [2 marks]

Circuit diagrams

G–E

1 Write down the name of a component:

(a) that measures current. _______________________________ [1 mark]

(b) that gives out light when the current flows in one direction only. _______________ [1 mark]

(c) whose resistance falls if the temperature increases. _______________________ [1 mark]

(d) that can protect equipment if the current is too large. _____________________ [1 mark]

D–C

2 The diagram shows a series circuit and a parallel circuit, which use identical bulbs and cells. The ammeter reading in the series circuit is 0.4 A when the switch is closed.

Series circuit

A

V

Parallel circuit

A

V

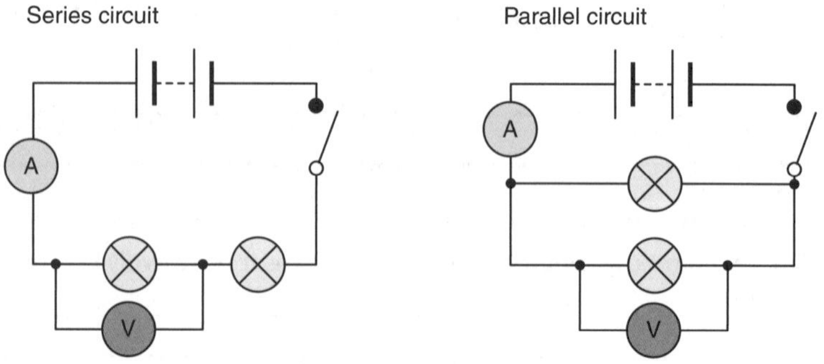

(a) What is the ammeter reading in the parallel circuit when the switch is closed?

_______________________________ [1 mark]

(b) The two bulbs are identical in the circuit, and each cell supplies 1.5 V.
What is the voltage supplied to each circuit?

____ [1 mark]

(c) Suggest a voltage reading for the voltmeter in the series circuit, giving a reason.

____

____ [2 marks]

Ohm's law

1 (a) Draw a circuit that could be used to measure the resistance of a piece of wire

[4 marks]

(b) When the circuit was built, it was used to measure the resistance of some wire. When the potential difference was 6 V, the current through the wire was 0.1 A. What is the resistance of the wire?

_______________ [3 marks]

2 Alex measured the resistance of a piece of wire.

(a) Explain what is meant by resistance.

_______________ [1 mark]

(b) The resistance of the piece of wire was 10 ohms.
Suggest a value for the resistance of a longer piece of the same wire. _______________ [1 mark]

D–C

(c) Write down two other factors that affect the resistance of a piece of wire.

_______________ [2 marks]

Non-ohmic devices

1 The diagram shows how the current varies with potential difference for a bulb.

(a) How does the temperature of the wire change when the current increases?

_______________ [1 mark]

Current

Potential difference

G–E

(b) How does the resistance of the wire change as it heats up?

_______________ [1 mark]

(c) Use the graph to explain when the resistance of the wire is lowest.

_______________ [2 marks]

2

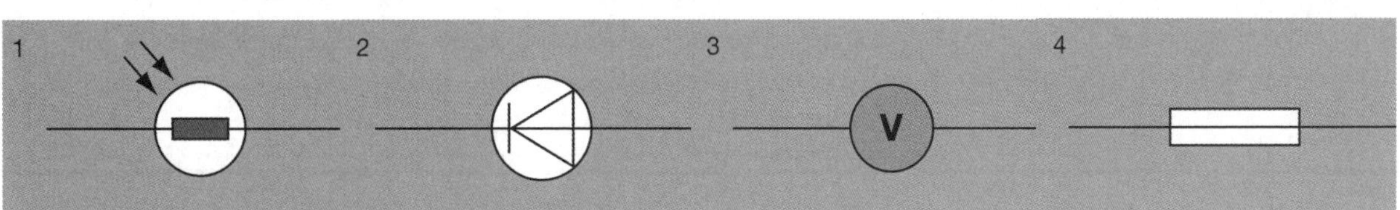

(a) How could you change the resistance of component 1?

_______________ [1 mark]

D–C

(b) Write down one use for a circuit that contains component 1. _______________ [1 mark]

(c) A diode only allows the current to flow in one direction. Which symbol shows a diode?

_______________ [1 mark]

Components in series

1 Peter has made a circuit with several lights wired in series.

G–E

(a) How does the brightness of each light change as he adds more lights in series?

______________________ [1 mark]

(b) What happens when one bulb in the circuit breaks? Explain your answer.

______________________ [2 marks]

(c) How does the current change in the circuit as Peter adds more bulbs?

______________________ [1 mark]

2 The circuit shows three bulbs wired in series.

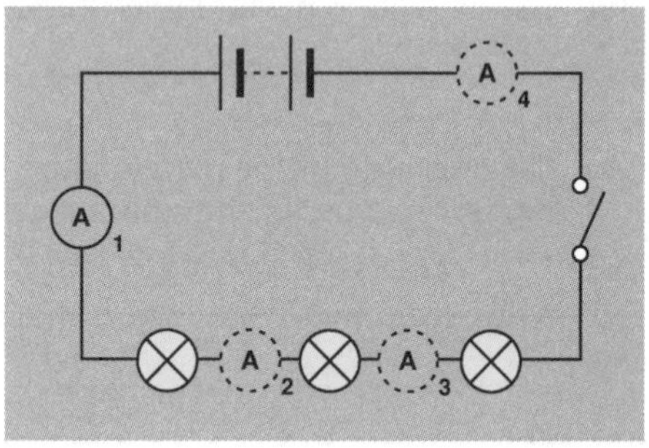

D–C

(a) Fred measures the current in four places in the circuit. What can you say about the readings?

______________________ [1 mark]

(b) There are two cells each supplying 1.5 V. What is the value of the potential difference supplied to the circuit? ____V [1 mark]

(c) The bulbs are all identical. Fred measures the potential difference across two of the bulbs. What is the reading on the voltmeter? _____ V [2 marks]

(d) An extra bulb, identical to the others, is added to the circuit in series. Explain how the resistance of the circuit changes in as much detail as possible.

______________________ [2 marks]

Components in parallel

1 Peter made a circuit with several lights wired in series. He changed the circuit so that each light was wired in parallel.

G–E

(a) How does the brightness of each light change as he adds more lights in parallel?

______________________ [1 mark]

(b) What happens when one bulb breaks in the circuit? Explain your answer.

______________________ [2 marks]

(c) How does the current leaving the battery change as Peter adds more bulbs?

______________________ [1 mark]

2 The circuit shows three bulbs wired in parallel. The circuit is switched on.

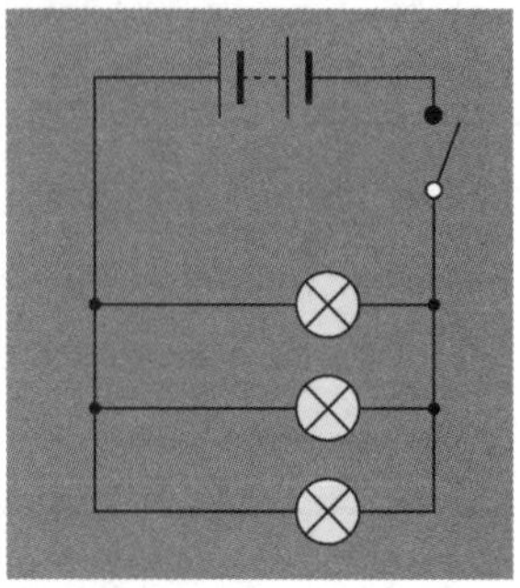

D–C

(a) If the current through each bulb is 0.5 A, calculate the current through the battery.

______________________ [1 mark]

(b) The voltage supplied by the battery is 6 V. Write down the voltage across each of the three bulbs. ______________________ [1 mark]

(c) Write down two advantages of wiring bulbs in parallel rather than in series.

______________________ [2 marks]

Household electricity

1 The diagram shows two traces from an oscilloscope.

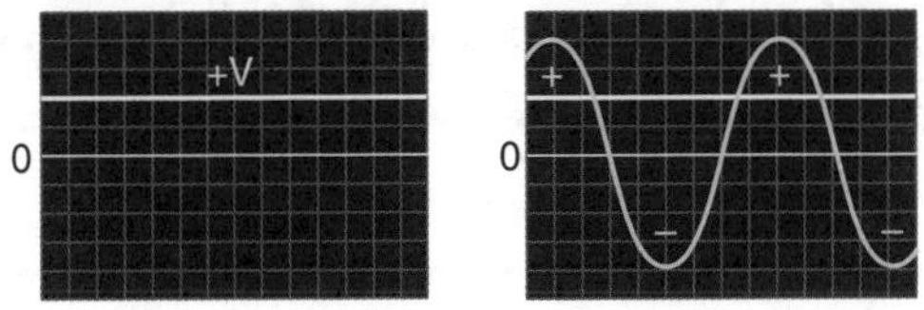

A B

(a) Which diagram shows the trace you would get from a battery?

______________________________ [1 mark]

(b) Explain what is meant by "the frequency of the current is 50 Hz".

____

______________________________ [2 marks]

G–E

2 Each square on the y-axis represents a voltage of 1 volt. Calculate the voltage of the wave shown on trace A.

____

______________________________ [2 marks]

D–C

Plugs and cables

1 Alf has finished wiring an electric plug. Describe three checks that he should make before using the equipment.

____

____

______________________________ [3 marks]

G–E

2 Explain why a radio does not use three-core wire.

____

____

______________________________ [3 marks]

D–C

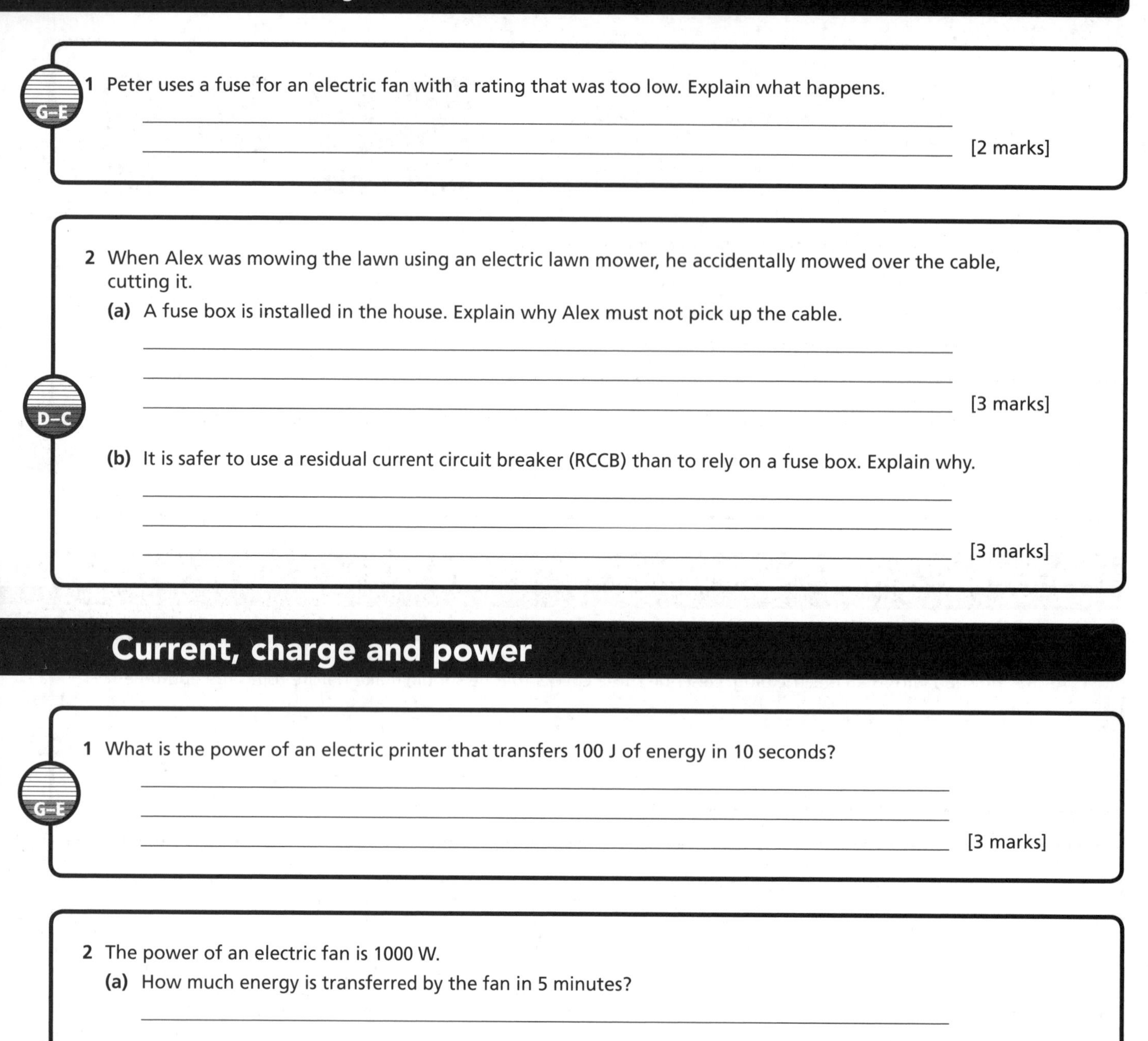

Electrical safety

G–E

1 Peter uses a fuse for an electric fan with a rating that was too low. Explain what happens.

____

____ [2 marks]

D–C

2 When Alex was mowing the lawn using an electric lawn mower, he accidentally mowed over the cable, cutting it.

(a) A fuse box is installed in the house. Explain why Alex must not pick up the cable.

____

____

____ [3 marks]

(b) It is safer to use a residual current circuit breaker (RCCB) than to rely on a fuse box. Explain why.

____

____

____ [3 marks]

Current, charge and power

G–E

1 What is the power of an electric printer that transfers 100 J of energy in 10 seconds?

____

____

____ [3 marks]

D–C

2 The power of an electric fan is 1000 W.

(a) How much energy is transferred by the fan in 5 minutes?

____

____

____ [3 marks]

(b) The current through the fan is 4.3 A. Calculate the charge transferred by the fan in 5 minutes.

____

____

____ [4 marks]

Structure of atoms

1 Complete the table to show the relative charges and masses of particles in the nucleus.

	Mass	Charge
electron	(a)	-1
proton	1	(b)
(c)	1	0

2 Carbon has several isotopes.

(a) What is different for different isotopes of carbon?

______________________________ [1 mark]

(b) What is the same for all atoms of carbon?

____

______________________________ [2 marks]

D–C

Radioactivity

1 **(a)** What is meant by the term "radioactive"?

______________________________ [1 mark]

(b) Which form of radiation is most ionising: alpha, beta or gamma?

______________________________ [1 mark]

G–E

(c) Which types of radiation can penetrate through the skin?

____

______________________________ [2 marks]

2 **(a)** Americium-241 emits alpha particles. Explain why smoke detectors containing Americium-241 are not a danger to people.

____

______________________________ [2 marks]

(b) Why can't a source of gamma rays be safely stored in a cardboard box?

______________________________ [1 mark]

More about nuclear radiation

G–E

1 Describe the plum pudding model of the atom.

____

____

____ [3 marks]

D–C

2 The diagram shows one of the first models of the atom.

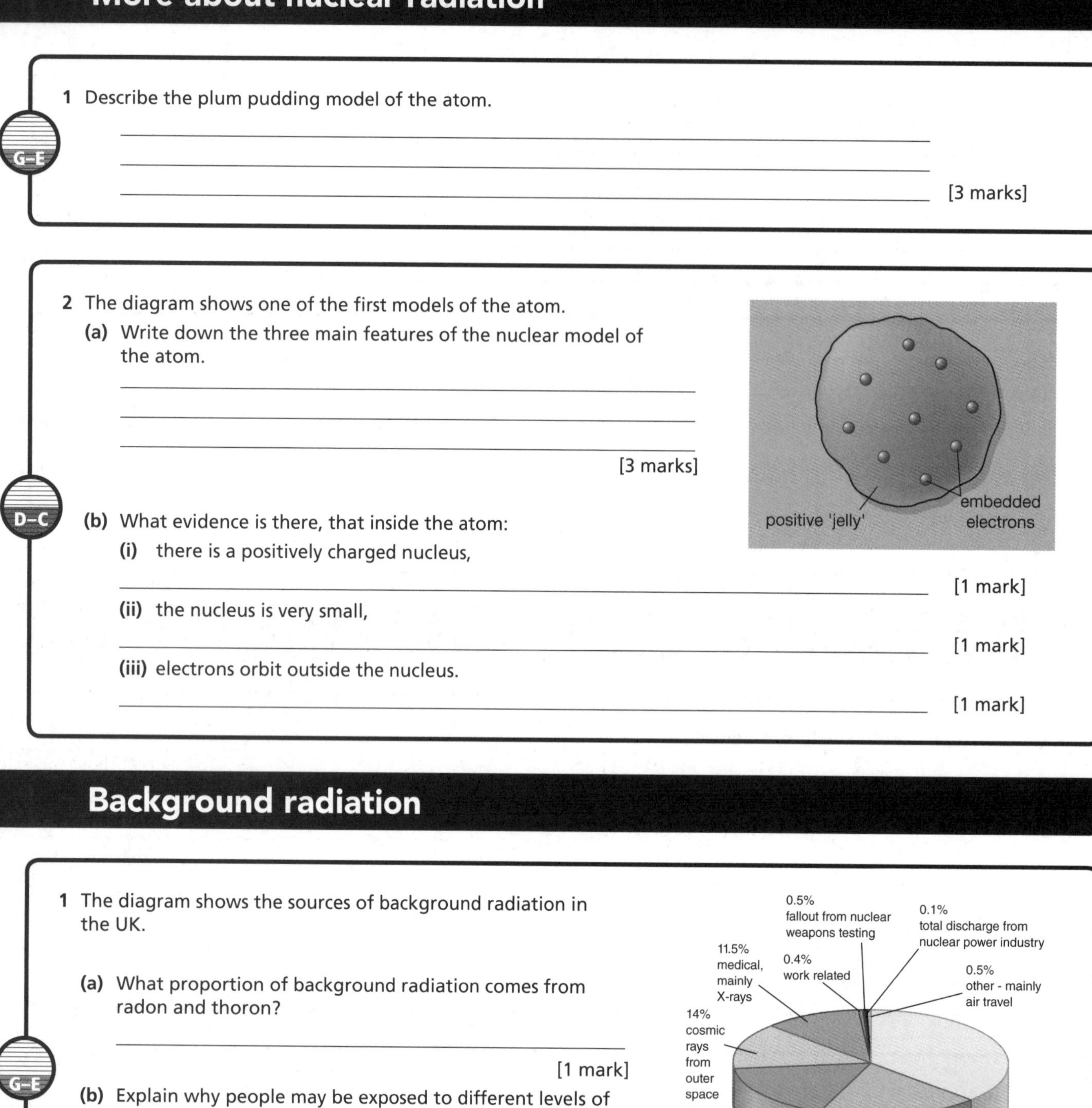

(a) Write down the three main features of the nuclear model of the atom.

____

____

____ [3 marks]

(b) What evidence is there, that inside the atom:

(i) there is a positively charged nucleus,

____ [1 mark]

(ii) the nucleus is very small,

____ [1 mark]

(iii) electrons orbit outside the nucleus.

____ [1 mark]

Background radiation

G–E

1 The diagram shows the sources of background radiation in the UK.

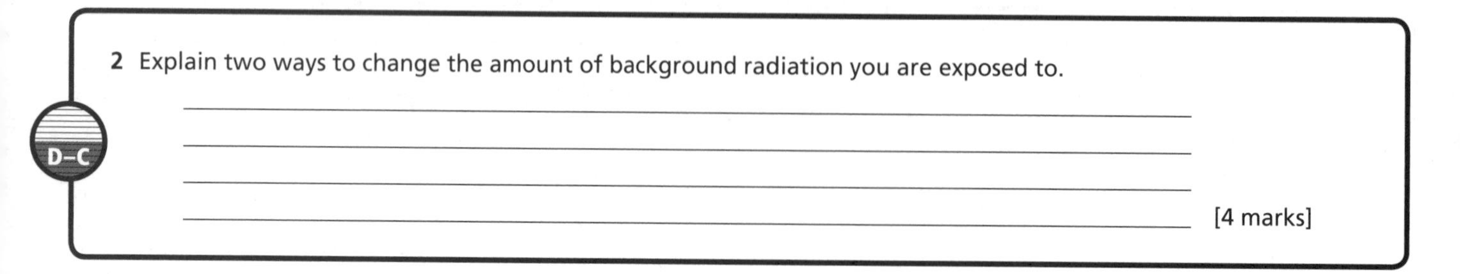

(a) What proportion of background radiation comes from radon and thoron?

____ [1 mark]

(b) Explain why people may be exposed to different levels of background radiation.

____

____

____ [3 marks]

D–C

2 Explain two ways to change the amount of background radiation you are exposed to.

____

____

____

____ [4 marks]

Half-life

1 (a) Explain what is meant by half-life.

[3 marks]

(b) The proportion of radioactive carbon in a wooden arrow has been measured. Explain whether the proportion of radioactive carbon in an arrow 10 000 years old will be more or less than that found in an arrow made from modern wood.

[2 marks]

2 A sample of cobalt has a half-life of 5 years. Its count rate is measured as 1200 counts per second.

(a) 5 years later, what will the count rate be?

[1 mark]

(b) How long will it take for the count rate to fall from 1200 counts per second to 150 counts per second?

[3 marks]

Using nuclear radiation

1 The diagram shows one method of controlling thickness of cardboard in a factory.

(a) Write down one reason why it is important to control the thickness of cardboard.

[1 mark]

(b) Why can't alpha radiation be used to monitor the thickness of cardboard?

[1 mark]

(c) Explain what adjustments the machine should make to the separation of the rollers if the amount of radiation reaching the detector decreases.

[2 marks]

2 Explain why a radioactive tracer used to investigate an underground leaking water pipe should emit gamma rays

[3 marks]

Nuclear fission

G–E

1 **(a)** What is meant by nuclear fission?

____ [2 marks]

(b) Write down one use for nuclear fission in the UK.

____ [1 mark]

D–C

2 The diagram shows a chain reaction.

(a) How many neutrons are produced from the first stage in this chain reaction?

____ [1 mark]

(b) Explain why a chain reaction must be controlled.

____ [3 marks]

Nuclear fusion

G–E

1 **(a)** Describe two differences between nuclear fission and nuclear fusion.

____ [2 marks]

(b) Write down one place in our solar system where nuclear fission takes place.

____ [1 mark]

D–C

2 Explain how nuclear fusion can explain the presence of different elements in the Universe.

____ [4 marks]

Life cycle of stars

1 Describe how the forces acting in a star control how it moves between different stages in its life cycle.

G–E

[5 marks]

2 Explain why the stable phase of a star lasts so long.

D–C

[3 marks]

Extended response question

Modern cars are designed to protect people in a crash. Describe the different safety features in a modern car. Explain how they help to protect people from death or injury.

The quality of written communication will be assessed in your answer to this question.

[6 marks]

B1 Biology Checklist

1 I can describe how lifestyle can affect health ☐

2 I can explain how white blood cells defend us against pathogens ☐

3 I can describe how Semmelweiss reduced the number of deaths in hospitals and explain the key evidence that led him to his conclusions ☐

4 I can explain how vaccination produces immunity against a disease ☐

5 I can explain the limitations of antibiotics and the implications of their overuse ☐

6 I can describe how to safely grow cultures of microorganisms in the laboratory ☐

7 I can describe the pathway taken by a nerve impulse during a reflex action, including how the impulse crosses a junction between two neurones ☐

8 I can explain why it is important to keep temperature and blood sugar level constant ☐

9 I can describe how FSH, LH and oestrogen control the menstrual cycle ☐

10 I understand plant growth responses (tropisms) ☐

11 I can explain how drugs are tested in the laboratory and in clinical trials ☐

12 I can use the terms legal, illegal, dependency and addiction when discussing drugs ☐

13 I can describe adaptations of plants and animals and explain why they have these ☐

14 I can describe how energy is moved along a food chain ☐

15 I can describe what a pyramid of biomass shows us ☐

16 I can explain how carbon is constantly cycled ☐

17 I understand that different characteristics depend on different genes and the environment ☐

18 I can describe different cloning techniques ☐

19 I can describe the advantages and disadvantages of GM crops ☐

20 I can explain how natural selection works ☐

B2 Biology Checklist

1 I can describe the main features of animal, plant, bacterial and yeast cells ☐

2 I can describe the function of the different parts (organelles) of a cell ☐

3 I can describe how oxygen needed for respiration passes through cell membranes by diffusion ☐

4 I can explain that, in multicellular organisms, cells may differentiate and become specialised ☐

5 I can describe the function of a number of animal and plant tissues, organs and systems ☐

6 I can write down the word equation for photosynthesis ☐

7 I can describe the three factors that limit the rate of photosynthesis and how they affect the process ☐

8 I can list the range of substances, together with their uses, that plants can make from glucose ☐

9 I can describe how and why you would use quadrats and transects when investigating the distribution of organisms ☐

10 I can list some examples of the different types of protein found in the human body ☐

11 I can explain that enzymes act as catalysts, increasing the rate of chemical reactions ☐

12 I can describe at least one domestic and one industrial use of enzymes, including its advantages and disadvantages ☐

13 I can describe differences between aerobic and anaerobic respiration ☐

14 I can list the differences between mitosis and meiosis ☐

15 I can describe the results of simple genetic crosses by interpreting genetic diagrams ☐

16 I can describe the chromosome combinations that determine sex ☐

17 I can describe some of the ways in which fossils are formed ☐

18 I can give some of the reasons that organisms become extinct ☐

19 I can describe how a new species can be formed by isolation ☐

C1 Chemistry Checklist

1. I can name and position the three subatomic particles in an atom of any of the first 20 elements using the periodic table, including writing the electronic structures ☐
2. I can describe how non-metal compounds share electrons to form molecules, and how the atoms of metal and non-metal compounds transfer electrons to form ions ☐
3. I can understand and use a symbol equation to determine the number of atoms in a reaction, and calculate the mass of a reactant or product from the masses of the other reactants and products ☐
4. I can explain how limestone can be used to produce quicklime, limewater, cement, mortar and concrete ☐
5. I can evaluate the social, economic and environmental impacts of quarrying limestone and metal ores ☐
6. I can describe how metals – such as iron, copper and aluminium – are extracted from their ores by reduction ☐
7. I can use the reactivity series to identify whether electrolysis or heating with carbon is the best method of extraction for a metal ☐
8. I can explain why electrolysis is an expensive method of extraction, and list the benefits of recycling metals ☐
9. I can describe the differences between iron, low- and high-carbon steel, and how making alloys produces metals that are more useful than the pure elements ☐
10. I can describe that crude oil is a mixture of a large number of compounds that can be separated into fractions by fractional distillation, using the differences in boiling points between molecules ☐
11. I can state the trends in boiling points, viscosity and flammability, as the mean molecule size in each fraction gets larger ☐
12. I can recognise that a molecule is an alkane (C_nH_{2n+2}) from its structural formula; and draw the structural formulae of methane (CH_4), ethane (C_2H_6), propane (C_3H_8) and butane (C_4H_{10}) ☐
13. I can describe the process of combustion, and describe the polluting effects of carbon, nitrogen and sulfur compounds ☐
14. I can evaluate the economic, ethical and environmental issues surrounding the use of both crude-oil-based fuels and biofuels ☐
15. I can describe how hydrocarbons can be broken down (cracked) to produce alkanes and unsaturated hydrocarbons called alkenes (C_nH_{2n}) that contain a double carbon–carbon bond ☐
16. I can describe how alkenes can be used to make polymers such as poly(ethene) and poly(propene) ☐
17. I can describe some useful applications of polymers; and new uses that are being developed, including biodegradable plastic bags ☐
18. I can describe two ways of making ethanol ☐
19. I can explain how plant material can be processed to produce plant oils ☐
20. I can describe how mixtures of oil and water can be emulsified, and explain how an emulsifier works ☐
21. I can describe how to test if an alkene or a plant oil contains carbon=carbon bonds using bromine water ☐
22. I can describe the structure of the Earth, and explain how tectonic plates cause earthquakes and volcanic eruptions ☐
23. I can describe the chemical content of the Earth's first atmosphere, how it formed, and how it has changed to the atmosphere today ☐
24. I can explain how carbon is stored in rocks, fossil fuels and the sea, and the consequences of increasing the concentration of carbon dioxide in the air ☐

C2 Chemistry Checklist

1. I can describe how atoms lose or gain electrons to gain noble gas structures ☐
2. I can describe how atoms join using ionic and covalent bonds to make compounds ☐
3. I can describe the following structures: sodium chloride crystal lattice; metallic lattices; and macromolecules, such as diamond, graphite and silicon dioxide ☐
4. I can explain the different properties of compounds, including electrical conductivity, in terms of their structure ☐
5. I can describe the differences between thermosetting and thermosoftening polymers; how shape memory alloys work; and the benefits and uses of nano science ☐
6. I can describe the atomic structure of the first 20 elements using the periodic table, and use the group number to find the number of outer electrons of the other elements ☐
7. I can calculate the relative formula mass of a compound, and know that the relative atomic mass in grams is known as a mole of substance and has identical numbers of particles ☐
8. I can calculate the percentage composition of an element in a compound ☐
9. I can explain why the theoretical yield in a reaction is not always achieved, and explain that many chemical reactions are reversible ☐
10. I can explain why using instrumental methods, such as gas chromatography and mass spectrometry, provide a quick and reliable analysis of substances ☐
11. I can calculate the rate of a chemical reaction, and know that concentration (and pressure in gases), temperature, particle size, and catalysts all affect the rate of a reaction ☐
12. I can use collision theory to explain how and why the rate of a chemical reaction will change when one variable is altered ☐
13. I can plot, use and interpret rate-of-reaction graphs ☐
14. I can identify exothermic and endothermic reactions and describe their use in hand warmers, sports injury packs and self-heating food applications ☐
15. I can choose the correct acid for making a salt, and describe how to make soluble and insoluble salts ☐
16. I can describe the difference between an alkali and a base ☐
17. I can describe the process of electrolysis of both solutions and molten liquids in terms of electron transfers; and the uses of the products, including electroplating and obtaining sodium and chlorine from molten sodium chloride, and sodium hydroxide, chlorine and hydrogen from a solution of sodium chloride ☐

P1 Physics Checklist

1. I can compare how heat is transferred by conduction, convection and radiation ☐
2. I know what factors have an effect on the way we heat and insulate buildings ☐
3. I can use kinetic theory to explain the properties of different states of matter ☐
4. I know what is meant by efficient use of energy, and how to compare the efficiency of energy transfers ☐
5. I can calculate the energy transferred by electrical appliances ☐
6. I can describe different methods of generating electricity including those using renewable and non-renewable energy sources ☐
7. I can evaluate the most suitable ways of generating electricity in different circumstances ☐
8. I can explain why electricity is transmitted at high voltages in the National Grid ☐
9. I can describe properties of transverse waves, including reflection and refraction ☐
10. I can describe the uses and hazards of electromagnetic waves ☐
11. I can calculate the speed of waves ☐
12. I can describe the properties of longitudinal waves such as sound waves ☐
13. I know what is meant by red-shift ☐
14. I can describe the Big Bang theory and explain why scientists believe the Universe is still expanding ☐

P2 Physics Checklist

1. I can calculate a resultant force, and describe its effect on objects including elastic objects ☐
2. I can calculate the acceleration of an object ☐
3. I can draw distance–time graphs and velocity–time graphs, and describe what they show ☐
4. I can describe different factors that affect the stopping distance of a vehicle ☐
5. I can explain why a moving object reaches a terminal velocity ☐
6. I can carry out calculations involving power and energy ☐
7. I can explain momentum changes during collisions and explosions using calculations ☐
8. I can describe how a static electric charge builds up and its effects ☐
9. I know standard circuit symbols and can draw and interpret voltage–current graphs for different components ☐
10. I can use Ohm's law ☐
11. I can calculate current, voltage and resistance in series and parallel circuits ☐
12. I can describe safe practices in the use of electricity ☐
13. I can describe the structure of atoms ☐
14. I can compare different properties of ionising radiation ☐
15. I know the origins of background radioactivity ☐
16. I can carry out half-life calculations ☐
17. I know what nuclear fission is and what is meant by a chain reaction ☐
18. I know what nuclear fusion is and how it is involved in the formation of different elements ☐
19. I can describe the life cycle of stars ☐

B1 Answers

Pages 148–149

Diet and energy

1a food [1]

b Any sensible suggestion, e.g. moving, chemical reactions inside cells, heart beating [1]

c It is stored as fat. [1]

2a Richard is exercising as well (more) [1]. Richard has a higher metabolic rate [1].

b If he exercises then he will use up stored fat to release energy [1]. If his metabolic rate is higher than Susan's, then chemical reactions in his cells will use up more energy than hers [1].

Diet, exercise and health

1a liver [1]

b It can lead to heart disease. [1]

2 A high level of cholesterol in the blood increases the risk of developing plaques [1] in the walls of the arteries that supply the heart with blood [1]. Plaques make it more likely that a blood clot will form in the artery. The clot can block the artery [1]. This will stop blood carrying oxygen getting to the heart muscle so the heart cells die [1]. This area of the heart no longer functions [1]. This is a heart attack.

Pathogens and infections

1 A microscope [1] because bacteria are too small to be seen with the naked eye / they have to be magnified [1].

2 Bacteria produce toxins [1]. Viruses reproduce inside our body cells [1]. When the cell gets filled up with viruses it bursts open, killing the cell [1].

Fighting infection

1a to kill pathogens [1]

b white blood cells [1]

2 By producing antibodies [1] and by producing antitoxins [1].

Pages 150–151

Drugs against disease

1a to reduce pain [1]

b The paracetamol has stopped the pain [1] for a while, but it cannot get rid of the sore throat as this is caused by a microorganism [1].

2a A drug that kills bacteria. [1]

b They spread the mucus onto a jelly [1] and add paper discs soaked in different antibiotics [1]. After a while they inspect the dish. The paper disc with the biggest clear area around it is the most effective antibiotic [1] because it stopped the growth of the bacteria in the mucus [1].

Antibiotic resistance

1a It cannot be killed by (most) antibiotics. [1]

b They are already ill and so do not have strong defences against disease / have a weakened immune system. [1]

2 One bacteria in a person undergoes a random mutation which means it is resistant to an antibiotic [1]. The person takes antibiotics to kill the bacteria. It works, except for the resistant one [1]. The bacterium now has no competitors and grows rapidly [1]. It divides and makes lots of identical copies of itself [1]. There is now a population of antibiotic-resistant bacteria.

Vaccination

1 When you have it the first time your white blood cells take time to make the antibody that can fight it [1]. During this time the chicken pox pathogen multiplies and makes you feel ill [1]. The next time the pathogen enters your body your white blood cells can make the antibody very quickly, so the pathogen gets destroyed before it can multiply and make you feel ill [1].

2a It is to stop children getting measles, mumps and rubella. [1]

b A small amount of the dead or inactive viruses that cause the diseases is injected into the blood [1]. The white blood cells attack them, just as they would attack living pathogens [1]. They remember how to make the antibody [1], so the child is now immune to the diseases.

c There was a study published that linked the vaccine to autism. [1]

d It was discovered that there was never any evidence to prove the link. [1]

Growing bacteria

1a **i** *E. coli* [1], **ii** agar jelly [1]

b It had no microorganisms on it.

2 If it was 37 °C or more, it would encourage the growth of pathogens that live in the body [1]. If the dish is then opened, the pathogens could escape and cause illness [1].

Pages 152–153

Co-ordination, nerves and hormones

1 Information is passed along nerves [1] from my eyes to my brain [1] and then to my hand [1].

2 The brain and spinal cord. [1]

Receptors

1 receptors [1] chemicals [1] effector [1]

2a sensory [1]

b arrow drawn to the right [1]

c (electrical) impulses [1]

Reflex actions

1a One example, such as: blinking/knee jerk/sneezing/coughing. [1]

b to protect us from harm [1]

2 Receptors in the skin detect the pin/pain [1], impulse sent along sensory neurone to spinal cord [1], impulse travels across synapse to relay neurone [1], impulse travels across synapse to motor neurone

[1], impulse travels along motor neurone to arm muscle [1], muscle contracts pulling hand away [1].

Controlling the body

1a Any one of: water content, ion (salt) content, temperature, concentration of sugar in the blood. [1]

b So chemical reactions can take place inside the cells. [1]

2 This is the temperature that enzymes work best at [1]. Without enzymes, important chemical reactions in the body would not take place [1].

Pages 154–155

Reproductive hormones

1a It is there so that a fertilised egg can sink into it [1] and grow into a baby [1].

b it breaks down / menstruation [1]

2 High levels of oestrogen stops the production of FSH [1]. Without FSH, eggs cannot mature [1] so no eggs are released to be fertilised [1].

Controlling fertility

1a Either one of: FSH, LH [1]

b Either one of: FSH stimulates a woman's eggs to mature [1] LH stimulates an egg's release [1].

2 Fertility drugs are given to stimulate the maturation of several eggs [1]. Eggs are collected and fertilised by sperm from the father [1]. The embryos formed are inserted into the mother's uterus (womb) [1].

Plant responses and hormones

1a enables photosynthesis in the leaves [1]

b helps to firmly anchor the plant into the ground so it doesn't fall over/to absorb water [1]

2a positive [1]

b negative [1]

Drugs

1 an unwanted effect [1]

2 i One mark for any legal drug: e.g. alcohol or nicotine.

ii One mark for any illegal drug: e.g. cannabis or heroin.

b Nicotine is addictive [1], so a person will suffer withdrawal symptoms if they stop smoking [1].

Pages 156–157

Developing new drugs

1 It takes many years, many people involved. [1]

2a Some of the volunteers are given a placebo [1], which does not contain the drug [1], but neither they nor their doctor knows whether they have been given it [1].

b the age of the soldiers [1]

c One reason, for example: the recovery time for the soldiers who took the drug was less than for those who took the placebo. [1]

Legal and illegal drugs

1a Cannabis: illegal [1], alcohol: legal [1], heroin: illegal [1], nicotine: legal [1], ecstasy: illegal [1].

b It can cause mental illness in young people [1], it can lead on to other more harmful drugs / drug-taking [1].

2a The number of deaths from drinking alcohol has risen (between the years 2000 and 2008) [1]. More men die from drinking alcohol then women [1].

b The number of people who died from alcohol is much higher than those who died from heroin (9031, which is around 10 times higher) [1]. This is because many more people drink alcohol than take heroin [1].

Competition

1 B (light, water, space, nutrients) [1]

2 The weeds will compete with the vegetable plants (for light, water, space and nutrients) [1], so the vegetable plants will not be able to grow as well [1].

Adaptations for survival

1a Any two from: thick fur to insulate it (keep it warm) [1], white fur for camouflage to catch prey [1], small ears to reduce heat loss [1].

b They will die/move away to a colder region. [1]

2a It deters predators from eating it. [1]

b The yellow and black are warning colours [1] so predators of the hoverfly think it is dangerous and leave it alone [1].

Pages 158–159

Environmental change

1 It will no longer be able to survive [1] because it is adapted to live in cold conditions [1].

2 Numbers will decrease [1] because plants will die through lack of water [1] so there will be less food for them [1].

Pollution indicators

1a It is a waste released to the environment. [1]

b Any one suitable example, e.g. sulfur dioxide, carbon dioxide, sewage. [1]

2a oxygen meter [1]

b It is low. [1]

c Mayfly larvae / freshwater shrimp / stonefly larva [1]. They can only live in unpolluted water [1].

Food chains and energy flow

1a i seaweed [1] ii limpet/octopus/seal [1]

b It only eats plants. [1]

c predator [1] octopus [1]

2 One reason from: some of the light misses the leaves altogether / hits the leaf and reflects / hits the leaf but goes all the way through without hitting any chlorophyll / hits the chlorophyll but is not absorbed because it is of the wrong wavelength (colour). [1]

Biomass

1 the mass of living material [1]

2 The sketch should show: a pyramid with the levels labelled with the name of the organism, starting with grass at the bottom [1], each level should then get progressively smaller [1].

Pages 160–161

Decay

1 bacteria [1] fungi [1]

2a To increase the moisture content [1] so the numbers of microorganisms in the compost increase [1], which speeds up decay and the production of compost [1].

b To let in oxygen [1] so the microorganisms responsible for decay can live (respire) [1].

c It contains nutrients [1] that increase their growth [1].

Recycling

1 It releases nutrients into the soil [1] that plants need in order to grow [1].

2 Added to food chain diagram: a label saying 'microorganisms', with arrows from all grass, antelope and lion pointing towards it.

The carbon cycle

1a decay/respiration (by microorganisms) [1]

b eating [1]

2a i combustion/burning [1]

ii photosynthesis [1]

b from eating plants or other animals [1]

c The animal will die [1] and the tissue will be decayed by microorganisms [1] which will carry out respiration [1] and release carbon dioxide [1].

Genes and chromosomes

1a genes [1]

b nucleus [1]

c fertilisation [1] sperm [1]

2 He inherited one from his mother in her egg [1] and one from his father in his sperm [1].

Pages 162–163

Reproduction

1a sperm [1]

b eggs [1]

c fertilisation [1]

d sexual [1]

2a asexual [1]

b It has the same genes. [1]

c They reproduce sexually [1] so have a different mix of genes [1].

Cloning plants and animals

1 genetically identical organisms [1]

2a So she has many identical copies of it. [1]

b Either: use tissue culture [1], take a piece of the tissue from the plant [1] and grow it in a sterile nutrient liquid or gel [1].
Or: take a cutting [1], cut off a small piece of the stem [1] dip it in hormone rooting powder and place it in soil [1].

Genetic engineering

1a It means adding a gene to an organism [1] that has come from another [1].

b Any one from: for making bacteria to make human insulin/herbicide-resistant crops/crops that produce a toxin that kills pests. [1]

2a They have genes from another organism.

b The gene for insulin is cut from a human chromosome; and a chromosome from bacterium is cut open using enzymes [1]. The insulin gene is inserted into the bacterium's chromosome [1]. The chromosome is put back into the bacterium [1].

c The bacteria grow best at this temperature [1] so insulin is produced more quickly [1].

Evolution

1 living things changing over time [1]

2 Giraffes will have a gene for long necks [1] which would be passed on to their offspring [1]; and reaching for leaves cannot change the gene [1].

Page 164

Natural selection

1a He will not hear predators [1] so will be eaten [1].

b The rabbit cannot reproduce [1] and pass on the gene for bad hearing to its offspring [1].

2a It's the same colour as the tree trunk (camouflaged) [1]. Makes it difficult to be spotted by predators [1].

b a change in a gene [1]

c The black moths are now camouflaged [1], so they do not get eaten [1] and they live to reproduce and pass on the gene to their offspring [1].

Evidence for evolution

1a It is the remains or impressions made by a dead organism. [1]

b They show how organisms have changed [1] over time [1].

2a Because each animal uses its limbs differently/they are adapted. [1]

b They all have the bones in the same place [1] which shows that birds, bats and humans all have a common ancestor [1] from which they evolved [1].

Page 165

Extended response question

5–6 marks

A detailed description of how a difference in concentration of auxins in the developing root will cause the roots to bend downwards. The answer will include information such as that in the roots a high concentration of auxins inhibits cell elongation. Auxin tends to accumulate on the lower side of a root. The lower side of the shoot grows more slowly than the upper surface. This causes the root to bend downwards. This is a tropism called gravitropism (or geotropism).

All information in answer is relevant, clear, organised and presented in a structured and coherent format. Specialist terms are used appropriately. Few, if any, errors in grammar, punctuation and spelling.

3–4 marks

Limited description of why a root bends downwards. The answer will state that auxins inhibit cell growth in roots. The upper side of the root will grow more, causing it to bend and this causes the root to bend downwards.

For the most part the information is relevant and presented in a structured and coherent format. Specialist terms are used for the most part appropriately. There are occasional errors in grammar, punctuation and spelling.

1–2 marks

An incomplete description, stating that the upper side of the root will grow more, causing it to bend downwards.

Answer may be simplistic. There may be limited use of specialist terms. Errors of grammar, punctuation and spelling prevent communication of the science.

C1 Answers

Pages 166–167

Atoms, elements, and compounds

1a formula [1]

b element [1]

c compound [1]

d symbol [1]

2a copper [1]

b copper carbonate [1]

c It is made of more than one element. [1]

d Malachite contains both rock and copper carbonate [1], so it is a mixture [1].

Inside the atom

1a electron [1]

b neutron [1]

c proton [1]

d B and C [1]

2a An atom of phosphorus contains 15 protons [1], 15 electrons [1] and has a mass of 31. [1]
It has 16 neutrons. [1]

b **i** N, As, Sb, or Bi [1]

ii Cl [1]

iii Si [1]

c 3 concentric circles with two crosses on inner most circle, eight crosses on middle circle [1] and five crosses on outer circle. [1]

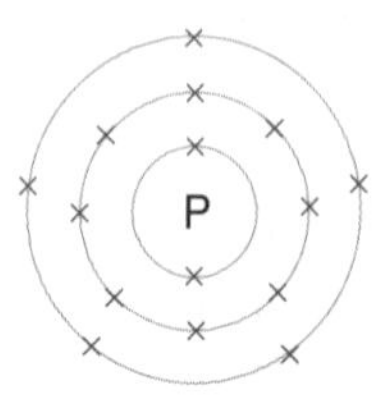

Element patterns

1a the number of protons [1]

b the number of protons and neutrons/how much the atom weighs [1]

c It is the same as the atomic number or number of protons. [1]

2a by proton number [1]

b periods [1]

c groups [1]

d They have similar properties/characteristics, [1] just like people in the same family. [1]

Combining atoms

1a a shared pair of electrons [1]

b

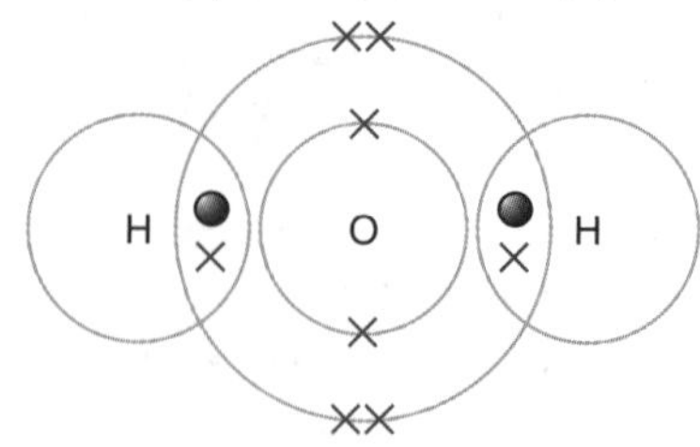

One mark for showing a pair of electrons between O and H atom, [1] for eight electrons on the outer shell of O atom.

c 2,8,1 [1]

d It loses its outer electron [1], and loses a negative charge, so it becomes a positively charged ion.[1]

e The chlorine gains the electron lost by the sodium[1], and becomes a negative ion [1], which is then attracted to the sodium ion.[1]

Pages 168–169

Chemical equations

1a Either of: $ZnCO_3$ or HCl [1]

b 3 [1]

c 2 [1]

2a Carbon dioxide. [1]

b **i** 3 [1]

ii 2 [1]

Building with limestone

1a calcium carbonate [1]

b cement [1], mortar [1] and concrete [1]

c The acid rain reacts with the limestone [1] slowly dissolving the stone/building [1].

2 Social advantages: more employment, more money to pay for community facilities. [1]

Social disadvantages: pollution effects on local population (not on wildlife). [1]

Economic advantages: more money in the local economy, better roads and railways. [1]

Economic disadvantages: reduction in tourism. [1]

To gain full marks you must mention both advantages and disadvantages.

Heating limestone

1a carbon dioxide [1]

b calcium hydroxide [1]

c Soil can be too acidic [1], the slaked lime neutralises the acidity [1].

2a Carbon dioxide. [1]

b Bubble the gas through limewater [1] a white precipitate appears in the solution, or it goes milky/cloudy white. [1]

c **i** Clay [1].

ii Mortar is sand and cement [1], concrete is sand, cement and gravel/stones/aggregate. [1]

d Mortar is good at sticking things together [1],

concrete has greater strength than mortar when made into a beam. [1]

Metals from ores

1a ores [1]

b Any two from: zinc, copper, iron, lead. (2)

c Electrolysis uses electricity [1] and this is very expensive/more expensive than heating with carbon [1].

2a i Haematite [1]

ii Zincite [1]

iii Bauxite. [1]

b Bauxite because aluminium is higher in the reactivity series than carbon [1], so it cannot be reduced by carbon. [1]

c Removal of oxygen [1] from a compound.

Pages 170–171

Extracting Iron

1a i Limestone [1], ii coke [1] accept carbon.

b The oxygen in the air reacts with the carbon in the coke [1] to make carbon dioxide. More carbon dioxide is produced by the iron oxide (haematite reacting with carbon monoxide [1] made from carbon dioxide reacting with more coke. [1]

c Slag is formed from the impurities in the ore used. [1]

Metals are useful

1a good conductor of electricity [1]

b shiny [1]

c good conductor of heat [1]

d unreactive [1]

2a The atoms are in a regular arrangement/layers, [1] that can slide past each other without braking apart. [1]

b The outer electrons are only loosely held to each atom [1], so they are free to move within the layers [1], allowing electricity to flow.

c Metals are made from a single type of atom [1], alloys are mixtures of different metals and have different types of atoms in them [1].

d The different atoms are often different sizes, distorting the structure [1], and making it harder for the atoms to slide past each other. [1]

Iron and steel

1a It makes it brittle/easy to snap or break. [1]

b carbon dioxide [1]

c It is less brittle [1], so it can be easily bent/shaped [1].

2a It does not rust [1], the metal looks attractive in use. [1]

b They can be used for different uses. [1]

c They are in the transition metals.[1]

Copper

1 One mark for each in this order: concentrate, sulfur, carbon, impure (4)

2a Impure copper. [1]

b Copper. [1]

c At the cathode. [1]

d Under the anode. [1]

e They gain an electron, become atoms [1] and stick to the cathode [1]

Pages 172–173

Aluminium and titanium

1a i It is less dense than titanium. [1]

ii It has a higher melting point so is safe at higher speeds. [1]

iii It is cheaper than titanium [1], and has a lower melting point which means it can be recycled easier [1].

b Both are unreactive, so aluminium will not react with the drink or titanium with body fluids. [1]

2a Using electricity to break down a compound into simpler substances. [1]

b bauxite [1]

c i It needs electricity [1], which is expensive to produce [1].

ii Aluminium ore melts at a high temperature. [1] Adding cryolite reduces this temperature [1], saving energy. [1]

Metals and the environment

1a Aluminium melts at a lower temperature than is used to make aluminium. This makes recycling cheaper. [1]

b It uses less energy [1], and reduces demand for ores from rainforests that would be destroyed [1].

2a **Advantages**, two from: recycled metal uses less energy to purify, saves on world resources, there's no need to buy fresh ore, waste material is relatively cheap. [2]

Disadvantages, two from: an efficient collection system is needed that costs money, it's often mixed with other rubbish so needs to be separated, it needs to be transported for processing. [2]

Comparison of merits of recycling and using new metals [1] is given in answer, that justifies a clear preference [1].

b Any three from: causes dust, destroys land for previous use, increase of traffic in area, causes noise pollution, may produce environmentally damaging by-products/waste. [3]

A burning problem

1a oil, coal, and natural gas (methane) [1]

b A source of energy that can be replaced/won't run out. [1]

c hydrogen and carbon [1]

d water [1] carbon dioxide [1]

2a It remains in the air for some time [1], reducing the amount of sunlight that reaches the ground [1].

b Volcanic gases contain sulfur dioxide, which dissolves in rainwater [1] and makes the rain acidic [1].

c It may increase the quantity of carbon dioxide gas in the atmosphere. [1]

Reducing air pollution

1a Sulfur in the fuel reacts with oxygen to make sulfur dioxide, which makes acid rain [1], so removing the sulfur helps reduce acid rain [1].

b i sulfur dioxide [1]

ii limestone/ calcium carbonate [1]

2a To convert nitrogen oxides to nitrogen before the exhaust gases are released, or carbon monoxide to carbon dioxide. [1]

b It reduces acid rain. [1]

c Ethanol is renewable as it is grown [1]; and when burnt only produces CO_2 that was just removed from the air [1] by photosynthesis, so doesn't increase the amount of CO_2 in the air [1].

Pages 174–175

Crude oil

1a has many different sized molecules with different boiling points [1]

b diesel [1]

c liquid petroleum gas [1]

d The greater the mean number of carbon atoms, the higher the boiling range. [1]

2a a compound containing only carbon and hydrogen [1]

b i fractional distillation [1]

ii bitumen [1]

ii petroleum gas [1]

c The crude oil is heated up [1] so the liquids become gases/evaporate [1], as they rise up the column they cool and condense/become liquids [1], so they are separated by their boiling points [1].

Alkanes

1a CH_4 [1]

b 4 [1]

c a shared pair of electrons [1]

d

```
    H  H
    |  |
 H—C—C—H
    |  |
    H  H
```

2a If there are 8 C atoms, then the H atoms will be 2 × 8 = 16 [1] plus another 2, so 18 in total [1].

b i harder to burn/less flammable [1]

ii increases [1]

iii increases [1]

Cracking

1a We need more diesel and petrol than is present in crude oil. [1]

b i C_2H_4

ii It has a double C=C bond. [1]

iii polymers/plastics [1]

2a $C_{10}H_{22}$

b ethene [1]

c a double bond [1]

d It is very reactive [1] and can be used to make polymers [1].

Alkenes

1a A general formula can be used to work out the formula of a specific compound by substituting n for the number of carbon atoms. [1]

b C_4H_8, [1] if n = 4, then 2n = 8 [1]

c two single covalent bonds [1]

2a Alkenes have a double bond; alkanes do not. [1]

b the double bond can react with other substances [1], it is easy to break the double bond [1].

c There are double bonds that can be broken to allow more atoms to join into the molecule. [1]

Pages 176–177

Making ethanol

1 a)

```
    H  H    H
    |  |   /
 H—C—C—O
    |  |
    H  H
```

[1]

b i fermentation [1]

ii Yeast is living and needs to be warm to work quickly. [1]

2a water or steam [1]

b a fuel made from biological material [1]

c **Advantages**: the fuel is renewable/can be replaced, does not contribute to global warming/CO_2 in the air. [2]

Disadvantages: uses land that could grow crops for people/forces food prices up, hinders development of non-carbon based transport fuels. [2]

Polymers from alkenes

1a polymer [1]

b a large number [1]

c to make the reaction happen quickly [1]

d poly(styrene) [1]

2a a chain of repeating molecules bonded together [1]

b monomer [1]

c The double bond breaks apart [1], each end of the broken double bond joins with a different monomer molecule [1] to make a chain.

Designer polymers

1a to provide materials with a large number of different properties [1]

b i It allows the joint to move freely. [1]

ii It doesn't react with your body fluids. [1]

iii The Teflon has worn away so it is hard to move. [1]

2a Lightweight means cars and plane weigh less [1], use less fuel [1] and are easier to mould into the shapes required [1].

b It will stop you getting wet [1], but allow evaporated sweat to pass out of the coat [1].

c To make materials better suited to their use, or different applications. [1]

Polymers and waste

1a broken down by bacteria, fungi, etc., or by microbes [1]

b They don't decay/rot quickly [1]; it wastes valuable raw materials (as they are made from crude oil). [1]

c i The plastic can be recycled. [1]

ii the type of plastic, or recycling to be used [1]

2a i won't degrade quickly/will fill up landfill sites [1]

ii use up hydrocarbon resources/produce poisonous or obnoxious gases [1]

iii lots of different polymer types so sorting would be costly [1]

b Using landfill doesn't allow the resources to be easily used again [1], burning prevents the hydrocarbons being used again as they become CO_2, H_2O, and other gases [1], recycling allows the plastic to be re-used, saving crude oil resources [1].

Pages 178–179

Oils from plants

1 crushed [1], liquid [1], filtered [1], dissolved [1], distillation [1]

2a They are rich in energy. [1]

b seeds, nuts and fruits [1]

c The plant parts are crushed [1], and the oil filtered from the pulp [1].

d The oil is added to the potato [1], the potato is cooked at a higher temperature [1] causing different changes to the molecules than if cooked at a lower temperature [1].

Biofuels

1a They are easily replaced and so won't run out. [1]

b the Sun [1]

c On burning, it produces the same quantity of carbon dioxide as it absorbed when it was growing. [1]

d There is no sulfur to make sulfur dioxide which causes acid rain. [1]

2a **Advantages,** any two from: cheaper than fossil fuels, renewable resource, carbon neutral, could use agricultural waste products such as manure/waste from food crops. [2]

Disadvantages, any two from: reduction in land for food use, rainforest turned into land to grow fuel crops, increase in costs of food crops as reduced land available to grow food. [2]

Comparison of merits of biofuels against fossil fuels [1] is given in an answer, that justifies a clear stated preference [1].

b Plant oils can be reacted with alcohols such as methanol [1], to make less viscous esters [1] that can be used in diesel engines.

Oils and fats

1a Saturated fats have only single C–C bonds. [1]

b Animal fats have melting points above room temperature. [1]

c **i** It is a mixture of liquid oil and solid butter with different melting points. [1]

ii The vegetable oil has unsaturated molecules [1], reducing the proportion of saturated fat to unsaturated fat molecules. [1]

2a The orange colour disappears. [1]

b There's no mention of the mass of the spread used [1], and no mention of the volume of ethanol used [1].

c Butter [1]. The reaction is quickest [1].

d An unsaturated fat has some double bonds [1], a saturated fat has very few double bonds [1].

Emulsions

1a a mixture of oil and water [1]

b The emulsifier prevents the oil and water mixture from separating. [1]

c The oil and vinegar would separate [1], which might make it look unappealing to consumers who then wouldn't buy it [1].

2a immiscible [1]

b It's an emulsifying agent or emulsifier. [1]

c **i** water and vinegar [1]

ii to act as an emulsifying agent [1]

Pages 180–181

Earth

1

Part	Name	Its structure
A	crust [1]	solid rock
B	mantle [1]	plastic/jelly-like [1] (with convection currents)
C	core [1]	made from hot iron and nickel [1], (solid inner, liquid outer)

Continents on the move

1a The continents are on (tectonic) plates [1] that are moving or drifting on the surface of the Earth. [1]

b Pangea has separated [1] and the pieces have moved to become the continents we now have. [1]

2a There was no evidence for the theory [1], no one could explain how the plates would move [1].

b The rock types and structures were similar in Africa and South America [1]. Rocks containing similar fossils were found in both Africa and South America [1].

c The North American plate is moving west and the Eurasian plate is moving east [1], carried by convection currents [1].

Earthquakes and volcanoes

1a They are not very powerful or destructive. [1]

b tectonic plates suddenly slipping past each other [1] that release energy [1]

c The hardness of all the rocks is not known [1], the forces building up are not known [1]/scientists have insufficient information (1 mark only).

2a It's a piece of the Earth's crust. [1]

b at the edges of tectonic plates [1]

c The meeting of the plates creates pressure in the mantle [1], this is relieved by magma being forced to the surface [1].

d **i** a tidal wave [1]

ii An earthquake took place in the Earth's crust beneath the sea [1]. The movement of the crust caused a shockwave in the water [1].

The air we breathe

1a B [1]

b C [1]

c i It depends on the nearby activity, for example a classroom full of respiring students will have more carbon dioxide than an empty field. [1]

ii Air over water will have more evaporated water in it. [1]

2a volcanoes [1]

b i 4.7 billion years [1]

ii B [1]

iii E [1]

c Ultraviolet light from sunlight would kill it [1]. The ozone protected the life from the ultraviolet light [1].

Page 182

The atmosphere and life

1a i The mean daytime temperature is above the boiling point of water. [1]

ii They might be damaged by the sulfuric acid clouds. [1]

b i photosynthesis [1]

ii It provides food for animals/it provides oxygen for animals to breathe. [1]

2a There are no fossil records of early life/It was too small to leave traces [1].

b Seawater allows all the compounds to be able to meet each other/Ultraviolet light could not damage or kill the emerging life. [1]

c All living things/proteins are made from amino acids [1]. If we know how to make amino acids then it is a possible route to making life [1].

Carbon dioxide levels

1a Man has burnt lots of fossil fuels. [1]

b Any two from: ice caps will melt; sea levels will rise; climate change. (2)

2a i A = photosynthesis [1]

ii B = burning [1]

iii C = death and decay [1]

iv D = respiration [1]

b Both plants and animals carry out respiration. [1]

c The carbon dioxide produced by burning, and given off to the air [1], has recently been removed from the air by photosynthesis [1].

Page 183

Extended response question

5 or 6 marks:

There is a clear, balanced and detailed description of the positive and negative environmental impacts involved with increasing the percentage of biofuels, with 5–6 points from the examples given. The answer shows almost faultless spelling, punctuation and grammar. It is coherent and in an organised, logical sequence. It contains a range of appropriate or relevant specialist terms used accurately.

3 or 4 marks:

There is some description of the positive and negative environmental impacts involved with increasing the percentage of biofuels, with 3–4 points from the examples given. There are some errors in spelling, punctuation and grammar. The answer has some structure and organisation. The use of specialist terms has been attempted, but not always accurately.

1 or 2 marks:

There is a brief description of the environmental impacts involved with increasing the percentage of biofuels, with 1–2 points from the examples given. The spelling, punctuation and grammar are very weak. The answer is poorly organised with almost no specialist terms and/or their use demonstrates a general lack of understanding of their meaning.

Possible points to make:

positive

- carbon neutral
- growing plants will remove carbon dioxide from atmosphere
- less crude oil will be needed
- less risk of oil spills if less oil needed
- biofuels are natural, and spillages will be easier to deal with.

negative

- will not reduce carbon emissions as still burning carbon dioxide
- more land will be needed to grow crops
- rainforest may be destroyed
- animal habitats may be disrupted or destroyed.

P1 Answers

Pages 184–185

Energy

1a to the surrounding air and road [1]

b to the surrounding air [1]

c to the arrow and surrounding air [1].

2a Electrical energy [1]; changes to heat [1]; kinetic energy [1]; and sound [1].

b It is transferred to the surroundings. [1]

c Olympus 1000 [1], it took least time to dry the cloth [1].

d Any two valid points, for example: distance from hairdryer to cloth [1]; dampness of cloth initially [1]; dryness of cloth finally [1]; size of cloth [1].

Infrared radiation

1a radiation [1]

b Black is better at absorbing radiation so energy is transferred to the car more quickly. [2]

2a Black [1]. Solar panels absorb infrared radiation from the Sun [1]. Black is the best colour to absorb infrared radiation [1].

b Any two from: radiation from the Sun is more intense from the south [1], more infrared radiation will be absorbed [1], the water will heat up quicker [1].

c The house loses heat by radiation when it is warmer than the surroundings [1]. At night/during winter, it is cooler outside the house compared to inside [1]. Black surfaces emit infrared radiation well [1].

Kinetic theory

1a solid -> liquid [1]

b The liquid does not have a fixed shape but the solid has. [1]

2a Melting is when a solid changes to a liquid. [1]

b Particles are in fixed positions in a solid [1] but change places in a liquid [1]. In a solid, ice bonds hold the particles together [1]. When ice melts, bonds between particles start to break and reform as the ice melts to liquid [1].

c Energy is needed to break bonds [1]. Energy is absorbed from the surroundings [1].

d Energy is absorbed more quickly. [1]

Conduction and convection

1a conduction [1]

b Metal is a good heat conductor [1] heat travels more quickly through a good conductor. [1]

c Heat is transferred by convection [1] warm fluids rise because they have a lower density. [1]

2a Convection is when particles move through a substance, transferring energy [1]; particles in solids cannot change places and move through a solid [1].

b The flask reduces conduction [1] with the poly(ethene) base/vacuum/insulated stopper, which is an insulator [1]. Reduces convection [1]: through the top using the stopper [1] OR through the sides using the vacuum [1].Reduces radiation [1] with the silvered walls [1].

Pages 186–187

Evaporation and condensation

1a Condensation is when a gas changes to liquid [1]. Evaporation is when a liquid changes into a gas [1].

b condensation [1]

2a Evaporation is when a liquid changes to a gas. [1]

b Boiling takes place throughout a liquid; evaporation takes place at the surface. [1] Boiling takes place at the boiling temperature, evaporation takes place at any temperature. [1]

c Any two from: warmer weather – molecules have more energy and can break free of bonds more easily [2], windy or drier weather – surrounding air does not become saturated [2].

Rate of energy transfer

1 Colour (white is better), material (use an insulator like expanded polystyrene), and surface area (keep this as small as possible). [3]

2 If the potato is cut, the surface area is larger [1] and the distance to the centre is smaller [1]. More heat can be transferred by radiation at the surface [1]. Heat is not transferred as far by conduction through the potato [1].

Insulating buildings

1a It slows down heat losses [1] so the heating bills are lower/less energy is wasted [1].

b It traps a layer of air [1] air is a good insulator/has a very low U-value [1].

2a The report can suggest savings that are greater than the cost of the report. [1]

b Total savings are £275 per year [1] or £1375 in 5 years [1]. Overall savings are £1080 [1].

c Loft insulation has a low U-value [1] as it is a good insulator/reduces heat losses [1].

d Maximum of 2 marks for two suggestions, and maximum 2 marks for two explanations, for example: cavity wall insulation or double glazing [1] reduces heat losses by conduction [1]; installing draught excluders [1] reduces losses by convection [1]; putting foil behind radiators [1] reduces heat losses by radiation [1].

Specific heat capacity

1a Oil reaches a higher temperature. [1]

b Water is best [1] it heats up less/can absorb more heat without heating up much [1].

2a Specific heat capacity is the energy absorbed by 1 kg of a material when its temperature increases by 1 °C. [1]

b Temperature rise = energy/(mass x specific heat capacity) [1] = 1800/(0.25 x 900) [1] = 8 °C [1]

Pages 188–189

Energy transfer and waste

1a Energy before and after an energy transfer [1] is the same [1].

b The total width of the arrows in and out are the same. [1] It matches the amount of energy transferred in different forms [1].

2a chemical energy -> kinetic energy + heat energy (2 marks if all forms of energy are correct; 1 mark if any form of energy is incorrect or missing)

b As wasted heat energy. [1]

c Conservation of energy states that energy is not lost or created during an energy transfer. [1] The width of the arrows shows the relative proportion of each form of energy [1]. The total width of the output arrows equals the width of the input arrow [1].

Efficiency

1 Less energy is wasted [1], so less energy is needed to do the same work [1], so less money is spent paying for electricity/less of the greenhouse gases are emitted [1].

2a efficiency = useful output energy/input forms of energy [1] = 60% [1]

b Allow sensible answers, for example: Yes [1], more efficient equipment saves money on running costs [1] so people should have this information [1].

OR No [1], Sankey diagrams are hard to interpret/ contain too much information [1]; the efficiency could be stated as a percentage (%) [1].

Electrical appliances

1 1 C [1]

2 A [1]

3 D [1]

4 B [1]

2a chemical (from food) [1] -> kinetic (winding the recharger) [1] -> chemical (in battery) [1]

b i For example: The recharger does not use mains electricity [1], so it can be used where there is no mains supply [1].

ii For example: A battery is a portable supply of electrical energy [1], so the phone can be used on the move [1].

Energy and appliances

1a 2200 joules

b The same energy is needed to heat the water [1]. A more powerful kettle transfers energy more quickly [1].

2 energy = power x time [1] = 850 W
850 x 15 x 60 seconds [1] = 765 000 J [1]

Pages 190–191

The cost of electricity

1 2 x 3 = 6 [1] kWh [1]

2a 0.1 x 1500 [1] = 150 kWh [1]

b 150 kWh at 15p [1] = £22.50 or 2250p [1] x 5 years = £112.50 [1]

c For example: less electricity is used lighting homes [1]; generating less electricity reduces carbon emissions from power stations [1].

Power stations

1 1 B [1]
2 D [1]
3 C [1]
4 A [1]

2a chemical in gas [1] → thermal in water and kinetic in steam [1] → kinetic in turbine and generator [1] → electrical in generator [1]

b Unwanted heat from combined heat and power stations is used to directly heat homes [1]. In power stations which only generate electricity, electricity is transferred to homes then used for heating [1] there are more stages in the energy transfer [1] energy is lost at each stage. [1]

Renewable energy

1a It may be too windy/too calm. [1]

b The reservoirs may not refill fast enough. [1]

c They emit greenhouse gases. [1]

2 Iceland uses geothermal energy [1]. This uses heat in volcanic rocks to produce steam to spin turbines [1]. Norway uses hydroelectricity [1]. This uses falling water trapped behind dams to spin turbines [1].

Electricity and the environment

1a Any three from: solar [1], hydroelectric [1], wind [1], wave [1], tidal [1], biomass [1].

b Any one advantage, which must be relevant to the island setting, e.g. no need to transport fuel [1], no polluting gases [1].

c Any one disadvantage, which must be relevant to the island setting, e.g. may not be reliable [1], may not generate enough electricity for everyone [1]

2

Type of station	Advantage of the scheme	Disadvantage of the scheme
Coal fired	a single power station in a relatively small land area generates large amounts of energy [1]	greenhouse gas emissions; mining damages the environment; large quantities of waste to dispose of [1]
Wind turbines	no pollution is emitted to the atmosphere [1]	wind farms take up large land areas; noise pollution [1]

Pages 192–193

Making comparisons

1a One advantage, e.g. from: more electricity produced for the same amount of fuel used [1], less greenhouse gases emitted per kWh produced by more efficient power stations [1].
One disadvantage, e.g. from: timescale [1], cost [1], carbon dioxide emitted during production processes [1].

b Three things, e.g. from: cost of building/running costs [1]; cost of decommissioning [1]; if the energy source is renewable/non-renewable [1]; impact on environment [1].

2a Credit sensible reason and explanation, e.g. more demand for lighting at night because it is dark [1]; more demand for electricity for heating in winter because it is cold [1]; more demand for electricity at meal times for cooking [1]. (3 marks maximum)

b Waves, wind, tidal, gas. [3]

The National Grid

1a It increases the AC voltage. [1]

b High voltages reduce the amount of energy wasted [1] during transmission of electricity [1].

2a Smaller current [1] so less energy is wasted as heat. [1]

b current = power/voltage = 200 000 000 / 400 000 [1] = 500 [1] A [1].

What are waves?

1a Energy. [1]

b One difference, e.g. mechanical waves cannot travel through a vacuum but electromagnetic ones can [1].

c i mechanical, ii electromagnetic, iii electromagnetic, iv mechanical. [4]

2 Credit valid answers
Difference 1: vibrations are perpendicular to direction of energy travel (transverse) or parallel to direction of energy travel (longitudinal). [1]
Difference 2: longitudinal waves cannot transfer energy through a vacuum but transverse waves can. [1]
Similarity 1: both waves transfer energy. [1]
Similarity 2: both types of wave travel from a source. [1]

Changing direction

1a diffraction [1]

b The radio waves diffract round mountains/hills/

buildings [1]; signals reach more places/spread round corners or obstacles [1].

2 Diffraction is the spreading of waves through a gap [1]. Diffraction is most pronounced when the wavelength of the wave is similar to the gap [1]. Doorways are a similar size to the wavelength of many sound waves [1].

Pages 194–195

Sound

1a In this order: amplitude [1], frequency [1].

b In this order: amplitude [1], frequency [1].

2a Any number between 441 Hz and 480 Hz. [1]

b It gets shorter. [1]

c The amplitude [1] gets bigger [1].

Light and mirrors

1a The light reflects. [1]

b The angle of reflection is the same as the angle of incidence [1], it travels away from the mirror [1].

2 The image is upright, virtual and the same distance behind the mirror. [3]

Using waves

1 Any two from: travel at the same speed/speed of light [1], can travel in a vacuum [1], transfer energy. [1]

2

Type of electromagnetic radiation	Use
Radio wave	BBC TV broadcasts
Microwave	Satellite TV
Infrared radiation	Remote control
Visible light	Photography

(1 mark for each correct pair)

The electromagnetic spectrum

1 A = infrared; B = visible; C = X-rays. [3]

2 wavelength = wavespeed/ frequency [1] = 300 000 000/ 2 700 000 000 [1] = 0.111 [1] m [1] (maximum of 3 marks if units are not correctly converted)

Pages 196–197

Dangers of radiation

1 1 = A [1], 2 = D [1], 3 = C [1], 4 = B [1].

2a Ultraviolet. [1]

b Skin cancer [1]; sunburn [1], premature aging on the skin [1].

c Wear a hat/cover up [1] so radiation cannot reach the skin [1]; stay inside between the hours of 11 am and 3 pm [1] to avoid very intense sunlight [1].

Telecommunications

1 a – microwaves [1], b – microwaves [1], c – visible light or infrared [1].

2 Microwaves transmit images to satellites [1] satellites transmit the images to receiving stations in Australia [1]. Microwaves travel so fast (300 million m/s) that the signal arrives in less than a second [1].

Cable and digital

1a Digital signals are on/off; analogue signals can have any value. [1]

b Pulses of light/infrared radiation [1] travel using total internal reflection/repeatedly reflecting off the inner surface [1].

2a Analogue. [1]

b The signal needs to be amplified/it gets weaker when it is transmitted long distances [1]; when the signal is amplified, any interference is amplified too [1].

Searching space

1 Two reasons, e.g. to see further/fainter objects; to detect radiation that the eye can't see. [2]

2a Different objects emit different types of electromagnetic radiation [1]; the same detectors cannot detect all forms of electromagnetic radiation [1].

b It is larger [1] because radio waves are longer than visible waves [1].

Page 198

Waves and movement

1a higher; shorter [2]

b The pitch is closer to the original pitch/the pitch falls. [1]

2a Doppler effect. [1]

b The wave has a longer wavelength [1] but the same amplitude. [1]

Origins of the Universe

1a red shift [1]

b The distant stars are moving away from us; they are moving away faster than closer stars. [2]

2a All matter and energy in the Universe were in one place at one point in time [1]. About 14 billion years ago, a rapid expansion started to take place [1], which is still continuing [1].

b Cosmic microwave background radiation/the echo of the Big Bang [1]. The red shift which provides evidence that galaxies are all moving apart [1]. The larger red shift from more distant galaxies, which is evidence that more distant galaxies are moving apart faster [1].

Page 199

Extended Response Question

5 or 6 marks:

A detailed description of relevant properties of radio waves, microwave, infrared, and visible light (e.g. they all travel at the speed of light; some are absorbed by the atmosphere; diffraction effects are noticeable with radio waves). The communication use of these electromagnetic waves is linked to its properties; e.g. microwaves are not absorbed by the atmosphere, so can be used for satellite communication.

All information in answer is relevant, clear, organised, and presented in a structured and coherent format. Specialist terms are used appropriately. Few, if any, errors in grammar, punctuation and spelling.

3 or 4 marks:

A limited description of relevant properties of radio

waves, microwave, infrared, and visible light. Some communication uses of members of the electromagnetic spectrum are stated, but may not always be linked to its properties.

For the most part the information is relevant and presented in a structured and coherent format. Specialist terms are used for the most part appropriately. There are occasional errors in grammar, punctuation and spelling.

1 or 2 marks:

The answer includes an incomplete description of relevant properties, which may be linked to specific members of the electromagnetic spectrum. Some relevant uses of more than one type of electromagnetic wave should be stated.

Answer may be simplistic. There may be limited use of specialist terms. Errors of grammar, punctuation and spelling prevent communication of the science.

B2 Answers

Pages 200–201

Animal and plant cells

1a sperm/egg/zygote [1]

b nucleus [1]

2a plant [1]

b A: cell wall [1] B: vacuole [1]

c It's where protein synthesis takes place. [1]

d Chloroplasts absorb light energy to make food [photosynthesis] [1]. Roots are underground where there is no light so have no need for chloroplasts [1]. Leaf cells do receive light so require many chloroplasts [for photosynthesis] [1].

Microbial cells

1a They are made up of one cell. [1]

b yeast/bacteria/algae [1]

2a Similarity: Both have cell wall / nucleus / cytoplasm / cell membrane / vacuole / ribosomes. [1] Difference: Plant cell wall is made of cellulose, yeast cell wall isn't / Yeast does not contain chloroplasts. [1]

b Similarity: Both have cytoplasm/cell membrane. [1] Difference: Animal cell has a nucleus, bacteria do not / Bacteria has a cell wall, animal cells do not. [1]

Diffusion

1 molecules/particles [1], high [1], low [1].

2 The scent of the cake is caused by cake molecules [1]. The concentration of cake molecules will be high in the kitchen [1] but low in other parts of the house [1]. There will be a net movement of molecules from where there is a high concentration to a lower one [1] so the cake molecules will spread away from the kitchen into the rest of the house.

Specialised cells

1a They have structures that make them good at their job. [1]

b They are only made up of one cell [1], so the cell has to carry out every job [1].

2 It has haemoglobin [1] to transport oxygen [1] /a disc shape [1] for taking in and letting out oxygen [1].

Pages 202–203

Tissues

1a tissue [1]

b cell [1]

c cell [1]

d organ [1]

2a muscle [1]

b to release energy [1] which the cells need to contract [1]

Animal tissues and organs

1 It contains many different tissues. [1]

2 One mark for each: 1A, 2D, 3E, 4B, 5C.

Plant tissues and organs

1a to carry out photosynthesis [1]

b to absorb water/anchor the plant in the ground [1]

c to hold up the leaves to the sunlight/hold up the flowers so they can be pollinated [1]

2a **i** It stops too much water leaving the leaf [1] and protects the underlying cells [1].

ii to carry out photosynthesis [1]

b xylem [1]

Photosynthesis

1a It makes food for the rose. [1]

b No [1] because the cells do not contain chlorophyll [1].

2a The leaves will not go black. [1]

b No starch was present [1] because the plants were in the dark so could not carry out photosynthesis [1] so no glucose was made which is stored as starch [1].

c The part that went black was green [1] and had carried out photosynthesis to produce starch [1] because the cells contain chlorophyll [1]. The bit that stayed orange was white so did not carry out photosynthesis and produce starch [1].

Pages 204–205

Limiting factors

1 It is warmer in summer [1] and photosynthesis is faster at warm temperatures [1]. The days in summer are longer so there's more light [1] and the more energy a plant receives, the faster it can photosynthesise [1].

2a oxygen [1]

b As the light intensity increases, the rate of photosynthesis increases. [1]

c The rate of photosynthesis stops increasing. [1]

d Another factor is limiting the rate of photosynthesis, e.g. concentration of carbon dioxide [1] so even if the light intensity increases the rate of photosynthesis cannot [1].

c a source [increased concentration] of carbon dioxide [1]

The products of photosynthesis

1a energy [1] respiration [1]

b i cellulose [1] ii starch/fats/oils [1] iii protein [1]

2 It adds nitrogen to glucose [1]. Nitrogen comes from nitrate ions [1] which are absorbed from the soil through the roots [1].

Distribution of organisms

1 it is too cold [1] there is little air (oxygen) [1]

2a There's not enough light [1] so plants cannot photosynthesise [1].

b It needs to be where it will be covered in water for some periods of the day [1] or it will get too hot and dry out [1].

Using quadrats to sample organisms

1a Habitats are usually too large to count the numbers directly. [1]

b quadrat / transect [1]

2a 4×8 [1 mark] = 32 [1 mark]

b Make sure that the quadrat is placed randomly. [1]

Pages 206–207

Proteins

1 Any two from: antibodies, hormones, enzymes, to build muscle tissue. [2]

2a amino acids [1]

b The order determines the shape of the protein [1]. The protein will only function correctly if the shape is right [1].

Enzymes

1 protein [1], speed up [1], catalysts [1], specific [1]

2a i starch [1]

ii sugars [1]

b temperature [1]

c Any one from: concentration of starch/amylase, volume of starch/amylase, room temperature. [1]

2 Between 0 °C and 40 °C, the rate of reaction increased [1] because the molecules of starch and amylase had increasing amounts of energy [1] and were colliding more frequently and with more energy so they reacted to break down the starch more often [1]. After 40 °C, the rate started to slow down [1] as the amylase started to change shape [denature] [1] so the starch no longer fitted into the active site so the reaction could not occur [1].

Enzymes and digestion

1 They break down large food molecules into smaller ones [1] that can enter the blood [1].

2a starch [1]

b salivary [1]

c protease [1]

d stomach [1]

e lipase [1]

f glycerol [1]

Enzymes at home

1 by using microorganisms [1]

2a It contains lipase [1] which will break down the fat [1] so it becomes soluble and washes away [1].

b The stain will not come out [1] because the lipase will not work at this temperature [1]. It will denature [1].

Pages 208–209

Enzymes in industry

1 Protease

2a by using carbohydrase enzymes [1]

b Starch solution is cheaper. [1]

c Fructose is sweeter than glucose [1] so less is needed [1], which lowers the energy [calorie] content of the food [1].

d isomerase [1]

Aerobic respiration

1a true [1]

b true [1]

c false [1]

2a For respiration [1], to release energy from glucose [1].

b Air enters the lungs [1] and oxygen travels into the blood [1].

Using energy

1 Any two from: to maintain body temperature/to build large molecules up from small ones/for muscles to contract. [2]

2a His muscles are contracting [1] and using up energy [1].

b His body will use up more energy [1] to maintain his body temperature [1].

Anaerobic respiration

1 energy /oxygen[1], heart [1], oxygen [1], anaerobic [1]

2 Aerobic respiration releases more energy per gram of glucose [1], anaerobic produces lactic acid which is toxic [1].

Pages 210–211

Cell division – mitosis

1 The zygote divided [1], each small cell grew [1] and divided again. This happened many times [1].

2a The parent cell divides into two. [1]

b They have the same chromosomes. [1]

c To replace old skin cells that die/to repair cuts in the skin. [1]

Cell division – meiosis

1a 23 [1]

b 23 [1]

c 46 [1]

2a ovaries [1]

b Cells: 1 = 64 [1], 2 and 3 = 64 [1], 4, 5, 6 and 7 = 32 [1]

c It forms gametes with half the normal number of chromosomes [1], so during fertilisation a zygote is formed with a full set of chromosomes [1].

Stem cells

1a unspecialised [1]

b differentiation [1]

2 Stem cells could be taken from an embryo or adult bone marrow [1] and placed into the brain of the patient [1] where they differentiate to form new

brain cells [1].

Genes, alleles and DNA

1a genes [1]

b chromosomes [1]

c DNA [1]

2a possible father is B [1]

b The child shares DNA fragments with him [1] which were transferred from the father in his sperm [1].

Pages 212–213

Mendel

1 Breeding together pea plants. [1]

2 His ideas were only accepted when scientists had discovered how characteristics were inherited [genes/chromosomes]. [1]

How genes affect characteristics

1a X [1]

b XY [1]

c XX [1]

2a One from each parent. [1]

b bb [1]

c In either order [but always a capital letter first]: Bb [1] BB [1].

Inheriting chromosomes and genes

1 Sperm from the father can be X or Y [1] but eggs all contain X [1]. If an X sperm fertilises the egg then the baby is a girl; if a Y sperm fertilises the egg, the baby is a boy [1].

2a The one for short hair. [1]

b Hh (accept any letter as long as there is an upper case and corresponding lower case) [1]

c

Parent	short-haired	long-haired	
	Hh	hh	[1]
Gametes	H and h	h and h	[1]
Offspring			

	H	h
h	Hh	hh
h	Hh	hh

[1 mark]

d 1:1 [1 mark]

How genes work

1a double helix [1]

b Genes [1], genetic [1], proteins [1]

2 The allele is a section of DNA [1]. The order of the bases in the DNA is a code [1] to make a brown protein [pigment] [1] that the cells in her eye produce [1].

Pages 214–215

Genetic disorders

1 From his parents [1]. It is inherited [1].

2a cc (or any answer with two lower-case letters) [1]

b Their child (E) has cystic fibrosis [1] so they must have each passed on the cystic fibrosis allele [1].

c 50 per cent/1 in 2 [1]

Fossils

1 It has changed (evolved) over time. [1]

2 There are not many fossils of jellyfish [1] as they just decay [1], so scientists do not have a lot of evidence of how they used to look millions of years ago [1].

Extinction

1a All members of that species have died. [1]

b Any two from: hunting/changing the habitat (or an example, such as global warming)/introducing alien species. [2]

2 Any two from: The trout are eating them/The trout are better adapted at catching food/The trout are carrying a disease which they pass onto the fish. [2]

New species

1 The mule is sterile, animals from the same species will breed to produce fertile offspring. [1]

2 On each island there is a different environment [1]. The finches show variation [1] and the ones better adapted to living on that island survive [1] to pass on these genes to their offspring [1]. The finches on each island change so much from the mainland species that they are a new species [1].

Page 216

Extended response question

5-6 marks

A detailed description of how prolonged exposure to temperatures above 37 °C results in the denaturation of enzymes and how this effects their ability to function. The answer should contain details about the importance of the shape of the enzyme and why the functioning of enzymes is vital for life.
All information in answer is relevant, clear, organised and presented in a structured and coherent format. Specialist terms are used appropriately. Few, if any, errors in grammar, punctuation and spelling.

3-4 marks

Limited description of how high temperatures affect enzymes. The answer will state that temperatures over 37 °C leads to a change in shape of the enzyme which inhibits its function.
For the most part the information is relevant and presented in a structured and coherent format. Specialist terms are used for the most part appropriately. There are occasional errors in grammar, punctuation and spelling.

1-2 marks

An incomplete description, stating that enzymes do not work over 37 °C.
Answer may be simplistic. There may be limited use of specialist terms. Errors of grammar, punctuation and spelling prevent communication of the science.

C2 Answers

Pages 217

Investigating atoms

1a

particle	Relative mass	Relative charge
proton	(i)1	+1
(ii) neutron	1	(iii) none
electron	(iv) very small	-1

(for 1 mark each)

b i 14 [1]
ii 14 [1]
iii 2,8,4 [1]

Mass number and isotopes

1a the number of protons in the atom [1]

b the number of protons and neutrons in an atom [1]

c Protons are positively charged, electrons are negatively charged [1], the charges have to be equal to make an uncharged atom. [1]

2a An isotope is an atom of an element that has a different mass to other atoms. [1]

b Potassium-40 has one [1] more neutron [1].

c $\frac{(90 \times 39) + (10 \times 40)}{100} = \frac{3910}{100} = 39.1$ [2]

Another answer, but showing correct method, [1].

Pages 218–219

Compounds and mixtures

1a iron and sulfur [1]

b iron sulfide [1]

c iron + sulfur → iron sulfide [1]

d Iron sulfide is a different substance with different chemical and physical properties. [1]

2a an aqueous or water solution [1]

b 137 + (35.5 × 2) [1 mark] = 208 [1 mark]

c 208 grams [1]

Electronic structure

1a Any one of: beryllium, strontium, barium, or radium [1]

b 2 [1]

c They all have the same number of outer electrons [1] available for reactions [1].

2a 2,7 [1] for diagram
[1] for written structure

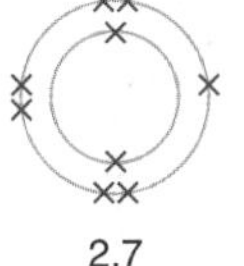

2,7

b 2 [1] for diagram
[1] for written structure

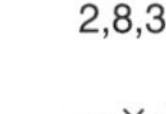

2

c 2,8,3 [1] for diagram
[1] for written structure

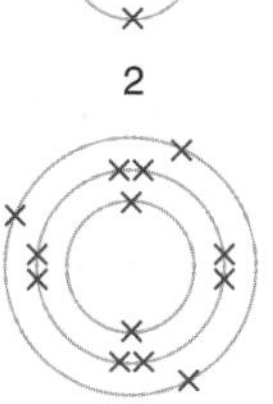

2,8,3

d 2,8,6 [1] for diagram
[1] for written structure

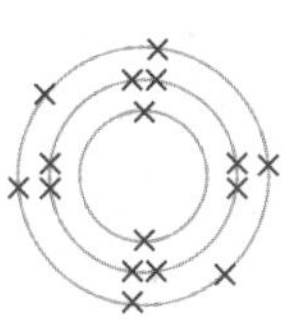

2,8,6

Ionic bonding

1 lose, [1], positively [1], negative ion [1], positive ion [1], a noble gas [1]

2a high melting or boiling point [1] and conducts in solution but not when solid [1]

b i Either a diagram showing 2,8,8 configuration or a statement that it is 2,8,8. [1]
ii Either a diagram showing 2,8 configuration or a statement that it is 2,8. [1]

c The potassium atom loses its outer electron to the fluorine atom, making two ions [1]. The oppositely charged ions then attract each other [1] [with an electrostatic attraction].

Alkali metals

1a it conducts electricity [1]

b so it can't react with water in the air [1]

c It is less dense than water. [1]

2a sodium + water → sodium hydroxide + hydrogen [1]

b They all have a single outer electron [1], which they lose to become 1+ ions. [1]

c It would react very violently [1] as it is lower in the group and so more reactive than potassium [1].

Pages 220–221

Halogens

1 fluorine boiling point; accept value greater than –100 °C, actual value is –188 °C [1]
chlorine Cl_2 [1] and green [1]
bromine [1] liquid [1]

2a

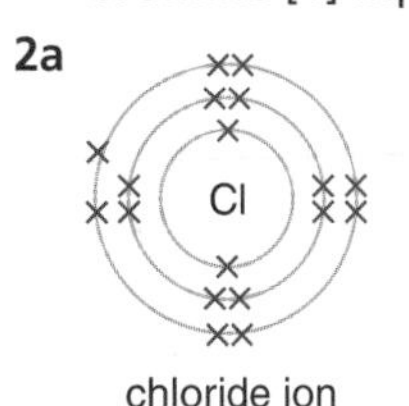

chloride ion

[1]

b 1– [1]

c argon [1]

Ionic lattices

1a R [1], it conducts electricity as a solid [1].

b S [1], it has a low melting point [1].

c P and Q [1], they have high melting points and conduct electricity in solution [1].

2a 6 [1]

b Sodium chloride has a crystal lattice but hydrogen chloride does not [1]. It is easy to separate hydrogen chloride molecules from each other, but much harder to separate the sodium and chloride ions of sodium chloride [1].

c Sodium goes to the negative electrode and chlorine goes to the positive electrode [1]. At the negative electrode, sodium forms [1]; at the positive electrode, chlorine gas molecules are made [1].

Covalent bonding

1 When two ***non-metal atoms*** join together they form covalent bonds. The outer shells of ***electrons*** overlap and form covalent bonds. A covalent bond is a shared ***pair*** of ***electrons***. [1 mark each]

2a

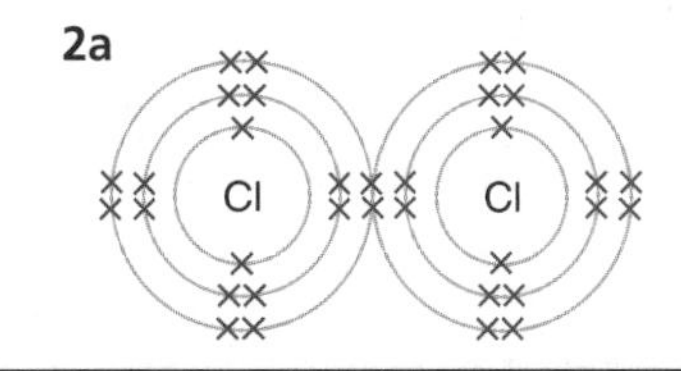

outer shells overlapping [1]

one pair of shared electrons drawn within the overlapping portion [1]

b a shared pair of electrons [1]

c Any two from: they only need to gain a few electrons [1] to achieve a noble gas structure [1], and sharing is the easiest way [1].

Covalent molecules

1a NH_3 [1]

b It has small molecules. [1]

c The molecules have no charge. [1]

2a fluorine and chlorine [1]

b 7 [1]

c As the molecule gets larger, the boiling point gets higher. [1]

d The boiling points get higher down the group [1], the higher the boiling point the greater the intermolecular forces [1], so the forces increase down the group [1].

Pages 222–223

Covalent lattices

1a i 4 [1]

ii 3 [1]

b They are joined to a large number of other atoms [1], so the atoms are difficult to separate from each other (1).

2a Each atom is joined to four others making a very strong structure [1] that is very hard to distort [1].

b The layers of graphite are held together by weak forces [1] so they slide easily past each other [1].

Polymer chains

1a a small molecule that can form part of a polymer [1]

b Polymer A has one monomer [1], but polymer B has two different monomers. [1]

c Polymer chains can have millions of monomer units to make a single chain. [1]

2a Thermosetting polymers harden on heating [1]. Thermosoftening polymers soften or melt on heating [1].

b It can be easily recycled after use. [1]

c Thermosetting polymers have bonds between different polymer chains [1]; thermosoftening polymers have no bonds between chains [1].

Metallic properties

1 One mark for each correct answer (shown in italics).

Metals	Non-metals
shiny	*dull*
good electrical conductor	*poor electrical conductor*
malleable or ductile (either)	brittle
good conductor of heat	poor conductor of heat

[4]

2a Diagram with most of the atoms the same size as in the printed diagram; the extra atoms are sized larger or smaller; and they distort the regular arrangement of layers. (All **three** for 2 marks, **two** for 1 mark.)

b An alloy that returns to its original shape when heated. [1]

c It is stretched to fit tightly at room temperature [1], then when warmed in the mouth it shrinks [1] pulling the teeth into position.

Modern materials

1a a material that can respond to changes in the environment [1]

b so they can remember their original shape [1]

c It can change colour at different temperatures. [1]

2a up to 300 atoms [1]

b A nanoparticle is from 1–100 nm in size. [1]

c The smaller sized particles will be colourless [1], so thick applications will not be visible [1].

Pages 224–225

Identifying food additives

1a three [1]

b five [1]

c Manufacturer C [1]

Instrumental methods

1a to keep the solvent and sample both as gases [1]

b so it can carry the sample through the tube [1]

c to record the different compounds in the sample as they reach the end of the long coiled tube [1]

2a using an instrument or machine to carry out the analysis [1]

b Any two from: they are quick to use [1], work with small samples of material [1], are easy to standardise [1].

c It allows the relative atomic or molecular mass to be found [1], which allows for identification of the substance [1].

Making chemicals

1 To make a chemical you need to know what **reactants** are needed. You need to **calculate** the masses needed of each one so the process is **economic**. You should consider the **hazards** of the reactants and **products** and their **environmental** problems. [6]

2a when an acid reacts with a base or alkali [1]

b a reaction where a solid is made out of two solutions [1]

c a reaction where a substance gains oxygen/loses electrons [1]

d When a substance gains oxygen/loses electrons, another substance will have lost oxygen/gained electrons [1], so one reaction is the opposite of the other. [1]

Chemical composition

1a Answer must be in this order: nitrogen, phosphorus and potassium. [1]

b It means 20% nitrogen, 15% phosphorus and 12% potassium [1] by mass [1].

c The farmer can select the best fertiliser combination

for the crop [1], as different crops need different combinations of NPK. [1]

2a 3.98 g [1]

b 3.18 g [1]

c 0.80 g [1]

d Cu = 3.18/63.5 = 0.05 [1 mark]

O = 0.80/16 = 0.05 [1 mark]

so 0.05 : 0.05 = 1 : 1 ratio, formula = CuO [1 mark]

Pages 226–227

Quantities

1a 40 + 12 + (16 × 3) = 100 [1 mark]

b 40 + 16 = 56 [1 mark]

How much product?

1a It is the maximum possible amount calculated that could be made. [1]

b It is the quantity actually produced by the reaction when carried out. [1]

c Not all the reactants were pure/not all the product was recovered from the solution. [1]

2a More than enough of a reactant to ensure that the other reactant completely reacts. [1]

b CuO = 63.5 + 16 = 79.5, 1.59 g/79.5 = 0.02 [1 mark], $CuSO_4$ = 63.5 + 32 +(16 × 4) = 159.5 [1 mark], so 159.5 × 0.02 = 3.19g [1 mark]

c (2.2/3.19) × 100 = 69% [1]

Reactions that go both ways

1a The reaction can go either way (in either direction), the reactants can become products, but the products can change back to reactants. [1]

b If the solution changes from acid to alkali, you don't need to add fresh indicator. [1]

2a the reaction is reversible / it is an equilibrium reaction [1]

b from blue to pink [1]

c The wet pink cobalt chloride will lose water as it dries and go blue [1]. Adding more water will make it go back to pink. [1]

Rates of reaction

1a decrease [1]

b no effect [1]

c increase [1]

d decrease [1]

e increase [1]

2a changes in mass [1] and time [1]

b At the start there are lots of reactants so it is quick [1], as the reaction continues the reactants are used up so the reaction slows down [1].

c 12.6 cm³ [1]

Pages 228–229

Collision theory

1a In a powder, there are more particles available to react. [1]

b There are fewer acid particles in the same volume of dilute acid as concentrated acid. [1]

c The particles move faster and collide more, giving more reacting collisions. [1]

2a It would increases the rate [1], as the higher the temperature the more reacting collisions will occur [1].

b It would increase the rate [1] as there are more particles on the surface able to react [1].

c It would decrease [1] as there are less acid particles to react with [1], so frequency of reacting collisions is reduced.

Adding energy

1a The particles move faster [1], so collide more frequently or there are more reacting collisions [1].

b If heated too strongly, the reaction would become too fast and pose a safety hazard. [1]

2a 9.6 cm³ [1]

b **i** curve to left of the two curves [1], that plateaus at 9.6 cm³[1]

ii The rate will be faster as the particles are moving faster so will produce more reacting collisions [1]; height will be the same as there are no more particles to react than before [1].

Concentration

1a the one in concentrated acid [1]

b There are more acid particles in the solution so there will be more collisions. [1]

2a A [1]

b C [1]. The curve is lowest/takes longest time to reach the end point/plateau [1].

c This is the maximum amount of hydrogen gas that can be produced from the reactants [1]. Changing the rate only changes the time it takes to make the products, not the quantity of products [1].

d 3.6 min [1]

Size matters

1 The dust will react far quicker than the lump if it catches fire/it is much easier to ignite the dust than the lump [1] causing an explosion in the mine [1].

2a iron filings [1]

b The iron filings have the smallest particles [1], so will react fastest as there are more particles available to react at the same time [1].

c The nail did not react [1]. There were no brown bits on the mat at the end/it was still the same size at the end [1].

Pages 230–231

Clever catalysis

1a transition metals [1]

b a substance that increase the rate of the reaction but is unchanged at the end [1]

c Iron is cheaper/more readily available. [1]

2a Reaction is five times faster [1] as 250/50 = 5 [1]

b The reaction rate was quicker with manganese(IV) oxide present [1] and there was no change in the mass of manganese(IV) oxide during the experiment [1].

c It could have lowered the activation energy; or it could have formed an intermediate that reacted easily, [1] so making it easier to release the oxygen [1].

Controlling important reactions

1a nitrogen + hydrogen ⇌ ammonia [1]

b It is very slow. [1]

2a Increasing the temperature increase the rate of a reaction [1], so to make the ammonia quickly 450 °C is used [1]

b Increasing the pressure forces the molecules closer together [1], this means there will be more collisions so a quicker reaction [1].

The ins and outs of energy

1a calcium oxide +6 [1] ammonium nitrate −7 [1]

b calcium oxide [1] the reaction produced heat energy [1]

2a endothermic [1]

b The products need more energy than the reactants [1], so energy is taken from the solution so the temperature falls [1].

c quicklime and water [1]

d The quicklime and water have more stored energy than the products they make [1], the surplus energy is released as heat energy [1].

Acid–base chemistry

1a hydrogen [1], hydroxide [1]

b The hydrogen and hydroxide ions react together to form a water molecule. [1]

2a An alkali can dissolve in water, a base cannot. [1]

b sodium nitrate and water [1]

c copper chloride [1] and carbon dioxide and water [1]

Pages 232–233

Making soluble salts

1a magnesium [1] hydrogen [1]

b potassium hydroxide [1]

c lithium sulfate [1]

2a sulfuric acid [1]

b i warm the mixture [1]

ii When no more cobalt oxide dissolves [1], check the pH of the solution [1].

c Filter the solution. [1]

d Evaporate the solution [1] until crystals appear[1].

Insoluble salts

1a does not dissolve in water [1]

b PbI_2 [1]

2a calcium carbonate and sodium chloride [1]

b Filter the solution. [1]

c Rinse the contents of the filter paper with water. [1]

Ionic liquids

1a They split up into charged ions [1] that are free to move and carry electric charge [1].

b when they are molten or liquid [1]

2a electrolysis [1]

b Dichromate ions are negatively charged [1] and will be attracted by the opposite charge of the positive electrode [1].

c The ion will gain an electron [1], becoming an atom of potassium [1].

Electrolysis

1a chlorine [1]

b Apply a lighted splint to the gas and it should go 'pop'. [1]

c sodium hydroxide [1]

2a the positive electrode [1]

b It loses two electrons [1] and dissolves in the solution [1].

c to allow copper to be deposited on the negative electrode [1]

d the impurities from the copper [1]

Page 234

Extended response question

5 or 6 marks:

There is a clear, balanced and detailed description of the positive and negative reasons for the use of brine rather than molten sodium chloride with 5–6 points from the examples given. The answer shows almost faultless spelling, punctuation and grammar. It is coherent and in an organised, logical sequence. It contains a range of appropriate or relevant specialist terms used accurately.

3 or 4 marks:

There is some description of the positive and negative reasons for the use of brine rather than molten sodium chloride with 3–4 points from the examples given. There are some errors in spelling, punctuation and grammar. The answer has some structure and organisation. The use of specialist terms has been attempted, but not always accurately.

1 or 2 marks:

There is a brief description of the reasons for the use of brine rather than molten sodium chloride with 1–2 points from the examples given. The spelling, punctuation and grammar are very weak. The answer is poorly organised with almost no specialist terms and/or their use demonstrating a general lack of understanding of their meaning.

possible points include:

positive

- it can be carried out at room temperature so saves energy
- hydrogen is produced at a low temperature, so it will be harder to ignite
- sodium hydroxide is produced directly, with molten reagents a further reaction is needed.
- with brine the sodium immediately reacts with the water solution to make NaOH.

negative

- with molten sodium chloride the sodium made has to react with water $2Na + 2H_2O \rightarrow 2NaOH + O_2$
- the more processes used the costlier it will be.

P2 Answers

Page 235

See how it moves

1a i travelling forward at a steady speed [1] ii stopped [1]

b Section C. [1]

c It is a curved line getting steeper. [2]

2a 80 km/h

b 40 km/h

Speed is not everything

1a It is accelerating.

b It is travelling at a steady speed. [2]

2a Velocity is speed in a certain direction. [2]

b Since velocity depends on speed and direction [1], although speed remains constant, the car's direction changes [1].

c Acceleration is change in speed/time [1]

$\frac{-40}{5}$ = –8 [1] m/s^2 [1]

Pages 236–237

Forcing it

1a 1000 – 400 – 400 = 200 N [2]

b accelerating [1]

2a The car and van pull in opposite directions [1] with equal forces [1].

b The tractor pulls harder / force is larger than the van [1].

Forces and acceleration

1a The resultant force is zero when the car is stopped [1] and forwards when the car starts moving. [1]

b The resultant force is zero when the car is travelling at a steady speed [1] and backwards when the car slows down [1].

2a As the force increases, the acceleration increases. [2]

b Yes, they have [1], you can see a clear pattern [1] OR No, they haven't [1], you can only say that the pattern applies up to forces of 5 N [1].

c To improve the reliability of the data by identifying anomalous results [1], allowing random errors to be identified [1].

Balanced forces

1a i In either order: e.g. gravity/reaction forces. [2]

ii In either order: tension in rope/force exerted by the person on the rope. [2]

b When the team moves, the weight/gravity stays fixed [1], the forces exerted by the right-hand team become larger than the forces exerted by the left-hand team [1].

2a air escaping from the balloon

b gravity pulls it down

c reaction force from the floor pushes up

Stop!

1a thinking distance [1], braking distance [1]

b She has less experience/may not recognise hazards [1].

c The stopping distance is longer

2a thinking distance increases with speed; or thinking distance doubles as speed doubles, etc [2]

b 9 m [1]

c The thinking distance is 6 m with no drink [1], so thinking distance has increased [1].

Pages 238–239

Terminal velocity

1 The car is travelling slower [1], the car has a smaller cross sectional area [1].

2a Terminal velocity is the top speed reached by a moving object. [1]

b i Weight is constant and greater than drag forces, so the resultant force is down [2].

ii Drag forces increase to match weight, which does not change. At terminal velocity, there is no resultant force. [2]

Forces and elasticity

1a yes

b no

c no

d yes

2a independent variable – force applied [1]

b dependent variable – extension of spring [1]

c He may make a mistake in his calculations/it is harder to compare his data with the other groups [1]

d He must control other variables, e.g. the spring used/how measurements are taken [1] so only one factor is changed at a time [1].

Energy to move

1a chemical energy in fuel [1]

b the atmosphere/surroundings as heat [1]

c the ground/surroundings as heat [1]

2a chemical [1] heat [1] sound [1]

b The rate of energy transferred depends on how quickly work is done against friction [1], friction is less on a smooth road [1].

Working hard

1a yes [1]

b no [1]

c no [1]

d yes [1]

2a 12 J [1]

b gravity [1]

c work = force x distance [1] = 2 x 25 = 50 [1] J [1]

Pages 240–241

Energy in quantity

1a Top of the diving board/top of his jump from the diving board – he is at the highest position. [2]

b When he enters the water – he is travelling fastest. [2]

2a mass x gravity x height = 6 x 10 x 0.8 [1] = 48 [1] J [1]

b 6 x 10 x 1.4 [1] = 84 [1] J [1]

Energy, work and power

1a 6000 J/10 s [1] = 600 W [1]

b The time is shorter. [1]

2a power = energy transferred/time [1] = 400 x 10 x 3/60 [1] = 200 [1] W [1]

b More power means more energy is transferred in the same time [1]; so the motor could lift a greater weight in the same time, or lift the same weight a higher distance in that time [1]. More power means the same work is done in less time [1]; so the weight is moved in a shorter time [1].

Momentum

1a 0 kg m/s [1]

b 0.5 × 3 [1] = 1.5 kg m/s [1]

2a mass x velocity [1] = 6 [1] kg m/s [1]

b 6 kgm/s [1]

c momentum/mass [1] = 6/4 = 1.5 [1] m/s [1]

Static electricity

1a positive charge [1]

b One way, e.g. hold it by small pieces of paper and see if it attracts them; see if it deflects a slowly running stream of tap water. [2]

2a Electrons [1] are rubbed off the cloth [1] and move onto the balloon [1].

b positive [1]

c Hold the balloon near the object [1], it attracts a positively charged object [1] and repels a negatively charged object [1].

Pages 242–243

Moving charges

1a same [1]

b They will repel each other [1] because they have the same charge and like charges repel [1].

2a electrostatic induction [1]

b Electrons in the wall are repelled by the negatively charged balloon [1], leaving an overall positive charge on the wall's surface [1].

Circuit diagrams

1a ammeter [1]

b light-emitting diode (LED) [1]

c thermistor [1]

d fuse [1]

2a 1.6 A [1]

b 3 V [1]

c 1.5 V [1] in the series circuit, voltage is shared between components [1]

Ohm's law

1a

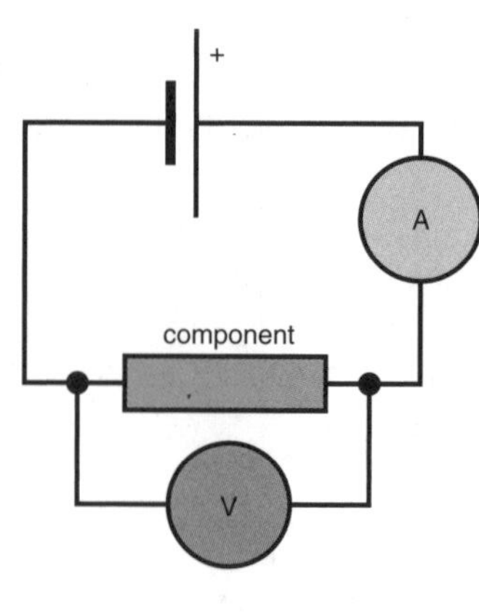

[4]

b resistance = voltage/current [1] = 6/0.1 = 60 [1] ohms [1]

2a resistance measures how easily electrons move through a material [1] OR resistance is voltage ÷ current [1]

b a number larger than 10 ohms [1]

c thickness of wire [1]; material the wire is made from [1]

Non-ohmic devices

1a It increases. [1]

b It increases. [1]

c The current changes with resistance [1]. Resistance is smallest (lowest) when the current is zero or very small (graph has a steep slope) [1].

2a Change the light intensity it is exposed to. [1]

b For example: security lighting. [1]

c symbol 2 [1]

Pages 244–245

Components in series

1a The lights get dimmer [1].

b None of the lights works because there is a break in the circuit. [2]

c gets less. [1]

2a They are all the same. [1]

b 3 V [1]

c 2 V [2]

d It increases [1]; OR it increases by a factor of 4/3 [2].

Components in parallel

1a The brightness does not change. [1]

b The other bulbs stay lit – each bulb has its own path for the current to flow through. [2]

c The current leaving the battery increases as more lights are added. [1]

2a 1.5 A [1]

b 6 V [1]

c e.g. brighter bulbs; if one bulb goes out the others stay on[1]; you can control individual bulbs[1]

Household electricity

1a A [1]

b The current repeatedly changes direction, [1] there are 50 cycles per second. [1].

2 2 Volts

Plugs and cables

1 Allow correct statements, e.g. any three from: the wires are connected to the correct pins [1](maximum 2 marks if colours and pins are correctly identified); the correct fuse is used [1]; the outer part of the cable is gripped under the cable grip [1]; the screws hold each wire tightly in place in each pins [1]; check no bare wires are visible [1].

2 The radio has a plastic case [1] which will not become live [1] so the earth wire is not needed [1].

Pages 246–247

Electrical safety

1 The fuse will melt and cut off the current / the fan stops working [1] even when there is no fault [1]

2a The fuse does not cut off the electric current [1]; the cable is still live [1]; there is a risk of electrocution [1].

b Credit correct answers: the RCCB cuts off the current very quickly [1] if there is a different current in different wires in the cable [1] so there is no risk of electrocution [1].

Current, charge and power

1 power = energy/time [1] = 100/10 [1] = 10 W [1]

2a energy = power x time [1] = 5 x 60 x 1000 [1] = 300 000 J [1] or 300 kJ

b charge = current x time [1] = 4.3 x 5 x 60 [1] = 1290 [1] C [1]

Structure of atoms

1a negligible or 1/2000 [1]

b 1 [1]

c neutron [1]

2a i negligible or 1/2000 [1]; ii +1 [1]; iii neutron [1].

b i number of neutrons;

ii number of protons and electrons.

Radioactivity

1a Radioactive – the nucleus emits ionising radiation, changing to a nucleus of a different element. [1]

b alpha [1]

c beta, gamma [2]

2a alpha particles cannot travel through the plastic case of the smoke detector

b gamma rays will penetrate through cardboard.

Pages 248–249

More about nuclear radiation

1 The positive charge is spread throughout the atom [1]; the negative charges are dotted throughout the atom [1]. The mass of the atom is spread throughout the atom [1].

2a positively charged [1]; central massive nucleus [1]; surrounded by negatively charged electrons [1]

b i deflection of some positively charged alpha particles [1]

ii most alpha particles not deflected [1]

iii electrons are relatively easy to remove from the atom [1]

Background radiation

1a 37% [1]

b Credit correct examples, e.g. some sources of background radiation increase with lifestyle choices, e.g. the number of flights taken [1] / medical history, e.g. X-rays [1] / where you live. [1]

2 move house [1] away from rocks that emit radon gas [1] take fewer flights [1] which reduces cosmic ray exposure [1].

Half-life

1a Half-life is the time [1] taken for the original count rate/activity/mass of radioactive atoms [1] to halve. [1]

b The proportions will fall, [1] the carbon atoms change into nuclei of other elements. [1]

2a 600 counts per second [1]

b Three [1] half-lives [1], which is 15 years [1].

Using nuclear radiation

1a e.g. reduce wastage/controls the quality [1]

b it is absorbed by cardboard [1]

c less radiation detected means the cardboard is too thick [1] so the rollers should move closer together [1]

2 It is traced using detector on the surface [1]. The radiation must be able to penetrate the ground [1] gamma rays are penetrating [1].

Pages 250–251

Nuclear fission

1a when a nucleus splits into two or more products [2]

b nuclear power / generating electricity [1]

2a 3 [1]

b More nuclei are involved at each stage of the chain reaction [1]; because each reaction produces more neutrons than are needed to cause fission [1]; too much heat/energy will be produced if too many nuclei are involved [1].

Nuclear fusion

1a in nuclear fission, nuclei split, but in nuclear fusion, nuclei join [1] nuclear fission involves large nuclei, but nuclear fusion involves light nuclei [1]

b the Sun [1]

2 Nuclear fusion creates different elements [1]; light elements form heavier elements [1]; in stars, elements up to iron are formed [1]; in supernovas heavier elements can form [1].

Life cycle of stars

1 Credit correct references to changes in forces at specific stages. Any five from: gravity forces the star to collapse / increases the pressure inside it [1] forces from heat generated in the star force it to expand [1] during each stage the forces balance [1] when the fuel runs out, less heat is generated and the star collapses [1] because gravity is greater than forces

from heat [1] in very massive stars, the heat generated can be enough to cause a rapid expansion [1]

2 The forces from fusion and gravity [1] are balanced [1]. There is a large supply of fuel for fusion reactions to take place [1]

Page 251

Extended response question

5 or 6 marks:

A detailed description of two safety features, e.g. airbags, crumple zones, seat belts, and side impact bars. The answer should include a discussion of the momentum and energy changes taking place, linking increased impact duration with a reduced risk of serious injury (change in speed takes longer, so the deceleration and force felt are smaller). The answer should explain how these safety measures increase the duration of the collision and/or absorb energy (e.g. seat belts stretch slightly, which extends the time a person takes to stop; crumple zones deform and absorb energy – this extends the time taken to come to a complete halt).

All information in answer is relevant, clear, organised, and presented in a structured and coherent format. Specialist terms are used appropriately. Few, if any, errors in grammar, punctuation and spelling.

3 or 4 marks:

A limited description of two safety features, e.g. airbags, crumple zones, seat belts, and side impact bars. There should be a discussion, which describes what these safety measures do, and some attempt to link this to increased impact duration or reduced energy transfer.

For the most part, the information is relevant and presented in a structured and coherent format. Specialist terms are used for the most part appropriately. There are occasional errors in grammar, punctuation and spelling.

1 or 2 marks:

An incomplete description of one or two safety features. There should be some description of energy changes or momentum changes during a collision.

Answer may be simplistic. There may be limited use of specialist terms. Errors of grammar, punctuation and spelling prevent communication of the science.